SPECTROSCOPY OF CONDENSED MEDIA

Dynamics of Molecular Interactions

SPECTROSCOPY OF CONDENSED MEDIA

Dynamics of Molecular Interactions

C. H. WANG

Department of Chemistry
University of Utah
Salt Lake City, Utah

1985

ACADEMIC PRESS, INC.

(Harcourt Brace Jovanovich, Publishers)

Orlando San Diego New York London
Toronto Montreal Sydney Tokyo

ACADEMIC PRESS, INC.
Orlando, Florida 32887

United Kingdom Edition published by
ACADEMIC PRESS, INC. (LONDON) LTD.
24/28 Oval Road, London NW1 7DX

Library of Congress Cataloging in Publication Data

Wang, C. H. (Chin Hsien), Date
Spectroscopy of condensed media.

Includes index.
1. Condensed matter--Spectra. 2. Molecular dynamics.
3. Nuclear magnetic resonance spectroscopy. I. Title.
QC173.4.C65W36 1984 541.2'8 84-6235
ISBN 0-12-734780-1 (alk. paper)

PRINTED IN THE UNITED STATES OF AMERICA

85 86 87 88 9 8 7 6 5 4 3 2 1

To My Children,
Nancy, Jeanny, Elaine, and Wilson Wang

Contents

5 NUCLEAR MAGNETIC RESONANCE SPECTROSCOPY

6 LINEAR AND NONLINEAR RAMAN SPECTROSCOPY OF THE CONDENSED PHASE

APPENDIX: LINEAR-RESPONSE THEORY AND FLUCTUATION–DISSIPATION THEOREM INCLUDING THE SPATIAL DEPENDENCE

Preface

This book has been developed from a seminar course on molecular spectroscopy of the condensed phase for advanced graduate students in physical chemistry. To date, most books on this subject have dealt mainly with the relation of spectroscopy to molecular structure. This book, however, stresses the use of molecular spectroscopy (including nuclear magnetic resonance and nonlinear optical spectroscopy) to study dynamic processes in condensed molecular systems.

Because of the availability of monochromatic radiation sources, high-resolution spectral instruments, and refined electronics and computers, the frequency, intensity, linewidth, and line shape of a spectrum in absorption and scattering experiments can now be measured with precision. The important task remaining in the study of matter is to unravel from these precise experimental results the fundamental properties of the molecules that constitute material systems.

It is natural that a detailed understanding of spectral features, especially in the condensed phase, depends on a knowledge of statistical mechanics. Spectroscopy of condensed matter cannot be separated from statistical mechanics; these two desciplines must therefore be presented together as an integrated subject. Since time-correlation functions occur naturally in the theory of spectroscopy, they are fundamentally important for the description of time-dependent processes inherent in the line shape of a spectrum. Much of the discussion in this volume deals with time-correlation functions; thus, this book may also be useful as an introduction to the subject of time-dependent statistical mechanics. However, because of the diversity and complexity of time-dependent statistical mechanics, only those subjects that are directly associated with spectral measurements are discussed. Moreover, due to the knowledge and interest of the author, many topics related to nonequilibrium statistical mechanics are treated only perfunctorily. It is hoped, however, that the discussion of spectroscopy will compensate for what is lacking in the treatment of statistical mechanics.

Chapter 1 begins with an elementary treatment of the relationship between transition probability and the time-correlation function of an isotropic electric dipole system. After the time-correlation function is defined, linear-response theory is introduced. Linear-response theory provides not only a useful description of the properties of time-correlation functions but also a unified approach to various types of absorption spectroscopy and theories of thermal fluctuations of dynamic variables involved in condensed-phase molecular spectroscopy. The theory of light scattering is discussed in Chapter 2, with special emphasis on dynamic molecular scattering that results from the translational motion of molecules in fluids. Chapter 3 deals with molecular rotation and theories of angular momentum. This provides a useful background for the discussion of the several selected topics in nuclear magnetic resonance presented in Chapter 5. Chapter 4 deals with the memory-function theory of the time-correlation function. Generalized hydrodynamic theory, as well as the connection between single-particle and collective (or many-particle) molecular reorientational dynamics, is also introduced in this chapter. Spontaneous and coherent Raman effects are treated in Chapter 6. Vibrational dephasing and relaxation processes and dynamics of phase-transition solids, which are closely connected in Raman and Brillouin scattering, are addressed. This chapter also discusses the use of nonlinear optical spectroscopy for the study of condensed matter.

The subject matter presented in this monograph has been developed by many workers. I have reorganized much of the material to make it more self-contained and useful for spectroscopic study of the condensed phase. This volume thus provides a comprehensive introduction to the fundamental principles governing the dynamic processes in molecular spectroscopy, emphasizing fundamentals of physical processes rather than descriptions of spectra that have been measured. Much rewriting was done each time I taught a graduate spectroscopy course at the University of Utah. I believe, therefore, that this book will be suitable for use as a text in advanced graduate courses on molecular spectroscopy. It will also be useful to scientists engaged in spectroscopic research of condensed matter.

I am much indebted to members of my family who have endured with me the hardship and the disruption of family life during preparation of this monograph. Special thanks are also given to Mrs. Donna Keyes, who has accurately typed the manuscript; to my students throughout the years; and to my daughter, Nancy Wang, all of whom have helped me correct errors throughout preparation of the manuscript.

1
Time-Correlation Functions and Spectroscopy

1.1 INTRODUCTION

Measurements of the absorption and scattering of electromagnetic radiation in materials are important in the study of molecular motion. This book stresses the use of these techniques to study the dynamics of molecular motion in dense molecular systems.

In a spectroscopy experiment, the investigator first disturbs the material system with an external force (used as a probe) and then monitors the effect of the perturbation on the material system. The induced effect can be expressed as a time-correlation function that describes the dynamic behavior of spontaneous fluctuations about the equilibrium state. Thus, the interpretation of a spectroscopic measurement is closely related to understanding the time-dependent spontaneous-fluctuation process. The experimental result is best expressed by the time-correlation function associated with the measured microscopic quantity.

Qualitatively, a time-correlation function describes how long a given property of the system persists until it is averaged out by the microscopic motion of the molecules in the system (Zwanzig, 1965). Mathematically, the time-correlation function $C_{AB}(t, t')$ is the product of the pair of dynamic variables $A(t)$ and $B(t')$ (at different times t and t', respectively) averaged over an equilibrium canonical ensemble,

$$C_{AB}(t, t') = \langle A(t)B(t')\rangle, \tag{1.1}$$

where the equilibrium canonical average is denoted by the angular brackets. The autocorrelation function $C_A(t)$ of quantity A is especially useful in spectroscopy. It is defined by the equation

$$C_A(t) = \langle A(t)A(0)\rangle, \tag{1.2}$$

where A can be any dynamic variable of the system. The time dependence is determined by the perturbations that arise from thermal motion and molecular interaction; these cause the positions and momenta of particles in the system to fluctuate over time. The time dependence of the dynamic variable A, which depends on the positions and momenta of all particles, is necessarily very complex; this is because the environment, which in the condensed phase consists of a great number of atoms and molecules, fluctuates in a complex fashion. The time dependence of $A(t)$ will generally resemble a noise pattern similar to the one depicted in Fig. 1.1. The noise pattern generated by the interaction of various degrees of freedom that determine A with the environment can be considered stochastic.

A simple example of treating a stochastic process is given by the equation

$$\partial/\partial t A(t) = i\Omega(t)A(t), \tag{1.3}$$

where the time dependence of the frequency $\Omega(t)$ is due to molecular interaction and thermal motion. Time dependence consists of two parts: an average part $\omega_0 = \langle\Omega(t)\rangle$ and a fluctuating part $\delta\Omega(t)$. If

$$\Omega(t) = \omega_0 + \delta\Omega(t), \tag{1.4a}$$

then clearly

$$\langle\delta\Omega(t)\rangle = 0. \tag{1.4b}$$

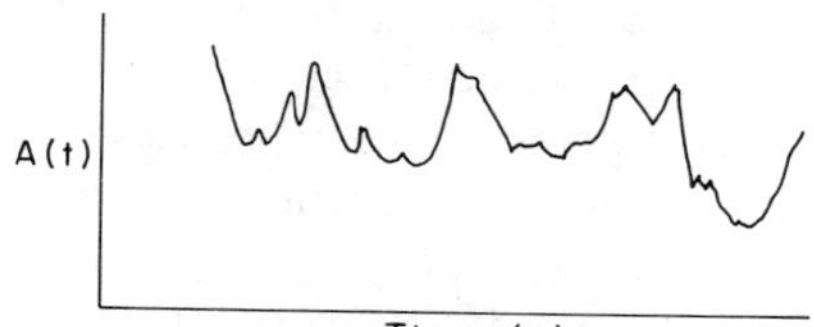

FIG. 1.1 Random fluctuations in time of the dynamic variable A.

Equation (1.3) is a stochastic equation that defines the time dependence of dynamic variable $Q(t)$ in terms of ω_0 and $\delta\Omega(t)$. The dynamics characterizing $A(t)$ are now incorporated in the random function $\delta\Omega(t)$. If knowledge of $\delta\Omega(t)$ could be obtained, then it would be possible to extract the dynamic information contained in $A(t)$, and vice versa.

It is useful to characterize the random variable $\delta\Omega(t)$ by the variance, defined as

$$\Delta = \langle(\delta\Omega)^2\rangle^{1/2}, \tag{1.5a}$$

and by the correlation time τ_c, given by

$$\tau_c = \frac{1}{\Delta^2}\int_0^\infty dt\,\langle\delta\Omega(t)\,\delta\Omega\rangle \tag{1.5b}$$

Later in this chapter, we shall provide an example relating the time-correlation function $C_A(t)$ to the line-shape function $I_A(\omega)$. A stochastic theory of the line-shape function will be discussed at the end of this chapter.

1.2 STATIONARY RANDOM PROCESS AND TIME-CORRELATION FUNCTIONS

In a measurement, the dynamic variable $A(t)$ is averaged over an interval of time from t_0 to $t_0 + T$, where the time interval T is chosen to be long, compared with the coherence time of the fluctuating variable $A(t)$, (i.e., the fluctuation period). The average quantity should correspond to the bulk property associated with the dynamic variable $A(t)$. In an equilibrium system, this bulk property is independent of the initial time t_0 at which the measurement is started. The random process for which an average value is calculated and which yields a result independent of the initial time t_0 is known as a stationary process (Wax, 1954). In other words, the values calculated in the stationary process do not depend on a shift of the origin of time. Thus, the equilibrium bulk property associated with $A(t)$ in a stationary random process can be calculated according to

$$\begin{aligned}\bar{A} &= \lim_{T\to\infty}\frac{1}{T}\int_{t_0}^{t_0+T} dt\,A(t)\\ &= \lim_{T\to\infty}\frac{1}{T}\int_0^T dt\,A(t).\end{aligned} \tag{1.6}$$

Although the dynamic variable $A(t)$ may appear to fluctuate randomly in time, temporal coherence exists if the value of $A(t + \tau)$ is separated from that of $A(t)$ by a small value of τ. The coherence is determined by the correlation time τ_c, to be described later. The correlation is lost, however, if τ is much longer than the correlation time.

A measure of this correlation for the property A separated by a time t is thus given by the time-autocorrelation function, which is defined in terms of the average as

$$\overline{A(t)A} = \lim_{T\to\infty} \frac{1}{T} \int_0^T d\tau \, A(t + \tau)A(\tau). \tag{1.7}$$

The time-correlation function given in Eq. (1.7) is defined in terms of the time average, which may be different from the ensemble-average time-correlation function given in Eq. (1.2). It will be shown in subsequent chapters that, in most theoretical calculations, one deals with the ensemble-average time-correlation functions, whereas in spectroscopic experiments the time-average time-correlation functions are measured. Although there exists no formal proof that the two time-correlation functions are equivalent in practice, the calculation of the ensemble-average time-correlation function has yielded results that are found to be consistent with the experiments (Uhlenbeck and Ford, 1963). Thus, we shall assume in the rest of this book that the two types of time-correlation functions are identical. Thus we set

$$C_A(t) = \langle A(t)A \rangle = \overline{A(t)A}, \tag{1.8}$$

where for simplicity we have written A for $A(0)$. The assumption that the time average is equal to the ensemble average is known as the *ergodic hypothesis*.

It is clear that the time-correlation function has the following properties:

$$\langle A^2 \rangle \geq \langle A(t)A \rangle \qquad \text{for} \quad t > 0, \tag{1.9a}$$

$$\lim_{T\to\infty} \langle A(t)A \rangle = \langle A \rangle^2. \tag{1.9b}$$

Equation (1.9b) follows because the values of the dynamic variable separated by a large value of time are not correlated, and the average of products becomes the product of averages.

It then follows from Eqs. (1.9) that the time-correlation function $C_A(t)$ decays in time from $\langle A^2 \rangle$ initially to $\langle A \rangle^2$ for large values of time. The detail of time decay depends on the dynamical nature of the physical system involved. For the case when the average $\langle A \rangle$ does not vanish, it is convenient to define the fluctuation in A by

$$\delta A = A - \langle A \rangle. \tag{1.10}$$

The time-correlation function of δA is then given by

$$\langle \delta A(t)\, \delta A \rangle = \langle A(t) A \rangle - \langle A \rangle^2. \tag{1.11}$$

It is now evident that the time-correlation function of δA always vanishes at large values of time.

The probable range of time with which the values of δA are correlated can be measured by the correlation time τ_c, which is defined as

$$\tau_c = \int_0^\infty dt\, \langle \delta A(t)\, \delta A \rangle / \langle |\delta A|^2 \rangle. \tag{1.12}$$

In many applications, it is useful to assume that the time-correlation function of δA decays as a single exponential with the relaxation time τ_r, so that

$$\langle \delta A(t)\, \delta A \rangle = \langle |\delta A|^2 \rangle e^{-t/\tau_r}. \tag{1.13}$$

Clearly, in this case $\tau_c = \tau_r$. However, in general, not all time-correlation functions decay as a single exponential, and Eq. (1.12) provides a proper method of determining the correlation time. The correlation function as given by Eq. (1.11) vanishes rapidly for $t > \tau_c$.

1.3 SPECTRAL POWER DENSITY AND THE WIENER–KHINTCHINE THEOREM

Spectral power density and the time-correlation function are two functions of central importance in the theory of spectroscopy of condensed media. The spectral (power) density for the stationary random variable $A(t)$ is defined as the Fourier transform of the time-correlation $C_A(t)$:

$$I_A(\omega) = \frac{1}{2\pi} \int_{-\infty}^{\infty} d\tau\, e^{-i\omega\tau} \langle A(\tau) A^* \rangle, \tag{1.14}$$

where, for the sake of generality, we allow A to be a complex quantity.

We shall now show that $I_A(\omega)$ corresponds to the spectrum of experimentally measurable quantity A. Consider that $A(t)$ is measured over a time interval from $-T/2$ to $T/2$ and that, over this time interval, $A(t)$ can be represented by a Fourier series as

$$A(t) = \frac{1}{\sqrt{T}} \sum_\omega A_\omega e^{i\omega t}. \tag{1.15}$$

The signal at the detector is proportional to $|A(t)|^2$. Using Eq. (1.15), we can express the signal averaged over the interval as

$$\begin{aligned}\overline{|A(t)|^2} &= \frac{1}{T}\int_{-T/2}^{T/2} dt\,|A(t)|^2 \\ &= \frac{1}{T}\sum_{\omega}\sum_{\omega'}\frac{1}{T}A_\omega A^*_{\omega'}\int_{-T/2}^{T/2} dt\,\exp[i(\omega-\omega')t] \\ &= \frac{1}{T}\sum_{\omega}\sum_{\omega'}A_\omega A^*_{\omega'}\frac{\sin(\omega-\omega')^{T/2}}{(\omega-\omega')^{T/2}}.\end{aligned} \tag{1.16}$$

For T large compared with the fluctuation period $(=1/\omega)$, we have

$$\lim_{T\to\infty}\frac{\sin(\omega-\omega')T/2}{(\omega-\omega')T/2} = \delta_{\omega,\omega'}, \tag{1.17}$$

where $\delta_{\omega,\omega'}$ is zero for $\omega \neq \omega'$ and one for $\omega = \omega'$. Thus, Eq. (1.16) reduces to

$$\overline{|A(t)|^2} = \langle |A(t)|^2\rangle = \frac{1}{T}\sum_{\omega}|A_\omega|^2. \tag{1.18}$$

Similarly, we can evaluate the time-correlation function

$$\begin{aligned}\langle A(\tau)A^*\rangle &= \overline{A(t+\tau)A^*(t)} \\ &= \frac{1}{T}\sum_{\omega,\omega'}\frac{1}{T}A_\omega A^*_{\omega'}\int_{-T/2}^{T/2} dt\,\exp[i(\omega-\omega')t]\exp(i\omega\tau).\end{aligned} \tag{1.19}$$

Performing the time integration and using Eq. (1.17), we obtain for $T \gg I/\omega$

$$\langle A(\tau)A^*\rangle = \frac{1}{T}\sum_{\omega}|A_\omega|^2 e^{i\omega\tau}. \tag{1.20}$$

We now carry out the Fourier inversion by multiplying both sides of Eq. (1.20) by $\exp(i\omega'\tau)$ and then performing the integration over τ from $-T/2$ to $T/2$ to obtain [with the help of Eq. (1.17)]

$$|A_\omega|^2 = \int_{-T/2}^{T/2} d\tau\,\langle A(\tau)A^*\rangle e^{-i\omega\tau}. \tag{1.21}$$

Substituting Eq. (1.21) into (1.18) gives

$$\langle |A(t)|^2\rangle = \langle |A|^2\rangle = \frac{2\pi}{T}\sum_{\omega}\left[\frac{1}{2\pi}\int_{-T/2}^{T/2} d\tau\,\langle A(\tau)A^*\rangle e^{-i\omega\tau}\right]. \tag{1.22}$$

For T greater than the fluctuation period, we can take the limit as $T \to \infty$. Then we can rewrite Eq. (1.22) as

$$\langle |A(t)|^2 \rangle = \frac{2\pi}{T} \sum_{\omega} I_A(\omega) = \sum_{\omega} \Delta\omega \, I_A(\omega), \tag{1.23}$$

where the last step follows because the frequency separation $\Delta\omega = 2\pi/T$. The term $I_A(\omega)$ is defined according to Eq. (1.14).

In the $T \to \infty$ limit, $\Delta\omega \to 0$; thus, Eq. (1.23) can be written as

$$\langle |A(t)|^2 \rangle = \int_{-\infty}^{\infty} d\omega \, I_A(\omega). \tag{1.24}$$

Thus we can conclude that the time average (or the ensemble average) of the signal at the detector is equal to the integral sum of the spectral power density $I_A(\omega)$ at all frequencies. In the $\Delta\omega \to 0$ limit, Eq. (1.20) can be written as

$$\langle A(\tau)A^* \rangle = \left(\frac{1}{2\pi}\right)\left(\frac{2\pi}{T}\right) \sum_{\omega} |A_\omega|^2 e^{i\omega\tau} \to \frac{1}{2\pi} \int_{-\infty}^{\infty} d\omega \, |A_\omega|^2 e^{i\omega\tau}.$$

In most of the spectroscopic experiments, a dispersive instrument (such as a grating monochromator, a Fabry–Perot interferometer, or a spectrum analyzer) is used to filter out the signal at a particular frequency interval, say, in the interval between ω_0 and $\omega_0 + \Delta\omega$. The signal that appears at the output terminal of the detector is a direct measure of the spectral power density at the frequency ω_0, as defined by the dispersive instrument. Thus, by tuning through different frequencies, that spectral density can be determined at a variety of frequencies. Knowing the spectral density at all frequencies, the time-correlation function can be determined through Fourier inversion, according to Eq. (1.14). The goal of any theoretical spectroscopic study is to model the physical system in terms of a few fundamental microscopic physical processes so that the calculated time-correlation function can be compared, refined, and finally made to agree with the experimentally observed time-correlation function. The experimental time-correlation function can be obtained by Fourier transforming the spectral power density. The Fourier transform relationship of the spectral density to the time-correlation function is known as the Wiener–Khintchine theorem.

The correlation-function description of spectral density is particularly useful whenever two physical systems are weakly coupled together, such as nuclear spins weakly coupled to the nonspin degrees of freedom (e.g., translational or rotational motions). To describe the effects of the weak interaction between the two systems, it is only necessary to know how the free motion of separate systems affects the weak coupling between them. In other words, all the relevant information is contained in the time-correlation function of the coupling Hamiltonian that deals with the interaction between the two separate systems.

For this reason, it is profitable to discuss the general properties of the time-correlation functions. However, before giving a detailed discussion of the general property of the correlation functions, we shall illustrate the relationship of the absorption-line shape to the associated time-correlation function. To do this we consider the absorption of light by a system of electric dipoles in a condensed medium.

1.4 ABSORPTION LINE SHAPE AND THE DIPOLE–DIPOLE TIME-CORRELATION FUNCTION§

Consider a many-body system collectively in the quantum state described by the vector $|i\rangle$. The system is perturbed by the electric field of an electromagnetic wave of frequency ω, and as a result it is promoted to other quantum states $|f\rangle$. The frequency of the radiation closely matches one of the Bohr frequencies:

$$\omega_{\mathrm{fi}} = (E_{\mathrm{f}} - E_{\mathrm{i}})/\hbar.$$

According to the well-known Fermi golden rule of time-dependent perturbation theory, the probability per unit time for making the transition from state $|i\rangle$ to state $|f\rangle$ is given by (Merzbacher, 1970)

$$P_{\mathrm{f}\leftarrow\mathrm{i}}(\omega) = (\pi/2\hbar^2)|\langle \mathrm{f}|\mathbf{E}_0\cdot\mathbf{P}|\mathrm{i}\rangle|^2[\delta(\omega_{\mathrm{fi}} - \omega) + \delta(\omega_{\mathrm{fi}} + \omega)], \qquad (1.25)$$

where $\mathbf{E}_0$ is the amplitude of the electric field of the radiation applied externally and $\mathbf{P}$ the total electric dipole moment operator for the molecules in the system.

The rate of energy loss from the radiation is given by

$$\sum_{\mathrm{i}}\sum_{\mathrm{f}} \hbar\omega_{\mathrm{fi}} P_{\mathrm{f}\leftarrow\mathrm{i}}\rho_{\mathrm{i}} = \dot{\varepsilon} = \frac{\pi}{2\hbar}\sum_{\mathrm{i}}\sum_{\mathrm{f}} \omega_{\mathrm{if}}\,\rho_{\mathrm{i}}|\langle \mathrm{f}|\mathbf{E}_0\cdot\mathbf{P}|\mathrm{i}\rangle|^2 \times [\delta(\omega_{\mathrm{if}} - \omega) + \delta(\omega_{\mathrm{if}} + \omega)], \qquad (1.26)$$

where ρ_{i} is the probability of finding the molecules in the initial state $|i\rangle$.

Interchange of i and f indices in Eq. (1.26) gives

$$\dot{\varepsilon} = \frac{\pi}{2\hbar}\sum_{\mathrm{i}}\sum_{\mathrm{f}} \omega_{\mathrm{fi}}(\rho_{\mathrm{i}} - \rho_{\mathrm{f}})|\langle \mathrm{f}|\mathbf{E}_0\cdot\mathbf{P}|\mathrm{i}\rangle|^2\,\delta(\omega_{\mathrm{if}} - \omega). \qquad (1.27)$$

§ This section is based on the review article by Gordon (1968).

Since ρ_{f} is related to ρ_{i} by $\rho_{\mathrm{f}} = \rho_{\mathrm{i}} \exp(-\beta\hbar\omega_{\mathrm{fi}})$, where $\beta = 1/kT$, Eq. (1.27) is further reduced to

$$\dot{\varepsilon} = \frac{\omega\pi}{2\hbar}[1 - \exp(-\beta\hbar\omega)] \sum_{\mathrm{i}} \sum_{\mathrm{f}} \rho_{\mathrm{i}} |\langle \mathrm{f} | \mathbf{E}_0 \cdot \mathbf{P} | \mathrm{i} \rangle|^2 \delta(\omega_{\mathrm{if}} - \omega), \quad (1.28)$$

where we have replaced ω_{fi} by ω because of the energy-conserving δ function.

The absorptive (imaginary) part of the electric susceptibility $\chi''(\omega)$ is equal to [compare with Eq. (1.162)]

$$\begin{aligned}\chi''(\omega) &= \dot{\varepsilon}/\omega I_0 \\ &= \frac{4\pi^2}{\hbar}[1 - \exp(-\beta\hbar\omega)] \sum_{\mathrm{i}} \sum_{\mathrm{f}} \rho_{\mathrm{i}} |\langle \mathrm{f} | \hat{e} \cdot \mathbf{P} | \mathrm{i} \rangle|^2 \delta(\omega_{\mathrm{fi}} - \omega), \quad (1.29)\end{aligned}$$

where I_0, the average energy density of the radiation field, is equal to $E_0^2/8\pi$ and $\hat{e}$ a unit vector indicating the polarization of the electric field of the electromagnetic radiation.

We next introduce the Fourier expansion of the Dirac δ function

$$\delta(\omega) = \frac{1}{2\pi} \int_{-\infty}^{\infty} dt \exp(-i\omega t) \quad (1.30)$$

and rewrite Eq. (1.29) as

$$\begin{aligned}\chi''(\omega) &= \frac{2\pi}{\hbar}\left[1 - \exp(-\beta\hbar\omega)\right] \int_{-\infty}^{\infty} dt \exp(-i\omega t) \\ &\quad \times \sum_{\mathrm{i}} \sum_{\mathrm{f}} \rho_{\mathrm{i}} |\langle \mathrm{f} | P_e | \mathrm{i} \rangle|^2 \exp\left[\frac{i(E_{\mathrm{f}} - E_{\mathrm{i}})t}{\hbar}\right] \\ &= \frac{2\pi}{\hbar}[1 - \exp(-\beta\hbar\omega)] \int_{-\infty}^{\infty} dt \exp(-i\omega t) \\ &\quad \times \sum_{\mathrm{i}} \sum_{\mathrm{f}} \rho_{\mathrm{i}} \langle \mathrm{i} | \exp(i\mathscr{H}t/\hbar) P_e \exp(-i\mathscr{H}t/\hbar) | \mathrm{f} \rangle \langle \mathrm{f} | P_e | \mathrm{i} \rangle, \quad (1.31)\end{aligned}$$

where the energy eigenvalues E_{f} and E_{i} have been expressed using the Hamiltonian of the material system and the quantity P_e is the component of $\mathbf{P}$ along $\hat{e}$; i.e., $P_e = \mathbf{P} \cdot \hat{e}$. The time-dependent Heisenberg operator $P_e(t)$ for the dipole at time t is given by

$$P_e(t) = e^{i\mathscr{H}t/\hbar} P_e e^{-i\mathscr{H}t/\hbar}. \quad (1.32)$$

The summation over the set of states f can be removed by the completeness relation

$$\mathbf{1} = \sum_{\mathrm{f}} |\mathrm{f}\rangle\langle\mathrm{f}|, \tag{1.33}$$

giving

$$\chi''(\omega) = \frac{2\pi}{\hbar}(1 - e^{-\beta\hbar\omega}) \int_{-\infty}^{\infty} dt\, e^{-i\omega t} \sum_{\mathrm{i}} \rho_{\mathrm{i}} \langle \mathrm{i}|P_e(t)P_e|\mathrm{i}\rangle$$

or

$$\chi''(\omega) = \frac{2\pi}{\hbar}(1 - e^{-\beta\hbar\omega}) \int_{-\infty}^{\infty} dt\, e^{-i\omega t} \langle P_e(t)P_e\rangle, \tag{1.34}$$

where the time-correlation function of the dipole-moment generator is given by

$$\langle P_e(t)P_e\rangle = \sum_{\mathrm{i}} \rho_{\mathrm{i}} \langle \mathrm{i}|P_e(t)P_e|\mathrm{i}\rangle = \mathrm{Tr}[\rho_0 P_e(t)P_e]. \tag{1.35}$$

Here ρ_0 is the equilibrium density operator given by $\rho_0 = e^{-\beta\mathscr{H}}/Z$, $\mathscr{H}$ being the Hamiltonian of the material system, and $Z = \mathrm{Tr}\, e^{-\beta\mathscr{H}}$.

The absorption line-shape function $I(\omega)$ is defined as

$$I(\omega) = \frac{\hbar\chi''(\omega)}{4\pi^2(1 - e^{-\beta\hbar\omega})}. \tag{1.36}$$

Using Eq. (1.34) we can write $I(\omega)$ as

$$I(\omega) = \frac{1}{2\pi}\int_{-\infty}^{\infty} dt\, e^{-i\omega t}\langle P_e(t)P_e\rangle. \tag{1.37}$$

This is the desired result, since the line-shape function is now expressed as the Fourier transform of the time-correlation function of the dipole-moment operator of the absorbing system. The dipole-correlation function describes the spontaneous fluctuations of the electric dipole moments in the absence of the external field.

The time-correlation function given in Eq. (1.37) has the dipole-moment operator $P_e(t)$ at time t occurring on the left-hand side. Such a correlation function is said to be *left sided*. In general, the left-sided function is not equal to the right-sided one, with the time-dependent operator $P_e(t)$ appearing on the right-hand side. Using the property that the trace is invariant with cyclic permutation, we find

$$\langle P_e(t)P_e\rangle = \langle P_e P_e(-t)\rangle. \tag{1.38}$$

Because the line-shape function is a measurable quantity and hence is real, we obtain

$$I(\omega) = I^*(\omega) = \frac{1}{2\pi}\int_{-\infty}^{\infty} dt \langle P_e(t)P_e\rangle^* e^{+i\omega t}$$
$$= \frac{1}{2\pi}\int_{-\infty}^{\infty} dt \langle P_e(-t)P_e\rangle^* e^{-i\omega t}, \tag{1.39}$$

where in the second line we have changed the integration variable from t to $-t$. Comparing Eq. (1.39) with (1.37) gives

$$\langle P_e(t)P_e\rangle = \langle P_e(-t)P_e\rangle^* = \langle P_e P_e(t)\rangle^*, \tag{1.40}$$

or

$$\langle P_e P_e(-t)\rangle = \langle P_e P_e(t)\rangle^*. \tag{1.41}$$

Therefore, in general $\langle P_e(t)P_e\rangle$ does not equal $\langle P_e(-t)P_e\rangle$ unless the one-sided time-correlation function is real. If it is complex, then it can be separated into a real and an imaginary part. The real part of a one-sided correlation function is an even function of time, and the imaginary part is an odd function. When the time-correlation function is complex, the spectral line-shape function is not symmetrical.

In this section we have shown that the incorporation of the Fermi golden rule of quantum mechanics with statistics yields the time-correlation function as the end result. However, the calculation carried out is lengthy and involves unnecessary approximations. Next we shall develop a general linear-response theory to analyze problems associated with a system that responds dynamically to a time-dependent external field $\mathbf{F}(t)$. We shall see how the time-correlation function comes about naturally in this theory.

1.5 THE LIOUVILLE EQUATION OF MOTION

To apply the linear-response theory, we need the Liouville equation, which we shall therefore derive. We shall begin with a classical system; a quantum system will be treated later.

The Classical-Mechanics Approach

In classical mechanics, although the exact motion of any system is completely determined by the initial conditions, it is impractical to obtain an exact solution for a complex system consisting of 10^{23} interacting molecules. In statistical mechanics, no attempt is made to obtain a complete solution for such a complex system; instead, an effort is made to predict certain average properties that are observed macroscopically. This is done by computing the values of the desired quantities by forming an average over all systems in an ensemble.

Each system in the ensemble is represented by a single point in the phase space. Thus an entire ensemble appears as a cloud of points in phase space. As time evolves, each phase point representing a system will trace out its trajectory. The trajectories of all systems are independent since each phase point represents an isolated system.

Because the points form a dense cloud, it is convenient to describe the behavior of such points in terms of a density function f such that the number of systems in the ensemble corresponding to phase points in $d\{\rho\}\, d\{q\}$ about the point $(\{p\}, \{q\})$ at time t is given by $f(\{p\}, \{q\}; t)\, d\{p\}\, d\{q\}$. Clearly, we must have

$$\int d\{p\}\, d\{q\}\; f(\{p\}, \{q\}; t) = S, \tag{1.42}$$

where S is the total number of systems in the ensemble. The terms $\{p\}$ and $\{q\}$ represent the sets of generalized momenta and conjugated coordinates that specify a phase point. For brevity of notation, we shall simply write p for $\{p\}$ and q for $\{q\}$. The density function is thus represented by $f(p, q; t)$.

The ensemble average of any dynamic function represented by $\phi(p, q)$ is given by

$$\bar{\phi}(t) = \frac{1}{S}\int dp\, dq\; \phi(p, q) f(p, q; t) = \int dp\, dq\; \rho(p, q)\phi(p, q), \tag{1.43}$$

where $\rho(p, q; t)$ is the normalized density function and is known as the distribution function. According to Eq. (1.42), it is clear that

$$\int dp\, dq\; \rho(p, q; t) = 1. \tag{1.44}$$

The distribution function $\rho(p, q; t)$ can vary with time as the phase points wander through phase space. Since the total number of phase points is conserved, we have

$$\int dp\, dq\; d\rho(p, q; t)/dt = 0. \tag{1.45}$$

Thus the distribution function $\rho(p, q; t)$ remains constant in time; that is, the integrand in Eq. (1.45) is equal to zero. We can write

$$\frac{d}{dt}\rho(p, q; t) = 0 = \frac{\partial\rho}{\partial t} + \sum_i \frac{\partial\rho}{\partial q_i}\dot{q}_i + \frac{\partial\rho}{\partial P_i}\dot{p}_i$$

or

$$\frac{\partial\rho}{\partial t} = -\sum_i \left(\frac{\partial\rho}{\partial q_i}\dot{q}_i + \frac{\partial\rho}{\partial p_i}\dot{p}_i\right). \tag{1.46}$$

Then, according to Hamilton's equations of motion,

$$\dot{q}_i = \partial\mathscr{H}/\partial p_i \qquad \text{and} \qquad \dot{p}_i = -\partial\mathscr{H}/\partial q_i, \tag{1.47}$$

with $i = 1, 2, \ldots, l$, where l is the number of generalized coordinates that completely specify the state of the body system, we can rewrite Eq. (1.46) as

$$\frac{\partial\rho}{\partial t} = -\sum_i \left(\frac{\partial\mathscr{H}}{\partial p_i}\frac{\partial\rho}{\partial q_i} - \frac{\partial\mathscr{H}}{\partial q_i}\frac{\partial\rho}{\partial p_i}\right) = -\{\mathscr{H}, \rho\}, \tag{1.48}$$

where the curly braces define the Poisson bracket.

Equation (1.48) is the Liouville equation. It is a linear partial-differential equation that governs the evolution of the distribution function $\rho(p, q; t)$. The solution of Eq. (1.48) is subject to the initial condition $\rho(t = 0) = \rho_0$. If at $t = 0$ the system is in thermal equilibrium, then we have ρ_0 equal to $\rho_e = (1/Z)e^{-\beta\mathscr{H}}$, where Z is the partition function of the canonical ensemble $\beta = 1/kT$.

The Liouville equation can be formally written as

$$\partial\rho/\partial t = -i\mathscr{L}\rho, \tag{1.49}$$

where the linear operator $\mathscr{L}$, defined as

$$\mathscr{L} = -i\sum_j \left(\frac{\partial\mathscr{H}}{\partial p_j}\frac{\partial}{\partial q_j} - \frac{\partial\mathscr{H}}{\partial q_j}\frac{\partial}{\partial p_j}\right) \equiv -i\left(\frac{\partial\mathscr{H}}{\partial p}\frac{\partial}{\partial q} - \frac{\partial\mathscr{H}}{\partial q}\frac{\partial}{\partial p}\right), \tag{1.50}$$

is known as the Liouville operator.

We now examine some important properties of the Liouville operator. The study is quite similar to that using quantum-mechanical operators and will thus be given in a minimum of detail. We first show that $\mathscr{L}$ is a Hermitian operator.

Consider the integral

$$\mathscr{L}_{mn} = \int dp\, dq\, g_m^*(\mathscr{L}g_n), \tag{1.51}$$

where g_m and g_n are arbitrary functions of p and q that can be integrated over phase space. The condition for hermiticity is

$$\mathscr{L}_{mn} = \mathscr{L}_{nm}^*, \tag{1.52}$$

where the asterisk indicates the complex conjugate.

According to the definition of the Liouville operator, Eq. (1.51) is given explicitly by

$$\mathscr{L}_{mn} = -i \int dp\, dq\, g_m^* \left(\frac{\partial \mathscr{H}}{\partial p} \frac{\partial}{\partial q} - \frac{\partial \mathscr{H}}{\partial q} \frac{\partial}{\partial p} \right) g_n. \tag{1.53}$$

We then integrate by parts, recalling that g_m and g_n vanish at the upper and lower limits, to obtain

$$\mathscr{L}_{mn} = i \int dp\, dq\, g_n \left(\frac{\partial \mathscr{H}}{\partial p} \frac{\partial}{\partial q} - \frac{\partial \mathscr{H}}{\partial q} \frac{\partial}{\partial p} \right) g_m^* = \mathscr{L}_{nm}^*. \tag{1.54}$$

This completes the proof. It is clear that the introduction of i in Eq. (1.50) makes $\mathscr{L}$ a Hermitian operator.

It then follows that the eigenvalues of a Hermitian operator are real, and the eigenfunctions associated with different eigenvalues are orthogonal. Moreover, the eigenfunctions of the Hermitian operator form a complete set. Thus the distribution function can be expanded in terms of the eigenfunction of the Liouville operator.

We shall next show that the eigenfunctions of the Liouville operator are complex. If ψ_κ is the eigenfunction of $\mathscr{L}$ with eigenvalue λ_κ, then $\mathscr{L}\psi_\kappa = \lambda_\kappa \psi_\kappa$, We take the complex conjugate of this equation and obtain

$$\mathscr{L}^* \psi_k^* = \lambda_k \psi_k^*. \tag{1.55}$$

But from Eq. (1.50) we have

$$\mathscr{L}^* = -\mathscr{L}; \tag{1.56}$$

thus Eq. (1.55) becomes

$$\mathscr{L} \psi_k^* = -\lambda_k \psi_k^*. \tag{1.57}$$

Therefore we have shown that, if ψ_k is an eigenfunction of $\mathscr{L}$ with eigenvalue λ_k, then ψ_k^* is also an eigenfunction of $\mathscr{L}$ with eigenvalue $-\lambda_k$. The fact that both λ_k and $-\lambda_k$ are eigenvalues of $\mathscr{L}$ is a direct consequence of the dynamic reversibility of classical mechanics. Indeed, as shown in Eq. (1.54), taking the complex conjugate of $\mathscr{L}$ is equivalent to a reversal of time. A time reversal changes p to $-p$ and thus changes the sign of $\mathscr{L}$.

DENSITY MATRIX AND THE QUANTUM-MECHANICAL LIOUVILLE EQUATION

We now turn to a discussion of general equations of a density matrix, which are very useful in treating a quantum-mechanical system. We begin by considering a system described by a wave function $\psi(t)$ at some instant of time, and ask for the expectation value $\langle \mathcal{O} \rangle$ of some operator. We then have

$$\langle \mathcal{O}(t) \rangle = \langle \psi(t) | \mathcal{O} | \psi(t) \rangle, \tag{1.58}$$

where $\psi(t)$ may not be an eigenstate of the Hamiltonian. Suppose we now expand $\psi(t)$ in a complete set of orthonormal functions U_n, which are independent of time, to obtain

$$\psi(t) = \sum_n C_n(t) U_n. \tag{1.59}$$

In terms of the functions U_n, $\langle \mathcal{O}(t) \rangle$ can be written as

$$\langle \mathcal{O}(t) \rangle = \sum_{n,m} C_m^*(t) C_n(t) \langle m | \mathcal{O} | n \rangle. \tag{1.60}$$

If the wave function is changed from $\psi(t)$ to some other wave function, then the expectation value $\langle \mathcal{O}(t) \rangle$ will differ because the coefficients $C_m^*(t)C_n(t)$ will change with time even though the matrix elements $\langle m | \mathcal{O} | n \rangle$ remain the same. We note that in calculating the expectation at some instant t, we need to know all the C_n coefficients. However, as seen in Eq. (1.60), the C coefficients always occur in the form of products, and thus it would be useful to calculate the product instead of the individual coefficients. Moreover, to compare with the experiment, we are interested in calculating the average expectation value for an ensemble of macroscopically identical systems. The product of coefficients $C_n(t)C_m^*(t)$ will vary from system to system. Thus, we need to perform an ensemble average over the C coefficients. We now denote the ensemble average by an overbar on the C coefficients, and Eq. (1.58) becomes

$$\overline{\langle \mathcal{O}(t) \rangle} = \sum_{m,n} \overline{C_n(t) C_m^*(t)} \langle m | \mathcal{O} | n \rangle. \tag{1.61}$$

The quantities $\overline{C_n(t)C_m^*(t)}$ form the density matrix. We shall consider it to be the matrix of an operator ρ, defined by the equation

$$\langle n | \rho | m \rangle = \overline{C_n C_m^*}. \tag{1.62}$$

Thus the ensemble average of the expectation value of $\mathcal{O}$ is given by

$$\begin{aligned} \overline{\langle \mathcal{O}(t) \rangle} &= \sum_{m,n} \langle n | \rho(t) | m \rangle \langle m | \mathcal{O} | n \rangle = \sum_n \langle n | \rho(t) \mathcal{O} | n \rangle \\ &= \mathrm{Tr}\, \rho(t)\mathcal{O} = \mathrm{Tr}\, \mathcal{O}\rho(t), \end{aligned} \tag{1.63}$$

where in obtaining Eq. (1.63) we used the closure relation

$$\sum_m |m\rangle\langle m| = \mathbf{I}. \tag{1.64}$$

In the future, we shall omit the bar indicating the ensemble average to simplify the notation, but note that whenever the density operator $\rho(t)$ is used, an ensemble average is intended.

The equation of motion governing $\rho(t)$ can be derived from the wave function $\psi(t)$, which develops in time according to the time-dependent Schrödinger equation

$$\mathscr{H}\psi(t) = i\hbar\, \partial\psi(t)/\partial t. \tag{1.65}$$

Using Eq. (1.60), we obtain

$$\sum_n C_n(t)\mathscr{H}U_n = i\hbar \sum_n \frac{\partial C_n(t)}{\partial t} U_n. \tag{1.66}$$

To find the equation that $C_n(t)$ obeys, we multiply both sides of Eq. (1.66) by U_m^* and then integrate. Using the fact that the U_n functions are orthonormal, we obtain

$$i\hbar \frac{\partial C_m(t)}{\partial t} = \sum_n C_n(t)\langle m|\mathscr{H}|n\rangle, \tag{1.67}$$

which is the basic equation for the time-dependent perturbation theory. We now use this equation to obtain the differential equation that the density matrix obeys. Since

$$\partial/\partial t\langle n|\rho|m\rangle = \langle n|\,\partial\rho/\partial t\,|m\rangle = \overline{\partial C_n C_m^*/\partial t} = \overline{C_n(\partial C_m^*/\partial t)} + \overline{(\partial C_n/\partial t)C_m^*}. \tag{1.68}$$

substituting Eq. (1.67) gives

$$\begin{aligned}\left\langle n\left|\frac{\partial\rho}{\partial t}\right|m\right\rangle &= \frac{1}{\hbar}\sum_k [\overline{C_n C_k^*}\langle m|\mathscr{H}|k\rangle^* - \overline{C_k C_m^*}\langle n|\mathscr{H}|k\rangle] \\ &= \frac{i}{\hbar}\sum_k [\langle n|\rho|k\rangle\langle k|\mathscr{H}|m\rangle - \langle k|\rho|m\rangle\langle n|\mathscr{H}|k\rangle] \\ &= \frac{i}{\hbar}\langle n|\rho\mathscr{H} - \mathscr{H}\rho|m\rangle. \end{aligned} \tag{1.69}$$

We have used the closure relation Eq. (1.64) to obtain the last step. In terms of the operator form, Eq. (1.69) becomes

$$\partial\rho(t)/\partial t = (1/\hbar)[\rho(t), \mathscr{H}] = -(i/\hbar)[\mathscr{H}, \rho(t)]. \tag{1.70}$$

This equation describes the rate of change of the density operator at a fixed point in space and is the quantum-mechanical form of Liouville's equation. The density operator is the quantum-mechanical equivalent of the distribution function for points in phase space [Eq. (1.48)].

The solution to Eq. (1.70) in the event that $\mathscr{H}$ is independent of time is

$$\rho(t) = e^{-i\mathscr{H}t/\hbar}\rho(0)e^{i\mathscr{H}t/\hbar}, \tag{1.71}$$

where $\rho(0)$ is the density operator at $t = 0$. However, if the Hamiltonian is dependent on time, then there is no simple solution to Eq. (1.70). In this case we must proceed with care to examine the detailed time-dependent behavior in $\mathscr{H}$ to obtain an approximate solution.

Consider now a system at time $t = 0$ to be in thermal equilibrium at temperature T. If we take U_n to be the eigenstates of the Hamiltonian $\mathscr{H}$, then the diagonal matrix element of the density operator $\rho_{nn}(0)$ would correspond to the Boltzmann population factor of this state, given by

$$\langle n|\rho(0)|n\rangle = \overline{C_n C_n^*} = e^{-\beta E_n}/Z, \tag{1.72}$$

where Z is the canonical partition function and $\beta = 1/kT$.

From Eq. (1.72) it follows immediately that $\rho(0)$ is given by

$$\rho(0) = e^{-\beta\mathscr{H}}/Z = \rho_e, \tag{1.73}$$

which is the density operator describing an equilibrium ensemble. Equation (1.73), together with the quantum-mechanical Liouville equation given by Eq. (1.70), completely specifies the time development of a statistical system.

The formal solution of the quantum-mechanical Liouville equation given in the form of Eq. (1.71) suggests that $\rho(t)$ may be regarded as a result of the transformation of the density operator at the initial time,

$$\rho(t) = G(t)\rho(0). \tag{1.74}$$

In this manner, the complex nature of the many-body problem is contained in $G(t)$, and we need only to focus our attention on the property of the time-evolution operator $G(t)$. Formulation of the problem in this manner is possible. This is begun by defining superoperator $a^{\times}$, which when acted on by b gives

$$a^{\times}b = [a, b]. \tag{1.75}$$

From Eq. (1.75) it then follows that

$$\exp(a^{\times})b = \exp(a)b\exp(-a). \tag{1.76}$$

Equation (1.76) can readily be proved by a series expansion of the operator

$$\exp(a^{\times}) = 1 + a^{\times} + \tfrac{1}{2}(a^{\times})^2 + \cdots.$$

Using Eq. (1.75) we can rewrite the quantum-mechanical Liouville equation as

$$\partial\rho(t)/\partial t = -(i/\hbar)\mathscr{H}^{\times}\rho(t) = -i\mathscr{L}\rho(t), \tag{1.77}$$

where $\mathscr{L}$ is called the von Neumann–Liouville operator, defined as

$$\mathscr{L} = \mathscr{H}^{\times}/\hbar. \tag{1.78}$$

The formal solution of Eq. (1.77) is

$$\rho(t) = \exp(-i\mathscr{L}t)\rho(0) = \exp(-i\mathscr{H}^{\times}t/\hbar)\rho(0) \tag{1.79}$$

which, according to Eq. (1.76), is equivalent to

$$\rho(t) = \exp(-i\mathscr{H}t/\hbar)\rho(0)\exp(i\mathscr{H}t/\hbar), \tag{1.80}$$

in complete agreement with Eq. (1.71).

It is clear that the operator $\exp(-i\mathscr{L}t)$ plays the role of the time-evolution operator $G(t)$. The introduction of the von Neumann–Liouville operator makes it possible to express the equation of motion for the density operator in a form identical to that for the classical distribution function. Further calculation with respect to classical and quantum systems can therefore proceed in a unified manner.

1.6 LINEAR-RESPONSE THEORY

We have formulated the equations of motion for the classical distribution function and for the quantum-mechanical density operator. We shall now analyze the dynamic response of the system to an external field $F(t)$ that varies over time in a prescribed fashion. Our calculation procedure corresponds closely to the procedure for a spectroscopic experiment. For example, in an absorption experiment, we apply an external radiation such as an infrared pulse to the system from time $t = 0$ to $t = t$, and we want to discover what happens to the system at time t by monitoring the intensity of the radiation. From this we can obtain information about the properties of the system.

Thus, we consider a system of N interacting molecules subject to an external time-dependent field $\mathbf{F}(t)$. Before the application of the external field, the Hamiltonian for the system is given by $\mathscr{H}$. The external field couples to the dynamic variable A and causes a perturbation represented by the perturbation Hamiltonian

$$\mathscr{H}_1(t) = -\mathbf{A}\cdot\mathbf{F}(t). \tag{1.81}$$

We are interested in calculating the response of the system when the perturbation is weak, so that the response calculated includes only terms linear in the external field. This linear approximation is known as the linear-response theory (Kubo, 1957). The calculation proceeds as follows.

The evolution of the system is described by the Liouville equation

$$\partial\rho(t)/\partial t = -i[\mathscr{L} + \mathscr{L}_1(t)]\rho(t), \tag{1.82}$$

subject to the initial condition

$$\rho(0) = e^{-\beta\mathscr{H}}/Z, \tag{1.83}$$

where, for a quantum-mechanical system, $\mathscr{L}$ is equal to $\mathscr{H}^{\times}/\hbar$ and $\mathscr{L}_1(t) = -F(t)A^{\times}/\hbar$. For simplicity, we shall consider $F(t)$ to be a scale field in the following derivation. For a classical mechanical system,

$$\mathscr{L} = -i\sum_j\left(\frac{\partial\mathscr{H}}{\partial p_j}\frac{\partial}{\partial q_j} - \frac{\partial\mathscr{H}}{\partial q_j}\frac{\partial}{\partial p_j}\right),$$

$$\mathscr{L}_1(t) = -i\sum_j\left(\frac{\partial\mathscr{H}_1(t)}{\partial p_j}\frac{\partial}{\partial q_j} - \frac{\partial\mathscr{H}_1(t)}{dq_j}\frac{\partial}{\partial p_j}\right).$$

If $\mathscr{L}_1(t)$ is zero, then the solution to Eq. (1.82) is equal to

$$\rho(t) = e^{-i\mathscr{L}t}\rho(0). \tag{1.84}$$

Thus, it is useful to define a quantity $\rho^*(t)$ (here the asterisk does not denote complex conjugate but rather, denotes a transformed density operator or a distribution function) by the equation

$$\rho(t) = e^{-i\mathscr{L}t}\rho^*(t). \tag{1.85}$$

Substituting Eq. (1.85) into (q.82) gives an equation for $\rho^*(t)$ as

$$e^{-i\mathscr{L}t}\,\partial\rho^*(t)/\partial t = -i\mathscr{L}_1(t)e^{-i\mathscr{L}t}\rho^*(t). \tag{1.86}$$

Multiplication of $e^{i\mathscr{L}t}$ from the left yields an equation for $\rho^*(t)$,

$$\partial\rho^*(t)/\partial t = -i\mathscr{L}_1^*(t)\rho^*(t), \tag{1.87}$$

where

$$\mathscr{L}_1^*(t) = e^{i\mathscr{L}t}\mathscr{L}_1(t)e^{-i\mathscr{L}t}. \tag{1.88}$$

The transformation to $\mathscr{L}_1^*(t)$ is canonical. The new representation (i.e., with the asterisk) is known as the interaction representation.

Equation (1.87) can be converted into the integral equation

$$\rho^*(t) = \rho^*(0) - i\int_0^t dt'\,\mathscr{L}_1^*(t')\rho^*(t'). \tag{1.89}$$

This integral equation can be solved by an iteration procedure. If $\rho^*(t')$ in the integral is replaced by $\rho^*(0)$, we obtain a first order solution:

$$\rho^*(t) = \rho^*(0) = i \int_0^t dt' \mathscr{L}_1^*(t')\rho^*(0); \tag{1.90}$$

and from Eq. (1.85) we obtain

$$\rho(t) = \rho_e - i \int_0^t dt' \exp[-i\mathscr{L}(t - t')]\mathscr{L}_1(t')\rho_e. \tag{1.91}$$

If the response of the system is observed through a physical quantity corresponding to a dynamic variable B, then

$$\begin{aligned} \langle \Delta B(t) \rangle &= \mathrm{Tr}\{B[\rho(t) - \rho_e]\} \\ &= -i \int_0^t dt' \, \mathrm{Tr}\{B \exp[-i\mathscr{L}(t - t^{i})]\mathscr{L}_1(t')\rho_e\}, \end{aligned} \tag{1.92}$$

where in the quantum-mechanical case the symbol Tr indicates taking the trace with respect to the quantity enclosed in the bracket. In the classical case, Tr denotes integration over the phase-space variables p and q. To be explicit, in the quantal case

$$\mathrm{Tr}\, \mathscr{O} = \sum_m \langle m | \mathscr{O} | m \rangle, \tag{1.93}$$

and in the classical case

$$\mathrm{Tr}\, \mathscr{O} = \int dp\, dq\, \mathscr{O}(p, q) \equiv \int d\Gamma\, \mathscr{O}(\Gamma), \tag{1.94}$$

where Γ denotes the phase-space variables $\{p, q\}$.

Because the trace is invariant to cyclic permutation, Eq. (1.92) can be rewritten as

$$\langle \Delta B(t) \rangle = -i \int_0^t dt' \, \mathrm{Tr}\{B(t - t')\mathscr{L}_1(t')\rho_e\}, \tag{1.95}$$

where $B(t)$ is the Heisenberg operator of B given by

$$B(t) = e^{i\mathscr{L}t}B = e^{i\mathscr{H}t/\hbar}Be^{-i\mathscr{H}t/\hbar}. \tag{1.96}$$

Equation (1.95), together with Eq. (1.96), is the basic result in the linear-response theory.

A few simple manipulations involving the property of the trace reduces Eq. (1.95) to

$$\langle \Delta B(t) \rangle = \int_0^t dt' \, \phi_{BA}(t - t')F(t') \tag{1.97}$$

where $\phi_{BA}(t)$ is known as the response function. In the quantal case it is given by

$$\phi_{BA}(t) = (i/\hbar)\,\mathrm{Tr}[B(t)(A, \rho_e)] = (i/\hbar)\langle[B(t), A]\rangle. \tag{1.98}$$

In the classical case it is given by

$$\phi_{BA}(t) = \langle\{B(t), A\}_{\mathrm{PB}}\rangle. \tag{1.99}$$

The subscript PB denotes Poisson's bracket.

The response function depends on time and varies according to the property of the medium. As will be shown later, $\phi_{BA}(t)$ is finite for all values of t. It goes to zero as $t \to \infty$. The range of values in which the response function differs appreciably from zero is the order of the relaxation time that characterizes the microscopic response $\langle\Delta B(t)\rangle$. The response function is proportional to the response $\langle\Delta B(t)\rangle$ when the system is subject to a δ-function type of an impulse.

The response function may also be expressed in different forms. Using the identity

$$[A, \rho_e] = \rho_e \int_0^\beta d\lambda\, e^{\lambda\mathscr{H}}[\mathscr{H}, A]e^{-\lambda\mathscr{H}} \tag{1.100}$$

and changing the variable from λ to τ by letting $\lambda = i\tau/\hbar$, Eq. (1.100) is reduced to

$$[A, \rho_e] = \frac{i}{\hbar}\rho_e \int_0^{\hbar\beta/i} d\tau\,[\mathscr{H}, A(\tau)] = \rho_e \int_0^{\hbar\beta/i} d\tau\,\frac{\partial}{\partial\tau}A(\tau) = \rho_e\left[A\left(\frac{\hbar\beta}{i}\right) - A\right]. \tag{1.101}$$

Substituting Eq. (1.101) into (1.98) gives

$$\phi_{BA}(t) = (i/\hbar)[\langle A(\hbar\beta/i)B(t)\rangle - \langle AB(t)\rangle]. \tag{1.102}$$

If the temperature is high, then we may expand $A(\hbar\beta/i)$ in a power series of $\hbar\beta/i$,

$$A(\hbar\beta/i) = A + (\hbar\beta/i)\dot{A} + \cdots, \tag{1.103}$$

and obtain

$$\phi_{BA}(t) = \beta\langle\dot{A}B(t)\rangle + \cdots. \tag{1.104}$$

This high-temperature limit corresponds to the classical result. Recall that the classical response function can be written as

$$\phi_{BA}(t) = \langle\{B(t), A\}_{\mathrm{PB}}\rangle = \int d\Gamma\, B(t)\{A, \rho_e\}_{\mathrm{PB}}. \tag{1.105}$$

The Poisson bracket $\{A, \rho_e\}_{\mathbf{PB}}$ is given by

$$\{A, \rho_e\}_{\mathbf{PB}} = \frac{\partial A}{\partial p}\frac{\partial \rho_e}{\partial q} - \frac{\partial A}{\partial q}\frac{\partial \rho_e}{\partial p} = -\beta\{A, \mathscr{H}\}_{\mathbf{PB}}\rho_e = \beta\dot{A}\rho_e. \quad (1.106)$$

Thus,

$$\phi_{BA}(t) = \beta \int d\Gamma\; B(t)\dot{A}\rho_e = \beta\langle B(t)\dot{A}\rangle = -\beta\langle \dot{B}(t)A\rangle. \quad (1.107)$$

The last step follows from the stationarity condition for an ensemble average.

As indicated in Eq. (1.97), the response induced by the external field lags behind the field. Values of the field applied at times earlier than t affect the response at time t. The response function represents the time lag of the system. In the case when the system could respond instantaneously to $F(t)$, the response function $\phi_{BA}(t)$ would be a δ function. Ordinarily, however, because of the finite time needed for molecules to respond to the external field, $\phi_{BA}(t)$ does not change with time instantaneously; it takes time for the system to respond to the field after it has been applied.

The integral as given in Eq. (1.97) is of a convolution form. Taking the complex Laplace transform of Eq. (1.97) we obtain

$$\langle \Delta B(z)\rangle = \phi_{BA}(z)F(z), \quad (1.108)$$

where

$$\langle \Delta B(z)\rangle = \int_0^\infty dt\; \langle \Delta B(t)\rangle e^{-izt} \quad (1.109)$$

and

$$F(z) = \int_0^\infty dt\; e^{-izt}F(t). \quad (1.110)$$

Most important, $\chi_{BA}(z)$ is given by

$$\chi_{BA}(z) = \int_0^\infty dt\; e^{-izt}\phi_{BA}(t). \quad (1.111)$$

The term $\chi_{BA}(z)$ is known as the generalized susceptibility of the system. The complex frequency z must lie in the lower half of the complex plane for the integrals in Eqs. (1.109)–(1.111) to converge.

It is convenient to introduce a function $R_{BA}(t)$ that is related to the response function $\phi_{BA}(t)$ by

$$R_{BA}(t) = \int_t^\infty dt'\; \phi_{BA}(t'), \qquad \text{or} \qquad \partial R_{BA}(t)/\partial t = -\phi_{BA}(t), \quad (1.112)$$

where $R_{BA}(t)$ is called the relaxation function. The relaxation function can be considered as the response $\langle \Delta B(t)\rangle$ when the system is subject to a step field $F(\tau)$, for which it is zero for $\tau \leq t$ and is constant for $\tau > t$. Using this definition to express $\phi_{BA}(t)$ in terms of $R_{BA}(t)$ and then substituting the result into Eq. (1.112), we obtain the generalized susceptibility $\chi_{BA}(z)$ as

$$\chi_{BA}(z) = -\int_0^\infty dt\, e^{-izt} \frac{\partial R_{BA}(t)}{\partial t}, \tag{1.113}$$

which, on integration by parts, gives

$$\chi_{BA}(z) = R_{BA}(0) - iz \int_0^\infty dt\, R_{BA}(t) e^{-izt}, \tag{1.114}$$

where $R_{BA}(0)$ is the area under the response function $\phi_{BA}(t)$ from $t = 0$ to $t = \infty$. Note $\chi_{BA}(0) = R_{BA}(0)$.

We now wish to prove a theorem relating the real part χ' and the imaginary part χ'' of the generalized susceptibility, namely, the Kramers–Kronig theorem.

Since z is a complex variable, the real part of z will be the frequency ω, and the imaginary part is introduced to guarantee the convergence of the integrals. For convenience, we use the symbol x for ω to make the formulas more familiar, and write $z = x + iy$, where $y \leq 0$. Thus,

$$\chi_{BA}(z) = R_{BA}(0) - iz \int_0^\infty dt\, R_{BA}(t) e^{-itx+yt}. \tag{1.115}$$

We want to show next that $\chi_{BA}(z) - R_{BA}(0)$ is an analytic function of z. To prove this, we apply the Cauchy derivative test, which says that if

$$\chi_{BA}(z) - R_{BA}(0) = u + iv, \tag{1.116}$$

where u and v are real, then u and v must satisfy the equations

$$\partial u/\partial x = \partial v/\partial y \qquad \text{and} \qquad \partial u/\partial y = -\partial v/\partial x. \tag{1.117}$$

From Eq. (1.115)

$$u = \int_0^\infty dt\, (y \cos xt - x \sin xt) R_{BA}(t) e^{yt}, \tag{1.118}$$

$$v = -\int_0^\infty dt\, (x \cos xt + y \sin xt) R_{BA}(t) e^{yt}, \tag{1.119}$$

which give

$$\frac{\partial u}{\partial x} = -\int_0^\infty dt\, [t(y \sin xt + x \cos xt) + \sin xt] e^{yt} R_{BA}(t) = \frac{\partial v}{\partial y}, \tag{1.120}$$

$$\frac{\partial u}{\partial y} = \int_0^\infty dt\, [\cos xt + t(y \cos xt - x \sin xt)] e^{yt} R_{BA}(t) = -\frac{\partial v}{\partial x}. \tag{1.121}$$

Therefore, the Cauchy relations are satisfied.

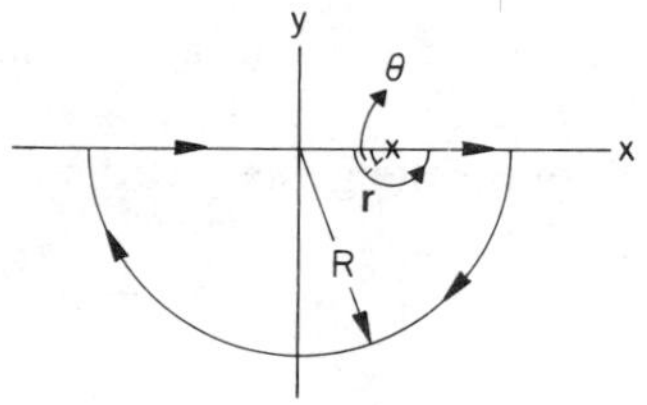

FIG. 1.2 Contour integration for the integral given by Eq. (1.222).

The presence of the term e^{yt} indicates that

$$|\chi_{BA}(z) - R_{BA}(0)| \to 0 \qquad \text{as} \quad y \to -\infty.$$

Furthermore, since $R_{BA}(t)$ is real and finite,

$$|\chi_{BA}(z) - R_{BA}(0)| \to 0 \qquad \text{as} \quad x \to \pm\infty.$$

Therefore, $\chi_{BA}(z) - R_{BA}(0)$ is a function that is analytic for $y \leq 0$ and goes to zero as $|z| \to \infty$ in the lower half of the complex plane.

Let us consider a contour integral along the path of Fig. 1.2 for the function $[\chi_{BA}(z') - R_{BA}(0)]/(z' - \omega)$. By Cauchy's theorem this integral vanishes, since the function $\chi_{BA}(z)$ has no pole inside the contour:

$$\oint dz' \frac{\chi_{BA}(z') - R_{BA}(0)}{z' - \omega} = 0. \tag{1.122}$$

Since $\chi_{BA}(z') - R_{BA}(0)$ goes to zero on the large circle of radius R, that part of the integral makes no contribution. The remaining contribution comes from the real axis and that which is on the semicircle $z' - \omega = re^{i\theta}$. Thus

$$\int_{-\infty}^{\omega - r} dx \frac{\chi_{BA}(x) - R_{BA}(0)}{x - \omega} + \int_{\pi}^{2\pi} d\theta \frac{\chi_{BA}(\omega) - R_{BA}(0)}{re^{i\theta}} rie^{i\theta}$$
$$+ \int_{\omega + r}^{\infty} dx \frac{\chi_{BA}(x) - R_{BA}(0)}{x - \omega} = 0.$$

Therefore,

$$\pi i[\chi_{BA}(\omega) - R_{BA}(0)] = \mathrm{P} \int_{-\infty}^{\infty} dx \frac{\chi_{BA}(x) - R_{BA}(0)}{\omega - x}, \tag{1.123}$$

where the symbol P stands for taking the principal part of the integral, i.e., taking the limit of the sum of the integrals $\int_{-\infty}^{\omega - r}$ and $\int_{\omega + r}^{\infty}$ as $r \to 0$.

Since the susceptibility $\chi_{BA}(\omega)$ at the real frequency ω has a real and an imaginary part, we can write

$$\chi_{BA}(\omega) = \chi'_{BA}(\omega) - i\chi''_{BA}(\omega). \tag{1.124}$$

Equating real and imaginary parts on both sides of Eq. (1.123), we find [note that $R_{BA}(0) = \chi_{BA}(0)$]

$$\chi'_{BA}(\omega) - \chi_{BA}(0) = \frac{1}{\pi} \mathrm{P} \int_{-\infty}^{\infty} dx \, \frac{\chi''_{BA}(x)}{x - \omega}, \tag{1.125}$$

$$\chi''_{BA}(\omega) = \frac{1}{\pi} \mathrm{P} \int_{-\infty}^{\infty} dx \, \frac{\chi'_{BA}(x) - \chi_{BA}(0)}{\omega - x}. \tag{1.126}$$

These two equations are the famous Kramers–Kronig relations. The significance of these equations is that there are restrictions placed on $\chi'(\omega)$ and $\chi''(\omega)$. The knowledge of $\chi''(\omega)$ at all values of ω enables us to compute χ' at any frequency. This result is important in reflection spectroscopy, which measures $\chi'(\omega)$. The Kramers–Kronig relations then allow the $\chi''(\omega)$ to be calculated so that it can be compared with the absorption spectrum, which measures $\chi''(\omega)$.

If in Eq. (1.111) we replace z by ω, then it is obvious that the generalized susceptibility $\chi_{BA}(\omega)$ follows the symmetry relation

$$\chi_{BA}(-\omega) = \chi^{*}_{BA}(\omega). \tag{1.127}$$

From Eq. (1.127) it follows that

$$\chi'_{BA}(-\omega) = \chi'_{BA}(\omega), \tag{1.128}$$

$$\chi''_{BA}(-\omega) = -\chi''_{BA}(\omega). \tag{1.129}$$

Thus, $\chi'_{BA}(\omega)$ is an even function of frequency ω, and $\chi''(\omega)$ is an odd function of ω. For small frequencies, $\chi_{BA}(\omega)$ can be expanded as a power series of ω. The expansion of $\chi'_{BA}(\omega)$ includes only even powers of ω and that of $\chi''_{BA}(\omega)$ includes only odd powers of ω. In the limit $\omega \to 0$, $\chi_{BA}(\omega)$ yields a real static susceptibility $\chi_{BA}(0)$.

From Eq. (1.115) we can write

$$\chi'_{BA}(\omega) = \chi_{BA}(0) - \omega \int_{0}^{\infty} dt \, R_{BA}(t) \sin \omega t, \tag{1.130}$$

$$\chi''_{BA}(\omega) = \omega \int_{0}^{\infty} dt \, R_{BA}(t) \cos \omega t. \tag{1.131}$$

1.7 FLUCTUATION–DISSIPATION THEOREM

We now want to establish a connection between $\chi''(\omega)$ and the time-correlation function. This connection is given by the celebrated fluctuation–dissipation theorem, originally discussed by Nyquist (1928) as a relation between noise and dissipation in electric resistors (Bernard and Callen, 1959; Kubo, 1957).

The fluctuation–dissipation theorem establishes a relationship between the response of a driven system and the correlation of its equilibrium fluctuations. This relationship provides a fundamental basis for analysis of irreversible processes.

Consider a simple irreversible process in which an external force $F(t)$ is applied to the system. The system responds to the external force, causing some macroscopic property to deviate from its equilibrium value, and becomes a function of time. The driving force $F(t)$ and the response of the driven system $\Delta A(t)$ are paired as each type of force tends to induce its particular type of response. For example, if F is pressure, then the response $\Delta A(t)$ is the volume change; if F is the magnetic field, then the response $\Delta A(t)$ is the induced magnetization. Since the experimentally observed response corresponds to the ensemble average of the response in the linear-response theory, the response of the system is denoted by $\langle \Delta A(t) \rangle$ and is given by Eq. (1.97), if B is replaced by A.

The noise of the equilibrium system, or the equilibrium fluctuation of the macroscopic property associated with the dynamic variable $A(t)$, is characterized by the equilibrium correlation function $\langle A(\tau)A(t+\tau)\rangle$. The relationship between the linear response $\langle \Delta A(t)\rangle$ and the equilibrium time-correlation function $\langle A(\tau)(t+\tau)\rangle$ is given by the fluctuation–dissipation theorem.

To establish the theorem, we return to Eq. (1.112) and write the relaxation function as

$$R_{BA}(t) = \frac{i}{\hbar}\int_t^\infty dt' \operatorname{Tr}[B(t')(A, \rho_e)]. \tag{1.132}$$

The identity given in Eq. (1.101) permits us to rewrite Eq. (1.132) as

$$R_{BA}(t) = \left(\frac{i}{\hbar}\right)^2 \int_t^\infty dt' \int_0^{\hbar\beta/i} d\tau \operatorname{Tr}\{\rho_e[\mathscr{H}, A(\tau)]B(t')\}. \tag{1.133}$$

Since the trace of the product of operations is invariant with respect to a cyclic permutation, i.e., $\operatorname{Tr}(ABC) = \operatorname{Tr}(CAB) = \operatorname{Tr}(BCA)$, it is easy to show that

$$\begin{aligned}\operatorname{Tr}\{\rho_e[\mathscr{H}, A(\tau)]B(t')\} &= \operatorname{Tr}\{\rho_e A(\tau)[B(t'), \mathscr{H}]\} = -(\hbar/i)\operatorname{Tr}[\rho_e A(\tau)\,\partial B(t')/\partial t'] \\ &= -(\hbar/i)\langle A(t)\dot{B}(t')\rangle.\end{aligned} \tag{1.134}$$

Using Eq. (1.134), the relaxation function $R_{BA}(t)$ can be written as

$$R_{BA}(t) = -\frac{i}{\hbar}\int_t^{\infty} dt' \int_0^{\hbar\beta/i} d\tau \, \langle A(\tau)\dot{B}(t')\rangle. \tag{1.135}$$

Exchanging the order of integration and carrying out the t' integration, we obtain

$$R_{BA}(t) = \frac{i}{\hbar}\int_0^{\hbar\beta/i} d\tau \, \langle A(\tau)B(t)\rangle - \frac{i}{\hbar}\lim_{t\to\infty}\int_0^{\hbar\beta/i} d\tau \, \langle A(\tau)B(t)\rangle. \tag{1.136}$$

Consider for the moment the second term on the right-hand side of Eq. (1.136):

$$\begin{aligned}\lim_{t\to\infty}\frac{i}{\hbar}\int_0^{\hbar\beta/i} d\tau \, \langle A(\tau)B(t)\rangle &= \frac{i}{\hbar}\lim_{t\to\infty}\int_0^{\hbar\beta/i} d\tau \, \langle AB(t-\tau)\rangle \\ &= \beta\langle A\rangle\langle B\rangle. \end{aligned} \tag{1.137}$$

The last step in Eq. (1.137) follows because the dynamic variables A and B become uncorrelated as t becomes large compared with the characteristic time of the fluctuations.

Although in some cases the relaxation may asymptotically approach a finite value at long times, there will be no loss of generality in the subsequent discussion by setting the asymptotic value to zero. In this case, the relaxation function is given simply by

$$R_{BA}(t) = \frac{i}{\hbar}\int_0^{\hbar\beta/i} d\tau \, \langle A(\tau)B(t)\rangle. \tag{1.138}$$

If the temperature is high (or τ small), we can expand $A(\tau)$ in a power series of τ:

$$A(\tau) = A(0) + \tau\dot{A}(0) + \cdots. \tag{1.139}$$

Then $R_{BA}(t)$ becomes

$$R_{BA}(t) = \beta\langle AB(t)\rangle + \mathcal{O}(\hbar B). \tag{1.140}$$

Therefore, at high temperature, $R_{BA}(t)$ is proportional to the time-correlation function. It is easy to show that this high-temperature limit corresponds to the classical result.

As previously discussed, in a quantum-mechanical system, the left-sided correlation function of the observable A and B given by $\langle B(t)A\rangle$ is not equal to the right-sided correlation function $\langle AB(t)\rangle$; it is more appropriate to define the time-correlation function according to

$$C_{BA}(t) = \mathrm{Tr}\{\rho_e[B(t)A]_+\}, \tag{1.141}$$

where the bracket $[AB(t)]_+$ denotes a symmetrized product. We have

$$[AB(t)]_+ = \tfrac{1}{2}[AB(t) + B(t)A]. \tag{1.142}$$

Defining the Fourier transforms of $R_{BA}(t)$ and $C_{BA}(t)$ as

$$\hat{R}_{BA}(\omega) = \frac{1}{2\pi}\int_{-\infty}^{\infty} dt\, e^{-i\omega t} R_{BA}(t) \tag{1.143}$$

and

$$\hat{C}_{BA}(\omega) = \frac{1}{2\pi}\int_{-\infty}^{\infty} dt\, e^{-i\omega t} C_{BA}(t), \tag{1.144}$$

we now show that

$$\hat{C}_{BA}(\omega) = E_T(\omega)\hat{R}_{BA}(\omega), \tag{1.145}$$

where $E_T(\omega)$ is the average energy of an oscillator with frequency ω at temperature $T = 1/k\beta$, given by

$$E_T(\omega) = \tfrac{1}{2}\hbar\omega \coth \tfrac{1}{2}\beta\hbar\omega. \tag{1.146}$$

To derive Eq. (1.145), we first take the Fourier transform of the relaxation function $R_{BA}(t)$ and use Eq. (1.138) to obtain

$$\hat{R}_{BA}(\omega) = \frac{i}{2\pi\hbar}\int_{-\infty}^{\infty} dt\, e^{-i\omega t}\int_{0}^{\hbar\beta/i} d\tau\, \langle AB(t-\tau)\rangle. \tag{1.147}$$

Changing the integration variable by setting $t' = t - \tau$ and then interchanging the order of integration, we obtain

$$\hat{R}_{BA}(\omega) = \frac{i}{2\pi\hbar}\int_{0}^{\hbar\beta/i} d\tau\, e^{-i\omega\tau}\int_{-\infty-\tau}^{\infty-\tau} dt'\, \langle AB(t')\rangle e^{-i\omega t'} \tag{1.148}$$

Because the τ integration is finite, the variable τ appearing in the lower and upper limits of the second integral in this equation is inconsequential and can be dropped. As a result, the τ integration in the first integral can be carried out independently of the second. We carry out the τ integration and obtain

$$\hat{R}_{BA}(\omega) = \frac{1}{2\pi\hbar\omega}(1 - e^{-\beta\hbar\omega})\int_{-\infty}^{\infty} dt\, \langle AB(t)\rangle e^{-i\omega t}. \tag{1.149}$$

The complex conjugate of Eq. (1.149) is equal to

$$\hat{R}^*_{BA}(\omega) = \frac{1}{2\pi\hbar\omega}(1 - e^{-\beta\hbar\omega})\int_{-\infty}^{\infty} dt\, \langle B(t)A\rangle e^{i\omega t}. \tag{1.150}$$

But since the relaxation function is real, it follows from Eq. (1.143) that

$$\hat{R}^*_{BA}(\omega) = \hat{R}_{BA}(-\omega). \tag{1.151}$$

Using this result, we obtain from Eq. (1.150) another expression for $\hat{R}_{BA}(\omega)$ by replacing ω with $-\omega$.

$$\hat{R}_{BA}(\omega) = \hat{R}^*_{BA}(-\omega) = \frac{1}{2\pi\hbar\omega}\,(e^{\beta\hbar\omega} - 1)\int_{-\infty}^{\infty} dt\; e^{-i\omega t}\langle B(t)A\rangle. \tag{1.152}$$

Equations (1.149) and (1.152) can now be combined to yield

$$\begin{aligned}\frac{1}{2\pi}\int_{-\infty}^{\infty} dt\; e^{-i\omega t}\langle[AB(t)]_+\rangle &= \frac{\hbar\omega}{2}\left(\frac{1}{e^{\beta\hbar\omega} - 1} + \frac{1}{1 - e^{-\beta\hbar\omega}}\right)\hat{R}_{BA}(\omega)\\ &= \frac{\hbar\omega}{2}\coth\left(\frac{\beta\hbar\omega}{2}\right)\hat{R}_{BA}(\omega)\end{aligned}$$

or

$$\hat{C}_{BA}(\omega) = E_T(\omega)\hat{R}_{BA}(\omega). \tag{1.153}$$

This completes the proof of Eq. (1.145).

With the availability of Eq. (1.153), we can now discuss the fluctuation–dissipation theorem. So far we have considered the calculation of the linear response resulting from the interaction of a material system with an external field. Attention will now be given to calculation of the rate of change of the energy of the system resulting from application of the external field. Application of the field to the system originally at equilibrium will do work on the system and cause energy to be absorbed by the system. This energy gain by the system will eventually be dissipated as heat because of the fluctuations of translational and rotational degrees of freedom caused by molecular motion that takes place within the material system. The fluctuation–dissipation theorem will relate χ'', which describes the energy absorption, to the correlation function, which describes the fluctuations.

In the presence of an interaction with the external field, the total energy of the system is equal to $\langle\mathscr{E}(t)\rangle = \mathrm{Tr}[\mathscr{H}(t)\rho(t)]$. The rate of absorption of the energy from the field is equal to

$$\begin{aligned}\partial\langle\mathscr{E}(t)\rangle/\partial t &= \mathrm{Tr}\{[\partial\mathscr{H}(t)/\partial t]\rho(t) + \mathscr{H}(t)[\partial\rho(t)/\partial t]\}\\ &= \mathrm{Tr}\{[\partial\mathscr{H}(t)/\partial t]\rho(t)\} + (i/\hbar)\mathrm{Tr}\{\mathscr{H}(t)[\mathscr{H}(t), \rho(t)]\}.\end{aligned} \tag{1.154}$$

The second term on the right-hand side of Eq. (1.154) vanishes identically. Furthermore, since the total Hamiltonian $\mathscr{H}(t)$ is given by

$$\mathscr{H}(t) = \mathscr{H} - AF(t), \tag{1.155}$$

Eq. (1.154) becomes

$$\partial\langle\mathscr{E}(t)\rangle/\partial t = -[\partial F(t)/\partial t]\,\mathrm{Tr}[A\rho(t)]. \tag{1.156}$$

The linear-response solution for $\rho(t)$ is given by Eq. (1.91). Substituting $\rho(t)$, as given by Eq. (1.91), into Eq. (1.156) we obtain, after some simple manipulation,

$$\partial\langle\mathscr{E}(t)\rangle/\partial t = -\partial F(t)/\partial t \int_0^t dt'\, \phi_{AA}(t-t')F(t'), \tag{1.157}$$

where we have assumed $\mathrm{Tr}(\rho_e A) = 0$, which is valid in most cases. The response function $\phi_{AA}(t)$ is given by

$$\phi_{AA}(t) = (i/\hbar)\langle[A(t), A]\rangle, \tag{1.158}$$

which is equivalent to the expression given in Eq. (1.98) with B set equal to A.

Consider next that $F(t)$ is a monochromatic field given by

$$F(t) = \tfrac{1}{2}(Fe^{i\omega t} + F^*e^{-i\omega t}). \tag{1.159}$$

Substituting the expression for $F(t)$ into Eq. (1.157) yields

$$\partial\langle\mathscr{E}(t)\rangle/\partial t = -\tfrac{1}{4}(i\omega Fe^{i\omega t} - i\omega F^*e^{-i\omega t}) \int_0^t d\tau\, \phi_{AA}(\tau)[Fe^{i\omega(t-\tau)} + F^*e^{-i\omega(t-\tau)}]. \tag{1.160}$$

Since the response function $\phi_{AA}(\tau)$ has an appreciable value only for the time of the order of the relaxation time τ_R and, in general, only the response of the system after $t \gg \tau_R$ is of interest, we can replace the upper limit in the integral in Eq. (1.160) by infinity, since for $t \gg \tau_R$, $\phi_{AA}(\tau)$ will become vanishingly small. Setting the upper limit t as infinity, we find that Eq. (1.160) reduces to

$$\frac{\partial\langle\mathscr{E}(t)\rangle}{\partial t} = \frac{\omega}{2}|F|^2 \int_0^\infty d\tau\, \phi_{AA}(\tau)\sin\omega\tau + (\text{terms varying with } e^{\pm i2\omega t}). \tag{1.161}$$

The time-dependent parts can be discarded because they vary with frequencies $\pm 2\omega$ and do not contribute to the steady-state energy dissipation.

It is obvious that the quantity expressed by the integral is equal to $\chi''_{AA}(\omega)$. Hence,

$$\partial\langle\mathscr{E}(t)\rangle/\partial t = (\omega/2)|F|^2\chi''_{AA}(\omega). \tag{1.162}$$

This is an important result. It shows that the energy dissipation is proportional to the imaginary part of susceptibility $\chi''_{AA}(\omega)$.

From Eq. (1.131), $\chi''_{AA}(\omega)$ is given by

$$\begin{aligned}\chi''_{AA}(\omega) &= \omega\int_0^\infty dt\, R_{AA}(t)\cos\omega t = \frac{\omega}{2}\int_{-\infty}^\infty dt\, R_{AA}(t)e^{-i\omega t}\\ &= \pi\omega\hat{R}_{AA}(\omega).\end{aligned} \tag{1.163}$$

The last step is possible because $R_{AA}(t)$ is an even function of time.

To prove that $R_{AA}(t)$ is even is simple. From Eq. (1.138) we have

$$R_{AA}(-t) = \frac{i}{\hbar}\int_0^{\hbar\beta/i} d\tau\, \langle A(\tau)A(-t)\rangle. \tag{1.164}$$

Changing the integration variable from τ to $\hbar\beta/i - \tau'$, we obtain

$$R_{AA}(-t) = -\frac{i}{\hbar}\int_{\hbar\beta/i}^{0} d\tau' \left\langle A\left(\frac{\hbar\beta}{i} - \tau'\right)A(-t)\right\rangle. \tag{1.165}$$

Several steps of manipulation reduce Eq. (1.165) to

$$\begin{aligned} R_{AA}(-t) &= \frac{i}{\hbar}\int_0^{\hbar\beta/i} d\tau' \operatorname{Tr}\left\{\rho_e \exp\left[\frac{i\mathscr{H}}{h}\left(\frac{\hbar\beta}{i} - \tau'\right)\right]A \right. \\ &\qquad \left. \times \exp\left[-\frac{i\mathscr{H}}{\hbar}\left(\frac{\hbar\beta}{i} - \tau'\right)\right]A(-t)\right\} \\ &= \frac{i}{\hbar}\int_0^{\hbar\beta/i} d\tau' \operatorname{Tr}[\rho_e A(-t)A(-\tau')] \\ &= \frac{i}{\hbar}\int_0^{\hbar\beta/i} d\tau' \langle A(\tau')A(t)\rangle = R_{AA}(t), \end{aligned} \tag{1.166}$$

where the permutation property of the trace was used to obtain the second line.

Combining Eq. (1.163) with (1.153), we obtain the celebrated fluctuation–dissipation theorem:

$$\chi''_{AA}(\omega) = (2\pi/\hbar)\tanh(\beta\hbar\omega/2)\hat{C}_{AA}(\omega). \tag{1.167}$$

The fluctuation–dissipation theorem can be interpreted as follows. A dissipation process is present because of interaction between a signal generator, which provides the external force, and a material system of complex structure. On interaction with the complex material system, the energy of the excited signal generator flows into the material system and is eventually distributed among the many modes of the material system. The dissipation of the energy of the signal generator can be regarded as being caused by the random molecular motion of the material system that acts on the signal generator. Fluctuations in the dynamic variable resulting from random molecular motion thus cause energy dissipation of the signal generator, and the resultant relationship between the loss in energy of the signal generator and the fluctuations in the material system is expressed by the fluctuation–dissipation theorem.

The above discussion does not include the spatial dependence. The spatial dependence is important when considering the scattering processes in the condensed phase. Inclusion of the spatial dependence in linear response theory is given in the appendix. In the next section, we shall provide an example of the application of the fluctuation–dissipation theorem by considering again the absorption of infrared radiation in a system of electric dipoles in the liquid state.

1.8 ABSORPTION OF RADIATION BY A SYSTEM OF ELECTRIC DIPOLES

We assume that the electric field $\mathbf{E}(t)$ is spatially uniform and therefore disregard its spatial dependence. The interaction between the field and the dipoles is given by $\mathscr{H}_1(t) = -\mathbf{P} \cdot \mathbf{E}(t) = -P_e E(t)$, where $P_e = \mathbf{P} \cdot \hat{e}$, $\hat{e}$ being the unit vector representing the polarization of the electric field. Assuming that the dipole-moment operator $\mathbf{P}(t)$ of the entire system at time t is given by the superposition of the dipole-moment operators of individual molecules, we obtain

$$\mathbf{P}(t) = \sum_{i=1}^{N} \mathbf{P}_i(t), \tag{1.168}$$

where $\mathbf{P}_i$ is the dipole moment associated with molecule i.

For an isotropic fluid, the time-correlation function is equal to

$$C(t) = \langle P_e(t) P_e \rangle = \tfrac{1}{3}\langle \mathbf{P}(t) \cdot \mathbf{P} \rangle, \tag{1.169}$$

which is the result obtained after averaging the dipole moment over all directions with respect to $\hat{e}$.

Substituting Eq. (1.168) into (1.169) gives

$$\langle \mathbf{P} \cdot \mathbf{P}(t) \rangle = \sum_{i} \langle \mathbf{P}_i(t) \mathbf{P}_i \rangle + \sum_{i,j}{}' \langle \mathbf{P}_i(t) \cdot \mathbf{P}_j \rangle, \tag{1.170}$$

where the prime appearing in the second term indicates that terms with $i = j$ are to be excluded. The first term on the right-hand side of Eq. (1.170) represents the self-correlation terms, and the second is the pair-correlation term which, in general, does not vanish in dense fluids. The paired correlated term is responsible for the intermolecular infrared absorption spectroscopy of liquids [Madden (1982)]. Thus, because of the presence of these cross terms, the interpretation of $C(t)$ cannot be perceived as depending only on single-particle dynamics. However, if the dipolar molecules are dilutely dissolved in some nonpolar solvent, then the pair-correlation terms become negligible, and the correlation function can be interpreted in terms of the dynamics of a single dipole.

In this case, P_i represents the dipole moment induced by the normal modes of vibration a molecule i:

$$\mathbf{P}^i = \sum_{k} (\partial \mathbf{P}_i / \partial Q_i^{(k)})_0 \, Q_i^{(k)} = \sum_{k} \mathbf{m}_i^{(k)} Q_i^{(k)}, \tag{1.171}$$

where $Q_i^{(k)}$ represents the normal coordinate of mode k of molecule i and $\mathbf{m}_i^{(k)} = (\partial \mathbf{P}_i/\partial Q_i^{(k)})_0$ is the transition dipole-moment induced by the vibration of mode k of molecule i.

Unless different modes of the same molecule interact through the intramolecular coupling, such as in the case of Fermi resonance, all normal modes can be considered independent, and we can treat the motion associated with each normal mode individually. Thus, for the specific normal mode designated by Q, the dipole time-correlation function $C(t)$ can be written as

$$C(t) = \tfrac{1}{3} \sum_{i,j} \langle [\mathbf{m}_i(t) Q_i(t) \cdot \mathbf{m}_j Q_j] \rangle. \tag{1.172}$$

If the vibration–rotation coupling is neglected, we can approximate $C(t)$ as

$$C(t) \simeq \tfrac{1}{3} \sum_{i,j} \langle \mathbf{m}_i(t) \mathbf{m}_j \rangle \langle Q_i(t) Q_j \rangle. \tag{1.173}$$

The correlation functions associated with $\mathbf{m}_i$ and $\mathbf{m}_j$ depend on the orientational motion of the induced dipole, and the correlation functions associated with Q_i and Q_j are owing to molecular vibrations.

For a weakly interacting molecular liquid, the N-body vibrational-wave function is approximately a product of the wave functions of individual molecules. This is because the vibrational-wave functions of individual molecules are not significantly perturbed by intermolecular interaction. As a result, by averaging over the phase of the vibrational-wave functions of different molecules, the cross terms with $i \neq j$ in Eq. (1.173) vanish identically. Therefore, in the absence of intermolecular coupling the vibration–rotation band can be approximated by the single-particle terms. We shall consider the dynamic equations that govern the single-particle and pair-correlation functions in Chapter 4.

Finally, it should be mentioned that $\chi''(\omega)$ is not measured directly in an absorption experiment; rather, it is the absorption coefficient $\alpha(\omega)$ that is directly measured. The absorption coefficient is defined as the ratio of the mean rate of energy dissipation per unit incident-energy flux S at the vicinity of the absorbing medium. According to classical electrodynamic theory, the steady-state energy flux S (or the energy flow per unit area per unit time) associated with the monochromatic electromagnetic field ($\mathscr{E}(t) = E_0 \hat{\mathbf{e}} \cos \omega t$) near the absorbing medium is given by

$$S = [\varepsilon'(\omega) c / 8\pi] E_0^2, \tag{1.174}$$

where c is the velocity of light and $\varepsilon'(\omega)$ the real part of the dielectric constant at frequency ω. According to Eq. (1.162), we obtain

$$\alpha(\omega) = \langle \mathscr{E}(t) \rangle / S = (4\pi\omega/c)[\chi''(\omega)/\varepsilon'(\omega)]. \tag{1.175}$$

Using the fluctuation–dissipation theorem, Eqs. (1.152) and (1.163), we can write $\alpha(\omega)$ as

$$\alpha(\omega) = \frac{2\pi\omega}{\hbar c\varepsilon'(\omega)}(e^{\beta\hbar\omega} - 1)\int_{-\infty}^{\infty} dt\, e^{-i\omega t}\langle P_e(t)P_e\rangle. \tag{1.176}$$

In connection with Eq. (1.175), it should be emphasized that at the frequency where absorption is appreciable, $\varepsilon'(\omega)$ also becomes a sensitive function of ω. To obtain the correlation function from the absorption experiment, it is essential to measure both $\alpha(\omega)$ and $\varepsilon'(\omega)$ accurately. Corrections to the correlation function owing to the dispersion of $\varepsilon'(\omega)$ have been found to be important.

1.9 SPECTRAL MOMENTS AND CUMULANTS

Consider now the series expansion of the time-correlation function. Although a calculation of the time-correlation function for all times is a difficult dynamic problem, at short times it is possible to expand the autocorrelation function $C_A(t)$ as a Taylor power series of t and evaluate expansion coefficients at the initial time. By Taylor's theorem, the expansion is given by

$$C_A(t) = \langle A(t)A^+\rangle = \sum_{n=0}^{\infty}\frac{t^n}{n!}\left[\frac{\partial^n}{\partial t^n}C_A(t)\right]_{t=0}. \tag{1.177}$$

If the autocorrelation function is an even function of time, then only the even powers of t are present in the expansion. The time derivative of $C_A(t)$ can be evaluated by a repeated application of Heisenberg's equation of motion for $A(t)$;

$$\partial A(t)/\partial t = (i/\hbar)[\mathscr{H}, A(t)]. \tag{1.178}$$

Continuous application of Heisenberg's equation gives

$$\partial^n A(t)/\partial t^n = (i/\hbar)^n[\mathscr{H}, [\mathscr{H}, [\mathscr{H}, \ldots, [\mathscr{H}, A(t)]\ldots,]]]. \tag{1.179}$$

This gives the general expansion formula for $C_A(t)$ as

$$C_A(t) = \sum_{n=0}^{\infty}\frac{t^n}{n!}\left(\frac{i}{\hbar}\right)^n\langle[\mathscr{H}, [\mathscr{H}, [\mathscr{H}, \ldots, [\mathscr{H}, A]\ldots,]]]A^+\rangle. \tag{1.180}$$

The coefficients in the time-power series can be identified as moments of the appropriate frequency spectrum. To illustrate the calculation, we define a normalized line-shape function $I(\omega)$, given by

$$I(\omega) = \frac{1}{2\pi} \int_{-\infty}^{\infty} dt\, e^{-i\omega t} \frac{\langle A(t)A^{+}\rangle}{\langle |A|^{2}\rangle}. \tag{1.181}$$

The Fourier inversion of this formula yields

$$\frac{\langle A(t)A^{+}\rangle}{\langle AA^{+}\rangle} = \int_{-\infty}^{\infty} d\omega\, e^{i\omega t} I(\omega) = \sum_{n=0}^{\infty} \frac{(it)^{n}}{n!} \int_{-\infty}^{\infty} \omega^{n} I(\omega)\, d\omega \tag{1.182}$$

For a normalized line-shape function with a reference frequency at $\omega = \omega_0$, the nth moment M_n with respect to the center-frequency point ω_0 is defined as

$$M_n = \int_{-\infty}^{\infty} d\omega\, (\omega - \omega_0)^{n} I(\omega). \tag{1.183}$$

Clearly, if $C_A(t)$ is an even function of t, then $I(\omega)$ is symmetrical with respect to $\omega_0 = 0$; consequently, all odd moments vanish.

Using Eq. (1.183), we rewrite Eq. (1.182) as

$$\frac{C_A(t)}{\langle A^{2}\rangle} = \sum_{n=0}^{\infty} \frac{(it)^{n}}{n!} M_n. \tag{1.184}$$

Comparing this series with that given in Eq. (1.180), we identify the coefficients as the spectral moments:

$$M_n = (1/\hbar)^{n} \langle [\mathcal{H}, [\mathcal{H}, [\mathcal{H}, \ldots, [\mathcal{H}, A] \ldots,]]] A^{+}\rangle / \langle A^{2}\rangle. \tag{1.185}$$

Given the Hamiltonian of the system and the dynamic variable associated with the specific problem, we can calculate the spectral moments and compare them with the experimental results. In practice, however, it is difficult to obtain accurate spectral moments because the ω^{2n} dependence tends to put emphasis on the spectral data in the wing region where the signal-to-noise ratio is poor; as a result, one can obtain only the first few moments experimentally.

The formulas for the first few moments are

$$\begin{aligned} M_0 &= 1, \\ M_1 &= (1/\hbar)\langle [\mathcal{H}, A] A^{+}\rangle / \langle |A|^{2}\rangle, \\ M_2 &= (1/\hbar^{2})\langle [\mathcal{H}, [\mathcal{H}, A]] A^{+}\rangle / \langle |A|^{2}\rangle \\ &= -(1/\hbar^{2})\langle [H, A][H, A^{+}]\rangle / \langle |A|^{2}\rangle. \end{aligned} \tag{1.186}$$

The moment expansion of the correlation function is valid only for small values of time. A better expansion of the time-correlation function, valid for larger values of time, is given in terms of the cumulants K_n, which are, by definition

$$\hat{C}_A(t) = \frac{\langle A(t)A^* \rangle}{\langle |A|^2 \rangle} = \exp\left[\sum_{n=1}^{\infty} \frac{K_n}{n!} (it)^n \right]. \tag{1.187}$$

Comparing the coefficients in the cumulant expansion with those in the moment expansion, we can obtain the cumulants in terms of moments. A more systematic way of obtaining the cumulants is to use the formula

$$K_n = (1/i^n)[\partial^n \ln \hat{C}_A(t)/\partial t^n]_{t=0}.$$

The first five cumulants are given explicitly as

$$\begin{aligned}
K_1 &= M_1, \\
K_2 &= M_2 - M_1^2, \\
K_3 &= M_3 - 3M_1M_2 + 2M_1^3, \\
K_4 &= M_4 - 4M_3M_1 - 3M_2^2 + 12M_2M_1^2 - 6M_1^4, \\
K_5 &= M_5 - 5M_4M_1 - 10M_3M_2 + 20M_3M_1^2 + 30M_2^2M_1 \\
&\quad - 60M_2M_1^3 + 24M_1^5.
\end{aligned} \tag{1.188}$$

Note that the cumulants, when expressed in terms of moments, do not contain moments of any order higher than the order of the cumulants. Further, for a Gaussian time-correlation function, only the second cumulant K_2 exists, and thus we have

$$\hat{C}_A(t) = \exp(-\tfrac{1}{2}t^2 K_2). \tag{1.189}$$

1.10 STOCHASTIC THEORY OF LINE SHAPE

A stochastic theory is the theoretical treatment of a problem in which the system is assumed to be perturbed by random forces. For example, when a heavy particle is suspended in a liquid, random collisions of the liquid molecules with the particle will cause it to undergo Brownian motion. The theory of Brownian motion is a stochastic theory. Similarly, if a particle is bound to a center by an elastic force, then the intermolecular interaction will cause random interruptions of the free oscillation and will broaden the spectral line as well as cause the relaxation of the oscillation from any initial state to the final equilibrium state.

Equation (1.3), the equation of motion for the dynamic variable $A(t)$, is stochastic because the frequency $\Omega(t)$ is a random function of time. In this section we shall consider this equation in detail and illustrate the essential points of the stochastic theory (Kubo, 1962).

We start by integrating Eq. (1.3) to obtain

$$\begin{aligned} A(t) &= A \exp\left[i \int_0^t dt'\, \Omega(t')\right] \\ &= A \exp\left[i\omega_0 t + i \int_0^t dt'\, \delta\Omega(t')\right] \end{aligned} \tag{1.190}$$

The time correlation $C_A(t)$ can then be written as

$$C_A(t) = \langle A(t)A^* \rangle = \langle |A|^2 \rangle e^{i\omega_0 t} F(t), \tag{1.191}$$

where the function $F(t)$, defined by

$$F(t) = \left\langle \exp i \int_0^t dt'\, \delta\Omega(t') \right\rangle, \tag{1.192}$$

is known as the relaxation function of A. The angular bracket is the ensemble average; $\delta\Omega(t)$ represents a random process, and we shall assume that the random process is ergodic and stationary in time.

By Eq. (1.181), the normalized line-shape function is given by

$$I(\omega) = \frac{1}{2\pi} \int_{-\infty}^{\infty} dt\, e^{-i(\omega-\omega_0)t} F(t). \tag{1.193}$$

Because of the function $F(t)$, the line-shape function is clearly broadened about the center frequency ω_0 by the random function $\delta\Omega(t)$.

We now write the relaxation function $F(t)$ in terms of cumulant averages

$$\begin{aligned} F(t) &= \left\langle \exp i \int_0^t dt'\, \delta\Omega(t') \right\rangle \\ &= \exp\Bigg\{ i \int_0^t dt_1 \langle \delta\Omega(t_1) \rangle + \frac{i^2}{2!} \int_0^t dt_1 \int_0^t dt_2 \langle \delta\Omega(t_1)\, \delta\Omega(t_2) \rangle_c \\ &\quad + \cdots + \frac{i^n}{n!} \int_0^t dt_1 \cdots \int_0^t dt_n \langle \delta\Omega(t_1)\, \delta\Omega(t_2) \cdots \delta\Omega(t_n) \rangle_c + \cdots \Bigg\}, \end{aligned} \tag{1.194}$$

where $\langle \delta\Omega(t_1)\, \delta\Omega(t_2) \cdots \delta\Omega(t_n) \rangle_c$ is known as the nth cumulant average, which is the generalization of the nth cumulant in the single-variable case given previously. For simplicity, we designate

$$\langle \delta\Omega(t_1)\, \delta\Omega(t_2) \cdots \delta\Omega(t_n) \rangle_c = K_n(\delta\Omega_1 \cdots \delta\Omega_n). \tag{1.195}$$

The explicit forms of some lower cumulant averages are given by

$$\begin{aligned}
K_1 &= \langle \delta\Omega_1 \rangle, \\
K_2 &= \langle \delta\Omega_1\, \delta\Omega_2 \rangle - \langle \delta\Omega_1 \rangle \langle \delta\Omega_2 \rangle, \\
K_3 &= \langle \delta\Omega_1\, \delta\Omega_2\, \delta\Omega_3 \rangle - \langle \delta\Omega_1 \rangle \langle \delta\Omega_2\, \delta\Omega_3 \rangle - \langle \delta\Omega_2 \rangle \langle \delta\Omega_1\, \delta\Omega_3 \rangle \\
&\quad - \langle \delta\Omega_3 \rangle \langle \delta\Omega_1\, \delta\Omega_2 \rangle + 2\langle \delta\Omega_1 \rangle \langle \delta\Omega_2 \rangle \langle \delta\Omega_3 \rangle.
\end{aligned} \tag{1.196}$$

The general form of the cumulant average can be obtained by first expanding $F(t)$ in terms of moments and then reorganizing the moments to compare with the expansion formula of $F(t)$ in a power series of cumulant averages.

The cumulant average has an important property in that if any component in the average is uncorrelated with the others, then the cumulant average vanishes. This property can easily be proven. To accomplish this, we divide the time period from 0 to t into the intervals $\Delta t_1, \Delta t_2, \ldots, \Delta t_n$. During each interval, the value of $\delta\Omega$ is changed to a new value because of intermolecular interactions:

$$F(t) = \left\langle \exp\left[i \int_0^t dt'\, \delta\Omega(t') \right] \right\rangle = \left\langle \exp\left(i \sum_{j=1}^{n} \delta\Omega_j\, \Delta t_j \right) \right\rangle. \tag{1.197}$$

We shall now assume that $\delta\Omega_1$ is uncorrelated with $\delta\Omega_2, \ldots, \delta\Omega_n, \ldots$. As a result,

$$\begin{aligned}
F(t) &= \langle \exp(i\, \delta\Omega_1\, \Delta t_1) \rangle \left\langle \exp\left(i \sum_{j=2}^{n} \delta\Omega_j\, \Delta t_j \right) \right\rangle \\
&= \exp[i \langle \delta\Omega_1 \rangle_c\, \Delta t_1 + (i^2/2) \langle \delta\Omega_1^2 \rangle_c (\Delta t_2)^2 + \cdots] \\
&\quad \times \exp\left(i \sum_{j=2}^{n} \langle \delta\Omega_j \rangle_c\, \Delta t_j + \tfrac{1}{2} \sum_{j,k=2}^{n} \langle \delta\Omega_j\, \delta\Omega_k \rangle_c\, \Delta t_j\, \Delta t_k + \cdots \right).
\end{aligned} \tag{1.198}$$

Thus, $\delta\Omega_1$ never mixes with $\delta\Omega_2, \ldots, \delta\Omega_n, \ldots$, i.e., if $\delta\Omega_1$ is uncorrelated with the others, there will be no cumulant average in which $\delta\Omega_1$ appears with the others. Therefore, we have shown that the cumulant average containing an uncorrelated variable vanishes identically.

Another important property of the cumulant expansion is that if the observation period t is long compared with the effective time interval of the intermolecular interaction Δt_j, or if $n \simeq t/\Delta t_j$ is large, then the modulation $\delta\Omega(t)$ is a Gaussian process, provided that each $\delta\Omega_j$ value is small. This is an application of the central-limit theorem, which we shall not prove. [See Kubo (1962) for this proof.] The Gaussian approximation has wide applicability. For example, when a molecule interacts with a few of its nearest neighbors (such as in crystals) it is sufficient to approximate the situation as a Gaussian process.

Thus, in the Gaussian approximation we have

$$F(t) = \left\langle \exp\left(i \sum_j \delta\Omega_j \Delta t_j \right) \right\rangle$$

$$= \exp\left(i \sum_j \langle \delta\Omega_j \rangle_c \Delta t_j + \frac{i^2}{2!} \sum_j \sum_n \langle \delta\Omega_j \Delta\Omega_n \rangle_c \Delta t_j \Delta t_n \right), \quad (1.199)$$

or in terms of continuous variables, the cumulant expansion for $F(t)$ is

$$F(t) = \exp\left[i \int_0^t dt_1 \langle \delta\Omega(t_1) \rangle_c + \frac{i^2}{2!} \int_0^t \int_0^t dt_1\, dt_2 \langle \delta\Omega(t_1)\, \delta\Omega(t_2) \rangle_c \right]. \quad (1.200)$$

In the case of $\langle \delta\Omega(t) \rangle = 0$, the first term on the right-hand side vanishes. Furthermore, according to Eq. (1.196), the second cumulant becomes

$$\langle \delta\Omega(t_1)\, \delta\Omega(t_2) \rangle_c = \langle \delta\Omega(t_1)\, \delta\Omega(t_2) \rangle,$$

and hence,

$$\frac{i^2}{2!} \int_0^t dt_1 \int_0^t dt_2 \langle \delta\Omega(t_1)\, \delta\Omega(t_2) \rangle = - \int_0^t d\tau\, (t - \tau) \langle \delta\Omega(\tau)\, \delta\Omega \rangle, \quad (1.201)$$

where the second step was obtained by utilizing the stationary property of the stochastic variable $\delta\Omega(t)$ [i.e., $\langle \delta\Omega(t_1)\, \delta\Omega(t_2) \rangle = \langle \delta\Omega(t_1 - t_2)\, \delta\Omega(0) \rangle$], and the double integral is converted into a single integral after changing to a new variable τ by letting $\tau = t_1 - t_2$.

Therefore, for a Gaussian modulation process, the relaxation function $F(t)$ is given by

$$F(t) = \exp\left[- \int_0^t d\tau\, (t - \tau) g(\tau) \right], \quad (1.202)$$

where $g(\tau)$ is the time-correlation function of the stochastic variable $\delta\Omega(\tau)$, given by

$$g(\tau) = \langle \delta\Omega(\tau)\, \delta\Omega \rangle. \quad (1.203)$$

The function $g(\tau)$ is a decreasing function of τ; it vanishes rapidly as τ increases. We assume a time τ_c' at which $g(\tau)$ has decreased appreciably. Or, to be precise, we define

$$\tau_c' = \frac{1}{\Delta^2} \int_0^\infty d\tau\, g(\tau) \quad (1.204)$$

to be the correlation time for $g(\tau)$. Here, Δ^2 is equal to $\langle \delta\Omega^2 \rangle$. The quantity Δ is known as the amplitude of modulation. It should be noted that τ_c' is not equal to the correlation time τ_c, defined previously in Eq. (1.56).

We now consider the behavior of $F(t)$ for two different regions of time. In one region, $t < \tau_c'$, $g(\tau)$ does not decrease appreciably, and we can set $g(\tau) \simeq g(0) = \Delta^2$. Thus we have

$$F(t) \simeq \exp\left[-g(0) \int_0^t d\tau\,(t - \tau)\right] = \exp\left(-\frac{\Delta^2 t^2}{2}\right). \tag{1.205}$$

Therefore, the initial decay of $F(t)$ is Gaussian, independent of the specific time dependence of the modulation.

In the other region, $t \gg \tau_c'$. Over such a long time, $g(\tau)$ has decayed to a small value, and without sacrificing the accuracy of the calculation, the upper limit of the integral appearing in Eq. (1.201) can be extended to ∞; thus

$$F(t) \simeq \exp\left[-\int_0^\infty d\tau\,(t - \tau)g(\tau)\right] = f \exp(-\Delta^2 \tau_c' t), \tag{1.206}$$

where the quantity f is given by

$$f = \exp\left[\int_0^\infty d\tau\,\tau g(\tau)\right]. \tag{1.207}$$

In this case, $F(t)$ decays exponentially with a relaxation time of $\tau_r = 1/(\Delta^2 \tau_c')$.

This discussion can be verified by assuming that $g(\tau)$ is given by an exponentially decreasing function, such as

$$g(\tau) = \Delta^2 \exp(-\tau/\tau_c'). \tag{1.208}$$

Substituting Eq. (1.208) into (1.202) and carrying out the calculation of the integral, we obtain

$$F(t) = \exp(-\Delta^2\{\tau_c' t - \tau_c^2[1 - \exp(-t/\tau_c')]\}). \tag{1.209}$$

Clearly, for $t < \tau_c'$ we can expand $\exp(-t/\tau_c')$ in a power series of t/τ_c', and the first nonvanishing term is

$$F(t) = \exp(-\Delta^2 t^2/2),$$

which is identical to Eq. (1.205). On the other hand, for $t \gg \tau_c'$, the term $\exp(-t/\tau_c')$ is practically zero, and $F(t)$ becomes

$$F(t) = \exp[-(\Delta\tau_c')^2] \exp(-\Delta^2 \tau_c' \tau),$$

which corresponds to Eq. (1.206), with f given by $\exp[-(\Delta\tau_c')^2]$.

In conclusion, we have shown that for small values of t, $F(t)$ decays as a gaussian, and for large values of time it decays exponentially. The behavior of $F(t)$ for $t < \tau_c'$ and for $t > \tau_c'$ is illustrated in Fig. 1.3.

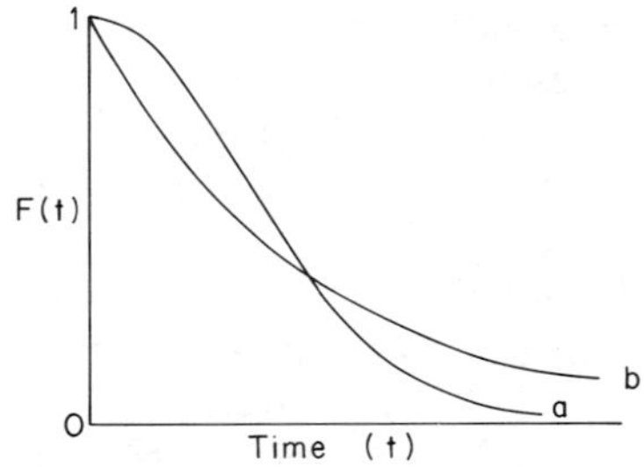

FIG. 1.3 The behavior of $F(t)$ for $t < \tau_c'$ (a, gaussian) and for $t > \tau_c'$ (b, exponential).

The spectral line shape has a distinctive feature according to the combination of the strength of the modulation (measured by Δ) and the speed of the modulation determined by τ_c'. Two situations can be distinguished: slow modulation ($\Delta\tau_c' \gg 1$) and fast modulation ($\Delta\tau_c' \ll 1$). We shall discuss these two separate situations by expanding $g(t)$ in power series of t as

$$g(t) = \langle \delta\Omega(t)\, \delta\Omega \rangle = \Delta^2 \left[\sum_{n=0}^{\infty} \frac{1}{(2n)!} \alpha_{2n} t^{2n} \right] \tag{1.210}$$

Note that the odd powers of t do not contribute. Clearly, $\alpha_0 = 1$ and

$$\alpha_2 = (1/\Delta^2)\langle [\partial^2\, \delta\Omega(t)/\partial t^2]\, \delta\Omega \rangle|_{t=0} = -\langle \dot{\Omega}^2 \rangle / \Delta^2, \tag{1.211}$$

where $\dot{\Omega} = \partial\, \delta\Omega(t)/\partial t\,|_{t=0}$ is the initial modulation speed. The last step in the α_2 expression follows from the stationary condition of the stochastic process.

Substituting Eq. (1.210) into Eq. (1.202) and carrying out the time integral, we obtain

$$\begin{aligned} F(t) &= \exp\left[-\Delta^2 \sum_{n=0}^{\infty} \frac{\alpha_{2n}}{(2n)!(2n+1)(2n+2)} t^{2n+2} \right] \\ &= \exp\left[-\left(\frac{\Delta^2 t^2}{2} - \frac{\langle \dot{\Omega}^2 \rangle t^4}{24} + \cdots \right) \right] \\ &= \exp\left[-\Delta^2 \left(\frac{t^2}{2} - \frac{\eta t^4}{24} + \cdots \right) \right], \end{aligned} \tag{1.212}$$

where τ_c' is given by $(\Delta^2/\langle \dot{\Omega}^2 \rangle)^{1/2}$ and corresponds approximately to the value given in Eq. (1.204). The parameter η is given by $1/(\Delta\tau_c')^2$, which is a measure of the modulation speed. Note that the t^2 term does not depend on the modulation speed; however, the higher terms do. This shows that for small values of time ($t \ll \tau_c'$), the oscillator is not affected by intermolecular interaction. With increasing values of time ($t \geq \tau_c'$), intermolecular interaction affects the modulation speed and thus influences the spectral line shape significantly. For slow modulation, we have $\Delta\tau_c' \gg 1$, and η is small. The $F(t)$ curve can be represented mainly by the Gaussian function as given by the first term on the right-hand side of Eq. (1.212). For fast modulation, $\Delta\tau_c' \ll 1$, and η is large. This means that the higher-order terms become important; i.e., the correlation function $F(t)$ does not initially decay as rapidly as in the slow-modulation case, but it decays faster at larger time values. The line-shape function corresponds to the Fourier transform of $F(t)$ [compare with Eq. (1.193)]; this means that in fast modulation the line shape has wider wings and a narrower central part.

The narrowing of the central part of the spectrum as a result of the increasing speed of modulation is a well-known effect in nuclear magnetic resonance (NMR). In NMR the dipolar linewidth is narrowed by the thermal motion of the nuclei. This phenomenon is called motional narrowing, which we shall discuss in detail in Chapter 5.

1.11 THE DISTRIBUTION-FUNCTION METHOD FOR THE RANDOM PROCESS

There are two approaches to the theory of random process. In the previous section, we focused our attention on actual random variation in time of the dynamic variable represented by $A(t)$. In that approach, we began by formally integrating the dynamic variable; then the relaxation function $F(t)$ [Eq. (1.192)] was introduced to calculate the time-correlation function of A. It was then shown that for the Gaussian random process, the relaxation function is related to the time-correlation function of the stochastic variable given by $g(\tau)$ [Eq. (1.202)]. The beauty of the stochastic approach is that it is often possible to obtain an adequate description of the time dependence of the relaxation function without carrying out a detailed analysis of $g(\tau)$. For example, if the random process is such that $g(\tau)$ decays rapidly as τ increases, then it is only necessary to know the area of $g(\tau)$ to describe the time evolution of $F(t)$. On the other hand, if the fluctuation period of the random process is slow, then only the initial value of $g(\tau)$ is needed.

Another method for developing the theory of random process is based on the probability-distribution function (Kubo, 1962; Wax, 1954). Since the stochastic variable does not depend on time completely, as in a *causal process*, the description of the stochastic process depends inherently on the probability-distribution function. The distribution-function method has an advantage over the previous approach because it can be generalized to allow the study of a non-Gaussian process as well as the discrete-jump process;

this is difficult to accomplish with the previous approach. The distribution method also permits formulation of the stochastic process in terms of the Fokker–Planck partial-differential equation. This makes possible a direct correspondence of the general theory of Brownian motion. The inertial effect can also be incorporated in the general theory of Brownian motion (Wax, 1954).

In the distribution-function approach, the dynamic variable A is a function of a stochastic variable, say x, which fluctuates rapidly in time. We shall need a set of distribution functions W_1, $W_2, \ldots, W_n$, to describe the statistical behavior of the dynamic system (Oppenheim *et al.*, 1977). These distribution functions are governed by the stochastic variable x and are defined as follows: $W_1(x, t)\,dx$ is the probability of finding x in the range $(x, x + dx)$ at time t; $W_2(x_2 t_1; x_1 t_1)\,dx_2\,dx_1$, the joint probability of finding x in the range $(x_1, x_1 + dx_1)$ at time t_1 and $(x_2, x_2 + dx_2)$ at time t_2; $W_3(x_3 t_3; x_2 t_2; x_1 t_1)\,dx_3\,dx_2\,dx_1$, the joint probability of finding x in the range $(x_1, x_1 + dx_1)$ at time t_1, $(x_2, x_2 + dx_2)$ at time t_2, and $(x_3, x_3 + dx_3)$ at time t_3, etc. Clearly, the distribution functions fulfill

$$W_n \geq 0, \qquad n = 1, 2, \ldots, \tag{1.213}$$

and

$$W_n = \int \cdots \int dx_{k+1}\, dx_{k+2} \cdots dx_k\, W_k(y_n t_n; \ldots; y_2 t_2; y_1 t_1). \tag{1.214}$$

W_2 is the (probability) distribution function of most interest to us. The time-correlation function is expressed in terms of W_2 as

$$\langle A(t) A^* \rangle = \iint dx_2\, dx_1\, A(x_2) A^*(x_1) W_2(x_2 t; x_1 0). \tag{1.215}$$

We shall next assume that all the statistical information can be described in terms of a Markov process, which is completely defined by W_2 and W_1 because the probability for finding x in the range $(x_n, x_n + dx_n)$ at time t_n depends on $x_n t_n$ and on the value of x at the previous time t_n. Not all dynamic processes are Markov processes. However, a process that is not Markov can still be considered a kind of projection of a more complicated Markov process. Therefore, we shall restrict ourselves to the study of the Markov process (Oppenheim *et al.*, 1977).

It is useful to define a conditional probability $P(x_2 t_2 | x_1 t_1)$, which gives the probability of finding x_2 at t_2 when the value of x is given by x_1 at t_1. Clearly, W_2 is related to P by

$$W_2(x_2 t_2; x_1 t_1) = P(x_2, t_2 | x_1, t_1) W(x_1 t_1). \tag{1.216}$$

We shall further assume that the Markov process is stationary and set the initial time t_1 as zero and $t = t_2 - t_1$. From the definition of the Markov process, we expect that

$$P(x_2, t|x_1) = \int dx\, P(x_2, t - t'|x, t')P(x, t'|x_1)$$

$$= \int dx\, P(x_2, t - t'|x)P(x, t'|x_1), \tag{1.217}$$

where the second equality follows from the stationary property of the Markov process. For brevity of notation, we have written $P(x_2, t|x_1)$ for $P(x_2, t|x_1, 0)$ and $P(x, t'|x_1)$ for $P(x, t'|x_1, 0)$. Equation (1.217) is known as the Smoluchowski equation and is the fundamental equation in the theory of random processes.

In terms of the conditional probability, the time-correlation function given in Eq. (1.215) can be written as

$$\langle A(t)A^*\rangle = \iint dx_2\, dx_1\, A(x_2)A^*(x_1)P(x_2 t|x_1)W(x_1). \tag{1.218}$$

To find a simple differential equation which governs the conditional probability P, we go back to Eq. (1.190) and treat the same problem in terms of the distribution-function approach. Equation (1.190) can be written as

$$A(t) = A \exp\{i\omega_0 t + i\theta(t)\}, \tag{1.219}$$

where

$$\theta(t) = \delta\Omega(t). \tag{1.220}$$

In Eq. (1.219) the stochastic variable x is θ. We shall also assume that initially the phase angle θ is equal to θ'. In terms of the stochastic variable θ, the time-correlation function of $A(t)$ can be written as

$$C_A(t) = \langle A(t)A^*\rangle = \langle |A|^2\rangle e^{i\omega_0 t}$$

$$= \langle |A|^2\rangle e^{i\omega_0 t} \iint d\theta\, d\theta'\, e^{i(\theta - \theta')}P(\theta, t|\theta')w_1(\theta'). \tag{1.221}$$

Equation (1.220) is similar to the Langevin equation for Brownian motion

$$\dot{u} + \xi u/m = f(t)/m, \tag{1.222}$$

where m is the mass of the Brownian particle, ξ the friction coefficient, and $f(t)$ the random force that modulates the Brownian particle. We shall study

Brownian motion in more detail in Chapter 4. Here we note, however, that Eq. (1.220) lacks the friction term that appears in Eq. (1.222).

In the following section we shall develop a general method for finding the conditional-probability function to calculate the correlation functions; this approach will result in a Fokker–Planck equation. To illustrate the application of the Fokker–Planck equation, we shall consider specifically the equation for $\theta(t)$ given in Eq. (1.220). Although the Fokker–Planck equation can predict a variety of dynamic processes with sufficient generality, we shall show that, for a Gaussian process, the conditional probability computed from Eq. (1.220) is a diffusion equation.

To derive the Fokker–Planck equation, we choose a time interval Δt that is long enough for the stochastic variable to undergo many fluctuations but short enough so that the state of the system as represented by the dynamic variable A does not change appreciably (Chandrasekhar, 1943). We assume that such a time interval exists. Given the value y at time 0, the probability of finding the value x at $t + \Delta t$ is given by the Smoluchowski equation [Eq. (1.217)] as

$$P(x, t + \Delta t \mid y) = \int dz\, P(x, \Delta t \mid z) P(z, t \mid y). \tag{1.223}$$

Next we consider the integral

$$R = \int dx\, g(x) \frac{\partial}{\partial t} P(x, t \mid y) = \lim_{\Delta t \to 0} \frac{1}{\Delta t} \int dx\, g(x) [P(x, t + \Delta t \mid y) - P(x, t \mid y)], \tag{1.224}$$

where $g(x)$ is an arbitrary function that is well behaved at all values of x and vanishes very rapidly as $x \to \pm\infty$.

Substituting Eq. (1.223) into the first term on the right-hand side of Eq. (1.224), we obtain

$$R = \lim_{\Delta t \to 0} \frac{1}{\Delta t} \left[\int dz \int dx\, g(x) P(x, \Delta t \mid z) P(z, t \mid y) - \int dx\, g(x) P(x, t \mid y) \right]. \tag{1.225}$$

For small Δt, the variable x is very close to z and we can expand the function $g(x)$ in a Taylor series about z as

$$g(x) = g(z) + g'(z)(z - x) + \tfrac{1}{2} g''(z)(z - x)^2 + \cdots. \tag{1.226}$$

Replacing $g(x)$ by its Taylor series, we find that Eq. (1.225) becomes

$$\int dx\, g(x) \frac{\partial}{\partial t} P(x, t\,|\,y) = \int dz\, [g'(z)A(z) + \tfrac{1}{2}g''(z)B(z)]P(z, t\,|\,y), \qquad (1.227)$$

where

$$A(z) = \lim_{\Delta t \to 0} \frac{1}{\Delta t} \int dx\, (z - x)P(x, \Delta t\,|\,z), \qquad (1.228)$$

$$B(z) = \lim_{\Delta t \to 0} \frac{1}{\Delta t} \int dx\, (z - x)^2 P(x, \Delta t\,|\,z). \qquad (1.229)$$

To obtain Eq. (1.227), we used the relation

$$\int dx\, P(x, \Delta t\,|\,z) = 1 \qquad (1.230)$$

because the conditional probability is normalized. The terms $A(z)$ and $B(z)$ are the first and second moments of the transition-probability function $P(x, \Delta t\,|\,z, t)$. We shall assume that higher moments vanish. This assumption becomes rigorous for the Gaussian process.

We next carry out partial integration with respect to z and then replace z with x at the end to obtain

$$\int dx\, g(x)\left[\frac{\partial P}{\partial t} + \frac{\partial}{\partial x}(AP) - \frac{1}{2}\frac{\partial^2}{\partial x^2}(BP)\right] = 0. \qquad (1.231)$$

Since we assumed $g(x)$ to be an arbitrary function, the expression in the square bracket must vanish. Thus, we obtain the Fokker–Planck equation

$$\frac{\partial P}{\partial t} = -\frac{\partial}{\partial x}(AP) + \frac{1}{2}\frac{\partial^2}{\partial x^2}(BP) \qquad (1.232)$$

This result can be generalized to an n-dimensional equation as

$$\frac{\partial P}{\partial t} = -\sum_i \frac{\partial}{\partial x_i}(A_i P) + \frac{1}{2}\sum_{i,j} \frac{\partial^2}{\partial x_i\, \partial x_j}(B_{ij}P) \qquad (1.233)$$

where A_i and B_{ij} are the first and second moments defined in the same manner as in Eqs. (1.228) and (1.229).

We now use the stochastic equation given by Eq. (1.220) to calculate the conditional probability. First we choose a time interval Δt and convert Eq. (1.220) into an integral equation:

$$\Delta\theta = \theta(t + \Delta t) - \theta(t) = \int_t^{t+\Delta t} dt'\, \delta\Omega(t'). \qquad (1.234)$$

The time interval Δt in Eq. (1.234) is chosen to be short enough so that θ does not change appreciably but long enough for $\delta\Omega(t)$ to undergo many fluctuations. We shall assume that such a time interval exists.

To obtain the moments, we set $\theta(t + \Delta t)$ to z and $\theta(t)$ to x in Eqs. (1.228) and (1.229). Therefore,

$$A(\theta) = \lim_{\Delta t \to 0} \frac{1}{\Delta t} \langle \Delta\theta \rangle = \lim_{\Delta t \to 0} \frac{1}{\Delta t} \int_t^{t+\Delta t} dt' \, \langle \delta\Omega(t') \rangle, \qquad (1.235)$$

which vanishes because $\langle \delta\Omega \rangle = 0$. Here the brackets indicate taking an average over the conditional probability $P(\theta, \Delta t \,|\, \theta)$. Since Δt is so short that θ does not undergo appreciable change, we consider the function P to be an equilibrium-distribution function. The angular brackets thus indicate an ensemble average. Further,

$$\begin{aligned} B(\theta) &= \lim_{\Delta t \to 0} \frac{1}{\Delta t} \langle (\Delta\theta)^2 \rangle \\ &= \lim_{\Delta t \to 0} \frac{1}{\Delta t} \int_t^{t+\Delta t} dt_2 \int_t^{t+\Delta t} dt_1 \, \langle \delta\Omega(t_2) \, \delta\Omega(t_1) \rangle \\ &= \lim_{\Delta t \to 0} \frac{2}{\Delta t} \int_0^{\Delta t} d\tau \, (t - \tau) \langle \delta\Omega(\tau) \, \delta\Omega \rangle. \end{aligned} \qquad (1.236)$$

We assume that the correlation function $\langle \delta\Omega(\tau) \, \delta\Omega \rangle$ decays very rapidly in the interval Δt. This is justified because of the assumption that $\delta\Omega(\tau)$ undergoes many fluctuations in Δt. As a result, Eq. (1.236) reduces to

$$B(\theta) = 2D, \qquad (1.237)$$

where

$$D = \Delta^2 \tau_c' = \int_0^\infty d\tau \, \langle \delta\Omega(\tau) \, \delta\Omega \rangle. \qquad (1.238)$$

If we go on to calculate the third and higher moments, then we shall need to calculate quantities of the form

$$\lim_{\Delta t \to 0} (1/\Delta t) \langle (\Delta\theta)^n \rangle.$$

This will involve calculating the multiple-time integral of a correlation function of the type $\langle \delta\Omega(t_n) \cdots \delta\Omega(t_2) \, \delta\Omega(t_1) \rangle$. To facilitate the calculation, we shall assume as before that $\delta\Omega(t)$ is a Gaussian process, and the central-limit theorem can then be used to reduce the multiple-time-correlation function to a product of pair-correlation functions

$$\langle \delta\Omega(t_n) \cdots \delta\Omega(t_2) \, \delta\Omega(t_1) \rangle = \sum_{\text{all pairs}} \langle \delta\Omega(t_i) \, \delta\Omega(t_j) \rangle \cdots \langle \delta\Omega(t_k) \, \delta\Omega(t_l) \rangle. \qquad (1.239)$$

The multiple-time integral of the time-correlation function of the form given here will vanish in the limit of $\Delta t \to 0$ because the value of the integral will contain a factor of $(\Delta t)^m$, with $m > 1$. Therefore, for the Gaussian process $\delta\Omega(t)$, no moments higher than the second will exist.

With the help of $A(\theta)$ and $B(\theta)$ given by Eqs. (1.235) and (1.237), respectively, the Fokker–Planck equation becomes

$$\partial P/\partial t = D\, \partial^2 P/\partial \theta^2, \tag{1.240}$$

which is the usual diffusion equation. Thus, with the distribution-function method, the random modulation owing to $\delta\Omega$ is treated as a diffusion of the phase angle to obtain the solution for the stochastic equation $\dot{\theta}(t) = \delta\Omega(t)$. In this limit, the problem of finding the conditional probability becomes that of finding the solution to a partial-differential equation of the diffusion type.

The solution to Eq. (1.240) with the initial condition at $t = 0$ given by

$$P(\theta, t \mid \theta') = \delta(\theta - \theta') \tag{1.241}$$

is

$$P(\theta, t \mid \theta') = (4\pi D t)^{-1/2} \exp[-(\theta - \theta')^2/4Dt]. \tag{1.242}$$

To obtain the time-correlation function $C_A(t)$ in accordance with Eq. (1.221), we shall also need the first probability-distribution function $W_1(\theta')$. In Markov processes W_1 can be obtained from the conditional probability P according to

$$\lim_{t \to \infty} P(\theta, t \mid \theta') = W_1(\theta). \tag{1.243}$$

This is, however, impossible to apply to Eq. (1.242) since this limiting process yields vanishing results for W_1. Note that the choice of the time interval Δt is inconsistent with the stochastic equation as given in Eq. (1.220). We shall show in Chapter 4 that, under some appropriate assumptions, the stochastic equation for $\theta(t)$ follows the Langevin type,

$$\dot{\theta}(t) = -\gamma\theta(t) + \delta\Omega(t), \tag{1.244}$$

rather than Eq. (1.220). Here γ is the kinetic coefficient that plays a role similar to the friction coefficient in the Langevin equation. If Eq. (1.244) were used to calculate the first moment, instead of a vanishing value, we would obtain

$$A(\theta) = -\gamma\theta. \tag{1.245}$$

The value of the second moment $B(\theta)$ is still given by Eq. (1.237). In this case, the Fokker–Plnack equation becomes

$$\partial P/\partial t = \gamma\, \partial(\theta P)/\partial\theta + D\, \partial^2 P/\partial\theta^2. \tag{1.246}$$

The initial condition for this equation for $t = 0$ is given by

$$P(t = 0) = \delta(\theta - \theta'). \tag{1.247}$$

Equation (1.246) can be solved by introducing a new variable s that is related to θ by

$$s = \theta e^{\gamma t}. \tag{1.248}$$

With respect to this new variable, Eq. (1.246) becomes

$$\partial P/\partial t = \gamma P + De^{2\gamma t}\, \partial^2 P/\partial s^2. \tag{1.249}$$

This equation is further simplified by introducing the function

$$X = Pe^{-\gamma t}. \tag{1.250}$$

In terms of X, Eq. (1.249) becomes

$$\partial X/\partial t = De^{2\gamma t}\, \partial^2 X/\partial s^2. \tag{1.251}$$

The solution to this equation with $s = s'$ at $t = 0$ is

$$X = \{2\pi D[\exp(2\gamma t) - 1]/\gamma\}^{-1/2} \exp\{-\gamma(s - s')^2/2D[\exp(2\gamma t) - 1]\}. \tag{1.252}$$

Transforming back to P, we obtain

$$P(\theta, t\,|\,\theta') = \{2\pi D[1 - \exp(-2\gamma t)]/\gamma\}^{-1/2} \times \exp\{-\gamma[\theta - \theta' \exp(-\gamma t)]^2/2D[1 - \exp(-2\gamma t)]\}. \tag{1.253}$$

For the conditional probability given here, we obtain for $t \to \infty$ a non-vanishing first-probability function for W_1 as

$$W_1(\theta) = \lim_{t\to\infty} P(\theta, t\,|\,\theta') = (\gamma/2\pi D)^{1/2} \exp(-\gamma\theta^2/2D). \tag{1.254}$$

This function vanishes if γ vanishes. Therefore, the diffusion equation derived from the stochastic equation given by Eq. (1.220) appears to be inconsistent with the conditional-probability approach.

To avoid this inconsistency, instead of carrying out the double integration involving θ and θ', the time correlation function $C_A(t)$ is calculated by taking the reference phase angle θ' to be zero in the conditional probability and in the dynamic variable A at $t = 0$. In this way, the function W_1 does not enter because the integration over the phase angle θ' is one (because W_1 is normalized).

Therefore, the time-correlation function is calculated according to

$$\begin{aligned} C_A(t) &= \langle |A|^2 \rangle e^{i\omega_0 t} \int_{-\infty}^{\infty} dt\, e^{i\theta} P(\theta, t\,|\,0) \\ &= \langle |A|^2 \rangle e^{i\omega_0 t} (4\pi Dt)^{-1/2} \int_{-\infty}^{\infty} dt\, e^{i\theta}\, e^{(-\theta^2/4Dt)}, \end{aligned} \tag{1.255}$$

which gives

$$C_A(t) = \langle |A|^2 \rangle e^{i\omega_0 t - Dt}. \tag{1.256}$$

This result is identical to that obtained previously by the direct calculation of the relaxation function $F(t)$.

In this section, we have derived the Fokker–Planck equation and have shown that it reduces to the ordinary diffusion equation under the appropriate approximations. Although the example used in the calculation is not sufficient to illustrate the power of Fokker–Planck equations, it should be emphasized that these equations are a very useful theoretical tool for studying the dynamic properties in nonequilibrium systems. The predictions by Fokker–Planck equations are generally valid for large as well as small values of time.

Furthermore, it should be emphasized that Fokker–Planck equations give the description of the physical system in terms of the distribution function in the phase space of the stochastic variables x_i. A statistical system is generally referred to as Fokker–Planck if the temporal evolution of the distribution function is given by a random diffusion of the ensemble's phase points superimposed on a systematic fluidlike drift of the phase points through the phase space. We can introduce a Fokker–Planck operator $\mathscr{L}$ by writing the Fokker–Planck equation as

$$\partial P/\partial t = -\mathscr{L}P, \tag{1.257}$$

where the operator $\mathscr{L}$ is given by

$$\mathscr{L} = \partial A_i/\partial x_i - \partial^2 B_{ij}/\partial x_i\,\partial x_j. \tag{1.258}$$

Here the repeated indices denote summation.

The Fokker–Planck operator is in general non-Hermitian, but it becomes Hermitian under the transformation

$$\mathscr{H} = W_1^{-1/2}\mathscr{L}W_1^{1/2} \tag{1.259}$$

where W_1 is the first probability-distribution function describing the initial statistical system.

Simple calculations show that the time-correlation function can be written as

$$\langle A[x(t)]A^*(x)\rangle = \int dx\, W_1^{1/2}(x)A(x)e^{-\mathscr{H}t}A^*(x)W_1^{1/2}(x). \tag{1.260}$$

In this formulation, the time-correlation function involves only a single integration with respect to the initial phase points represented by x. Another advantage of this formulation is the possibility of expanding $\mathscr{H}$ in terms of its eigenvalues and eigenfunctions and then evaluating Eq. (1.259) as a sum of terms associated with the eigenvalue problem of the operator $\mathscr{H}$. This method becomes very useful for problems possessing symmetry; it reduces the number of degrees of freedom and shortens the calculations.

2
Spontaneous Scattering of Light in Condensed Media

2.1 INTRODUCTION

The phenomenon of light scattering can be understood as the emission of radiation by the oscillation of dipole moments induced by an externally applied electromagnetic wave. The oscillation of the dipole moments generates radiation of low intensity having frequencies and directions of propogation different from those of the incident wave.

The basic scattering event may be viewed as a process in which a photon with energy $\hbar\omega_i$ interacts with a material system. The photon is absorbed by the medium, which simultaneously emits a photon with energy $\hbar\omega_s$. The frequency of the emitted photon ω_s can be either less or greater than the incident photon frequency ω_i. The polarization of the scattered photon can also be correlated with that of the incident photon. For $\omega_s < \omega_i$, the medium gains energy from the incident field in the amount $\hbar(\omega_i - \omega_s)$. This process is called Stokes scattering. On the other hand, if $\omega_s > \omega_i$, the medium loses energy in the amount $\hbar(\omega_s - \omega_i)$ to the incident field, and the process is called anti-Stokes scattering. The case in which the scattered photon has approximately the same frequency as the incident photon is known as Rayleigh scattering.

A detailed microscopic description of the light-scattering phenomenon must be provided by quantum theory; however, when the incident-radiation frequency is not close to the electronic-transition frequency, major results of the quantum theory can be obtained from the classical theory of light scattering. Thus, for most practical purposes, it suffices to develop a theory of light scattering from the principles of classical electrodynamics.

We shall now provide a formulation of the light-scattering process from a material system consisting of N interacting, nonspherical molecules. We shall assume that the interaction between light and matter can be described in terms of the polarizability tensor and derive an expression for the spectral power density of the scattered light in terms of the time correlation of the polarizability tensor. We shall show explicitly the effect of the intermolecular interaction on the intensity, polarization, and frequency of the scattered light.

2.2 INHOMOGENEOUS WAVE EQUATION FOR THE SCATTERED FIELD

The incident electromagnetic wave is characterized by an electric-field vector $\mathbf{E}_\mathrm{i}$ and a magnetic-field vector $\mathbf{H}_\mathrm{i}$. On introducing the electromagnetic wave that interacts with the medium consisting of N interacting molecules, the internal field at the position of each molecule is modified. In addition to the local field scattered fields are also present. We now denote the electric and magnetic fields of the scattered wave by $\mathbf{E}_\mathrm{s}$ and $\mathbf{H}_\mathrm{s}$, respectively.

Maxwell's equations provide a complete interrelationship of the electromagnetic fields of the scattered waves. Thus, they serve as a convenient starting point for the development of the basic theory of light scattering. Maxwell's equations for the various fields associated with the scattered waves in a nonconducting and noncharged medium are given by

$$\nabla \cdot \mathbf{D}_\mathrm{s} = 0, \tag{2.1}$$

$$\nabla \cdot \mathbf{B}_\mathrm{s} = 0, \tag{2.2}$$

$$\nabla \times \mathbf{E}_\mathrm{s} = (-1/c)\partial \mathbf{B}_\mathrm{s}/\partial t, \tag{2.3}$$

$$\nabla \times \mathbf{H}_\mathrm{s} = (1/c)\partial \mathbf{D}_\mathrm{s}/\partial t, \tag{2.4}$$

where $\mathbf{D}_\mathrm{s}$ and $\mathbf{B}_\mathrm{s}$ are the displacement and magnetic induction vectors, respectively. For a nonmagnetic system, we assume that $\mathbf{B}_\mathrm{s}$ is equal to $\mathbf{H}_\mathrm{s}$. The relationship between $\mathbf{D}_\mathrm{s}$ and $\mathbf{E}_\mathrm{s}$, the electric field of the scattered wave, is more complicated because of the present lack of a solution to the local-field problem. However, for developing a workable light-scattering theory, it suffices to neglect the reaction of the local field to the induced polarization and to assume that the incident and scattered electric fields make their separate contributions to the total electric displacement vectors $\mathbf{D}(\mathbf{r}, t)$. Thus, we write

$$\mathbf{D}(\mathbf{r}, t) = \mathbf{D}_i(\mathbf{r}, t) + \mathbf{D}_s(\mathbf{r}, t). \tag{2.5}$$

The displacement vectors $\mathbf{D}_i$ and $\mathbf{D}_s$ depend on space and time because of the presence of molecular motion. In the absence of the scattered field, the displacement vector $\mathbf{D}_i(\mathbf{r}, t)$ can be taken as $\boldsymbol{\varepsilon} \cdot \mathbf{E}_i$, where $\mathbf{E}_i$ is the electric field of the incident light and $\boldsymbol{\varepsilon}$ the local dielectric-constant tensor of the medium. The displacement vector is also a function of space and time. If the interactions between the incident and the scattered fields are neglected, then one might simply take $\mathbf{D}_s$ to be equal to $\boldsymbol{\varepsilon} \cdot \mathbf{E}_s$. This, however, is undesirable because the contribution from the fluctuations of the dielectric-constant tensor resulting from molecular motion now appears as a second-order term in the $\mathbf{D}_s$ expression. Thus, in the development of a light-scattering theory, it is useful to divide $\mathbf{D}(\mathbf{r}, t)$ into two parts, taking $\mathbf{D}_i$ to be equal to $\boldsymbol{\varepsilon}_0 \cdot \mathbf{E}_i$, where $\boldsymbol{\varepsilon}_0$ is the equilibrium average of the dielectric-constant tensor, which is space and time independent, and to write

$$\mathbf{D}_s = \boldsymbol{\varepsilon}_0 \cdot \mathbf{E}_s + 4\pi \mathbf{P}_s, \tag{2.6}$$

where $\mathbf{P}_s$ is the induced-polarization vector that is a function of the incident and scattered fields and vanishes if the incident field disappears. In linear theory, $\mathbf{P}_s$ depends only linearly on the incident field and is related to the polarizability density tensor $\boldsymbol{\alpha}$ by

$$\mathbf{P}_S = \boldsymbol{\alpha} \cdot \mathbf{E}_i. \tag{2.7}$$

Other terms involving $\mathbf{E}_s$, as well as higher orders of $\mathbf{E}_i$, are present in nonlinear processes. These will be considered in Chapter 6.

In an isotropic medium such as a liquid or cubic crystal, $\boldsymbol{\varepsilon}_0, = \varepsilon_0 \mathbf{I}$, where ε_0 is the zero-frequency dielectric constant and is only a function of temperature: $\mathbf{I}$ is a unit tensor. We shall adapt the calculation to the isotropic medium. (Generalizing the calculation to an anisotropic medium is straightforward but more cumbersome.)

We now use Maxwell's equations, together with Eq. (2.6) for $\mathbf{D}_s$, to express the electric field of the scattered wave in terms of $\mathbf{P}_s$. The approximation of Eq. (2.7) will then be used to obtain the spectral power density in terms of the time-correlation function of the polarizability density.

To accomplish this, we shall first obtain a wave equation for $\mathbf{D}_s$. This is carried out by eliminating $\mathbf{B}_s$ from Eq. (2.3). To do this, we use Eq. (2.4) after taking the curl of Eq. (2.3):

$$\nabla \times (\nabla \times \mathbf{E}_s) = -(1/c)\,\partial(\nabla \times \mathbf{B}_s)/\partial t = -(1/c^2)\,\partial^2 \mathbf{D}_s/\partial t^2 \tag{2.8}$$

Solving Eq. (2.6) for $\mathbf{E}_s$ (assuming $\boldsymbol{\varepsilon}_0 = \varepsilon_0 \mathbf{I}$),

$$\mathbf{E}_s = (\mathbf{D}_s - 4\pi \mathbf{P}_s)/\varepsilon_0, \tag{2.9}$$

and substituting the result into Eq. (2.8), we obtain an inhomogeneous wave equation for $\mathbf{D}_s$ as

$$\nabla^2 \mathbf{D}_s - (\varepsilon_0 c^2)\,\partial^2 \mathbf{D}_s/\partial t^2 = -4\pi[\nabla \times (\nabla \times \mathbf{P}_s)]. \tag{2.10}$$

Equation (2.10) can be solved if it is assumed that $\mathbf{P}_s$ is given by Eq. (2.7). This simplification results in an inhomogeneous wave equation with the source term given by $-4\pi[\nabla \times (\nabla \times \boldsymbol{\alpha} \cdot \mathbf{E}_i)]$. Solving this inhomogeneous equation gives a basic equation for calculating the scattered field $\mathbf{E}_s$.

To calculate the scattered field, consider that $\mathbf{E}_i$ is given by a plane-wave equation of the form

$$\mathbf{E}_i(\mathbf{r}, t) = \mathbf{E}_i \exp[-i(\mathbf{K}_i \cdot \mathbf{r} - \omega_i t)] + \text{complex conjugate} \tag{2.11}$$

and that $\mathbf{D}_s(\mathbf{r}, t)$ can be Fourier decomposed into various frequency components:

$$\mathbf{D}_s(\mathbf{r}, t) = \frac{1}{2\pi} \int_{-\infty}^{\infty} d\omega\, \mathbf{D}_s(\mathbf{r}, \omega)\, e^{+i\omega t}. \tag{2.12}$$

The Fourier inversion gives the frequency component $\mathbf{D}_s(\mathbf{r}, \omega)$ as

$$\mathbf{D}_s(\mathbf{r}, \omega) = \int_{-\infty}^{\infty} dt\, \mathbf{D}_s(\mathbf{r}, t) e^{-i\omega t}. \tag{2.13}$$

Since $\mathbf{D}_s(\mathbf{r}, t)$ is real, we have

$$\mathbf{D}_s^*(\mathbf{r}, \omega) = \mathbf{D}_s(\mathbf{r}, -\omega). \tag{2.14}$$

Substituting Eqs. (2.11) and (2.12) into (2.10) gives the following equation for $\mathbf{D}_s(\mathbf{r}, \omega)$:

$$\nabla^2 \mathbf{D}_s(\mathbf{r}, \omega) + k_s^2 \mathbf{D}_s(\mathbf{r}, \omega) = -4\pi \nabla \times [\nabla \times \boldsymbol{\alpha}(\mathbf{r}, \omega - \omega_i) \cdot \mathbf{E}_i \exp(i\mathbf{K}_i \cdot \mathbf{r})] \tag{2.15}$$

where $K_s = \omega\sqrt{\varepsilon_0}/c = n\omega/c$ ($n = \sqrt{\varepsilon_0}$ being the index of refraction), and

$$\boldsymbol{\alpha}(\mathbf{r}, \Omega) = \int_{-\infty}^{\infty} dt\, \boldsymbol{\alpha}(\mathbf{r}, t) e^{-i\Omega t}, \tag{2.16}$$

where $\Omega = \omega_i - \omega$.

Equation (2.15) now contains the spatial coordinate **r** as a variable. From the theory of differential equations, the formal solution $\mathbf{D}_s(\mathbf{r}, \omega)$ of this equation is [see, for example, Landau and Lifshitz (1960)]

$$\mathbf{D}_s(\mathbf{r}, \omega) = \int_V d^3r' \frac{\exp(iK_s|\mathbf{r} - \mathbf{r}'|)}{|r - r'|} \nabla' \times [\nabla' \times \boldsymbol{\alpha}(\mathbf{r}', \omega - \omega_i) \cdot \mathbf{E}_i \exp(i\mathbf{K}_i \cdot \mathbf{r}')] \tag{2.17}$$

where the prime sign associated with ∇ indicates that the gradient operator is to operate on the coordinate $\mathbf{r}'$. The integration is to cover all regions in the scattering volume V.

In Eq. (2.17), if we choose the origin of the coordinate system to be somewhere in the interior of the scattering volume, then the vector **r** representing the point at which the scattered field is observed will generally be large compared with the linear dimension of the scattering volume. As a result, the quantity $|\mathbf{r} - \mathbf{r}'|$ can be approximated as

$$|\mathbf{r} - \mathbf{r}'| = (\mathbf{r}^2 + \mathbf{r}'^2 - 2\mathbf{r} \cdot \mathbf{r}')^{1/2} \simeq r - \hat{r} \cdot \mathbf{r}', \tag{2.18}$$

where r is the amplitude of the radius vector **r** and $\hat{r}$ is the unit vector in the direction **r**.

With this approximation, the expression for $\mathbf{D}_s(\mathbf{r}, \omega)$ as given by Eq. (2.17) becomes

$$\mathbf{D}_s(\mathbf{r}, \omega) = \exp(iK_s r)/r \int_V d^3r' \exp(-i\mathbf{K}_s \cdot \mathbf{r}')\nabla' \times [\nabla' \times \boldsymbol{\alpha}(\mathbf{r}', \omega - \omega_i) \cdot \mathbf{E}_i \exp(i\mathbf{K}_i \cdot \mathbf{r}')] \tag{2.19}$$

where $\mathbf{K}_s = \mathbf{K}_s \hat{r}$ is the wave vector of the scattered light.

In Eq. (2.19) the gradient operator can be moved to operate on the function $\exp(-i\mathbf{K}_s \mathbf{r}')$ on its left-hand side by integrating by parts. This manipulation reduces Eq. (2.19) to

$$\mathbf{D}_s(\mathbf{r}, \omega) = -[\exp(iK_s r)/r]\mathbf{K}_s \times [\mathbf{K}_s \times \boldsymbol{\alpha}(\mathbf{K}_i - \mathbf{K}_s, \omega_i - \omega) \cdot \mathbf{E}_i] \tag{2.20}$$

where the quantity $\boldsymbol{\alpha}(\mathbf{q}, \Omega)$ is given by the spatial Fourier transform of $\boldsymbol{\alpha}(\mathbf{r}', \Omega)$ as

$$\boldsymbol{\alpha}(\mathbf{q}, \Omega) = \int_V d^3r' \boldsymbol{\alpha}(\mathbf{r}', \Omega) \exp(-i\mathbf{q} \cdot \mathbf{r}'). \tag{2.21}$$

Note that in Eq. (2.20) the vector $\mathbf{K}_i - \mathbf{K}_s = \mathbf{q}$ is associated with frequency $\omega_i - \omega_s$. The vector **q** is the scattering-wave vector and plays an important role in the discussion of light scattering.

In the vicinity of the scattering volume, $\mathbf{D}_s$ is related to $\mathbf{E}_s$ by Eq. (2.9), but as the scattered wave propagates away from the scattering volume toward the detector, the scattered field $\mathbf{E}_s$ is simply related to $\mathbf{D}_s$ by $\mathbf{E}_s = (1/\varepsilon_0)\mathbf{D}_s$. Therefore, the amplitude of the scattered field with the polarization vector $\hat{n}_s$ at a large distance from the scattering volume at position $\mathbf{r}$ and frequency is given by

$$E_s(\mathbf{r}, \omega) = \hat{n}_s \cdot \mathbf{E}_s(\mathbf{r}, \omega) = -[\exp(iK_s r)/r\varepsilon_0]\hat{n}_s \cdot \{\mathbf{K}_s \times [\mathbf{K}_s \times \boldsymbol{\alpha}(\mathbf{K}_s - \mathbf{K}_i, \omega - \omega_i) \cdot \mathbf{E}_i]\} \quad (2.22)$$

After working out the vector product and using the fact that in an isotropic medium the electromagnetic wave is transverse (i.e., $\hat{n}_s \cdot \mathbf{K}_s = 0$), Eq. (2.22) can be simplified to

$$\begin{aligned} E_s(\mathbf{r}, \omega) &= \hat{n}_s \cdot \mathbf{E}_s(\mathbf{r}, \omega) \\ &= -[\exp(iK_s r)/r\varepsilon_0]K_s^2 E_i \alpha_{si}(\mathbf{q}, \Omega), \end{aligned} \quad (2.23)$$

where $\alpha_{si}(\mathbf{q}, \Omega) = \hat{n}_s \cdot \boldsymbol{\alpha}(\mathbf{q}, \Omega) \cdot \hat{n}_i$; $\mathbf{q} = \mathbf{K}_s - \mathbf{K}_i$; $\Omega = \omega - \omega_i$; and $\hat{n}_i$ and E_i are the polarization vector and the amplitude of the incident light, respectively.

2.3 SPECTRAL POWER DENSITY AND THE POLARIZABILITY DENSITY TIME-CORRELATION FUNCTION

In a light-scattering experiment, the spectral intensity is measured rather than the scattered field itself. With the scattered field described by Eq. (2.23), the scattering intensity at the frequency ω_s in the frequency range between $\omega + d\omega$ and ω detected at position $\mathbf{r}$ is given by (after averaging over an ensemble)

$$\begin{aligned} I(\mathbf{r}, \omega) &= \frac{c}{4} |E_s(\mathbf{r}, \omega)|^2 \\ &= I_0 \frac{4\pi K_s^4}{\varepsilon_0^2 r^2} |\alpha_{si}(\mathbf{q}, \Omega)|^2, \end{aligned} \quad (2.24)$$

where I_0 is the incident intensity and, for the incident field given by Eq. (2.11), corresponds to $I_0 = (c/16\pi)|E_i|^2$.

Note that the scattered intensity is proportional to K_s^4 (or ω_s^4). The K_s^4 dependence shows that blue light is scattered more intensely than red light. The r^{-2} dependence in Eq. (2.24) accounts for the intensity reduction for a spherical wave. The most relevant quantity that connects the light-scattering spectrum with the molecular properties of the medium is $|\alpha_{si}(\mathbf{q}, \Omega)|^2$.

According to Eq. (2.16), the frequency of the scattered light is shifted from that of ω_i by the amount $\Omega = \omega_i - \omega_s$. This is because the polarizability density fluctuates over time. Furthermore, as shown by Eq. (2.21), the scattered-light intensity that appears in the nonforward direction (i.e., in a direction different from K_i) results from the spatial fluctuation of $\boldsymbol{\alpha}$. Clearly, if $\boldsymbol{\alpha}$ is independent of the position vector $\mathbf{r}'$, then we can carry out the $\mathbf{r}'$ integration and obtain a result that is nonvanishing when $\mathbf{K}_s = \mathbf{K}_i$, or when $\mathbf{q} = 0$. Therefore, for light to scatter in the nonforward direction with an accompanying frequency shift, $\boldsymbol{\alpha}$ must vary with time and position; i.e., it must fluctuate both in space and in time. These spatial and temporal fluctuations clearly exist in any condensed media at finite temperature because of the presence of unceasing molecular motion.

We shall now use the result of Eq. (1.24), wherein we showed that the superposition of the spectral density at all frequencies is equal to the mean-square fluctuations.

Taking $A(t) = \alpha_{si}(\mathbf{q}, t)$, we rewrite Eq. (1.24) as

$$\langle |\alpha_{si}(\mathbf{q}, t)|^2 \rangle = \int_{-\infty}^{\infty} d\Omega \, I_{si}(\mathbf{q}, \Omega), \tag{2.25}$$

where $I_{si}(\mathbf{q}, \Omega)$ is the spectral power density of the fluctuations in the polarizability density and is given by

$$I_{si}(\mathbf{q}, \Omega) = |\alpha_{si}(\mathbf{q}, \Omega)|^2 = \int_{-\infty}^{\infty} d\tau \, e^{-i\Omega\tau} \langle \alpha_{si}(\mathbf{q}, \tau) \alpha_{si}^*(\mathbf{q}, 0) \rangle. \tag{2.26}$$

The last step is the result of Eq. (1.21) for $\tau \gg \Omega^{-1}$. The time-correlation function of the polarizability tensor can be obtained by taking the inverse Fourier transform of Eq. (2.26):

$$C_{si}(\mathbf{q}, t) = \langle \alpha_{si}(\mathbf{q}, t) \alpha_{si}^*(\mathbf{q}, 0) \rangle = \frac{1}{2\pi} \int_{-\infty}^{\infty} d\Omega \, \langle |\alpha_{si}(\mathbf{q}, \Omega)|^2 \rangle e^{i\Omega t}. \tag{2.27}$$

As discussed previously, the time-correlation function of the fluctuations of the polarizability density relates the light-scattering spectrum with the molecular properties of the medium. This relationship is made evident if the polarizability density $\alpha_{si}(\mathbf{q}, t)$ for an assembly of molecules in the scattering volume is expressed in terms of a superposition of individual molecular polarizability. This is possible only if the electronic states of the molecules are not greatly perturbed by the intermolecular interaction. In other words, if the collision-induced effect can be ignored, then the polarizability-density tensor can be written as

$$\boldsymbol{\alpha}(\mathbf{r}, t) = \sum_{m} \boldsymbol{\alpha}^{(m)}(t)\, \delta[\mathbf{r} - \mathbf{r}_m(t)] \tag{2.28}$$

where $\boldsymbol{\alpha}^{(m)}(t)$ is the polarizability tensor of molecule m at position $\mathbf{r}_m(t)$. Summation over m is to include all molecules inside the scattering volume. The term $\boldsymbol{\alpha}^{(m)}(t)$ is time dependent because of molecular vibration and rotation, whereas $\mathbf{r}_m(t)$ depends on time because of translational motion.

The spatial Fourier transformation of the local polarizability density [Eq. (2.28)] is thus equal to

$$\boldsymbol{\alpha}(\mathbf{q}, t) = \int d^3r \exp(-i\mathbf{q} \cdot \mathbf{r})\boldsymbol{\alpha}(\mathbf{r}, t) = \sum_{m} \boldsymbol{\alpha}^{(m)}(t) \exp[-i\mathbf{q} \cdot \mathbf{r}_m(t)]. \tag{2.29}$$

Using this expression, we can rewrite the time-correlation function

$$C_{si}(\mathbf{q}, t) = \sum_{m,n} \langle \alpha_{si}^{(m)}(t)\alpha_{si}^{(n)} \exp\{-i\mathbf{q} \cdot [\mathbf{r}_m(t) - \mathbf{r}_n]\}\rangle. \tag{2.30}$$

Equation (2.30) is the basic expression that relates the molecular properties of the medium to the spectrum of the scattered light. We shall use this expression to discuss the Rayleigh–Brillouin-scattering and Raman-scattering spectra.

It should be emphasized that the approximation introduced in Eq. (2.28) is not valid for interpreting the collision-induced light-scattering spectra of molecular fluids. In general, when two molecules undergo collision, the electronic-wave function of one molecule is distorted by the other; the distortion persists for a time on the order of the duration of the collision, which is about 10^{-13} sec in a dense fluid. During this time of collision, the total polarizability of the colliding molecules is not equal to the sum of the polarizabilities of the two individual molecules. The induced effect also makes an important contribution not included in Eq. (2.28). Collision-induced light scattering in monatomic fluids has been studied extensively, but there is no quantitative description at present for collision-induced light-scattering spectra of more complex molecular fluids.

2.4 LIGHT SCATTERING FROM TRANSLATIONAL MOTION

In the previous section, we considered the scattering of light by a system of molecules. The spectrum of the scattered light was shown to be proportional to the Fourier transform of the time-correlation function $C_{si}(\mathbf{q}, t)$. We shall now investigate the types of information that can be extracted from the study of the light-scattering spectrum of a dense fluid. The simplest case to consider is scattering from a system of spherical molecules, assuming that the anisotropic part of the molecular polarizability tensor resulting from the collision-induced mechanism is negligible. In this simplification, the molecular-polarizability tensor in the molecular-coordinate frame is given by

$$\boldsymbol{\alpha}^{(m)} = \alpha^{(m)}\mathbf{I}, \tag{2.31}$$

where $\mathbf{I}$ is the unit tensor.

The component of the molecular polarizability in the laboratory coordinate frame is given by

$$\alpha_{si}^{(m)} = \alpha^{(m)}(\hat{n}_s \cdot \mathbf{I} \cdot \hat{n}_i) = \alpha^{(m)}(\hat{n}_s \cdot \hat{n}_i). \tag{2.32}$$

The correlation function $C_{si}(\mathbf{q}, t)$ then becomes

$$C_{si}(\mathbf{q}, t) = (\hat{n}_s \cdot \hat{n}_i)^2 \sum_{m,n} \langle \alpha^{(m)}(t)\alpha^{(n)} \exp\{-i\mathbf{q} \cdot [\mathbf{r}_m(t) - r_n]\}\rangle. \tag{2.33}$$

The time dependence in $\alpha^{(m)}(t)$ results from molecular vibration and rotation. However, in the case of spherical molecules, rotational motion will not change the molecular polarizability. Furthermore, if we are interested in scattering phenomena resulting from the translational motion of the molecules, the time dependence of $\alpha^{(m)}(t)$ given in Eq. (2.33) can be suppressed. This is because the vibrational motion that causes $\alpha^{(m)}$ to depend on time occurs on a much faster time scale and its effect is canceled, as long as we focus on the spectral region associated with the much slower translational motion.

For a system of identical particles, the polarizabilities of all molecules are identical. Thus, in Eq. (2.33) we can set $\alpha^{(m)} = \alpha^{(n)} = \alpha$ and rewrite $C_{si}(\mathbf{q}, t)$ as

$$C_{si}(\mathbf{q}, t) = \alpha^2(\hat{n}_s \cdot \hat{n}_i)^2 \left\langle \sum_{m,n} \exp\{-i\mathbf{q} \cdot [\mathbf{r}_m(t) - \mathbf{r}_n]\} \right\rangle. \tag{2.34}$$

We now wish to calculate $S(\mathbf{q}, \Omega)$, which is defined as

$$S(\mathbf{q}, \Omega) = \frac{1}{2\pi} \int_{-\infty}^{\infty} dt \exp(-i\Omega t) \left\langle \sum_{m,n} \exp\{-i\mathbf{q} \cdot [\mathbf{r}_m(t) - \mathbf{r}_n]\} \right\rangle. \tag{2.35}$$

Here $S(\mathbf{q}, \Omega)$ is known as the dynamic-structure factor and is one of the most important quantities in the theory of a many-body system. This factor enters not only in inelastic light scattering but also in x-ray and neutron scattering experiments, to name a few.

In a typical light-scattering experiment, a laser beam of frequency ω_i and wave vector $\mathbf{K}_i$ is introduced to interact with a sample. The intensity of scattered light with frequency $\omega_s = \omega_i \pm \Omega$ and wave vector $\mathbf{K}_s = \mathbf{K}_i \pm \mathbf{q}$ is then measured at the detector (Berne and Pecora, 1976). The schematic diagram for light scattering is given in Fig. 2.1.

We can now calculate the amplitude of $\mathbf{q}$. Since

$$\pm\mathbf{q} = \mathbf{K}_s - \mathbf{K}_i, \tag{2.36}$$

the square of Eq. (2.36) is

$$q^2 = K_s^2 + K_i^2 - 2K_s K_i \cos\theta, \tag{2.37}$$

where θ is the scattering angle.

The frequency shift Ω involved in light scattering from translational motion is very small compared with ω_i or ω_s; it is an excellent approximation to write

$$K_s \simeq K_i = 2\pi/\lambda_i = 2\pi n/\lambda_0. \tag{2.38}$$

Substituting Eq. (2.38) into (2.37) gives

$$q = |\mathbf{q}| = 2K_i \sin\theta/2 = (4\pi n/\lambda_0)\sin\theta/2. \tag{2.39}$$

Here n is the index of refraction of the fluid that gives rise to light scattering and λ_0 the wavelength of the incident light in a vacuum. For $\lambda_0 = 4880$ Å, $\theta = 90°$, and $n = 1.3$, q is equal to 2.4×10^5 cm^{-1}. Thus, the physical process probed by light scattering is of long-wavelength excitation. This contrasts with neutron scattering, in which the q involved is on the order of 10^8 cm^{-1}; therefore, neutron scattering probes short-wavelength excitations. In other words, the spatial resolution in light scattering is only about 10^{-5} cm, whereas, in neutron scattering it is about 10^{-8} cm.

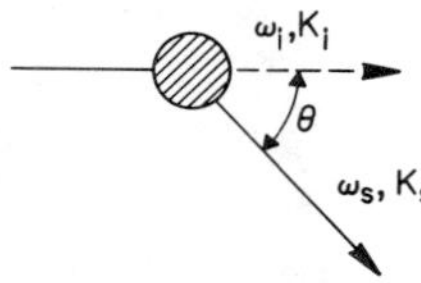

FIG. 2.1 Schematic diagram for light scattering.

Next we note that the phase function $\sum_m \exp[-i\mathbf{q}\cdot\mathbf{r}_m(t)]$ can be written as

$$\begin{aligned}\sum_m \exp[-i\mathbf{q}\cdot\mathbf{r}_m(t)] &= \int_V d^3r \exp(-i\mathbf{q}\cdot\mathbf{r}) \sum_m \delta[\mathbf{r}-\mathbf{r}_m(t)] \\ &= \int_V d^3r \exp(-i\mathbf{q}\cdot\mathbf{r})\rho(\mathbf{r}, t). \end{aligned} \tag{2.40}$$

where $\rho(\mathbf{r}, t)$ is the local-number density at time t and position $\mathbf{r}$ within the scattering volume.

We can resolve $\rho(\mathbf{r}, t)$ into an average and a fluctuating part as

$$\rho(\mathbf{r}, t) = \sum_m \delta[\mathbf{r}-\mathbf{r}_m(t)] = \delta\rho(\mathbf{r}, t) + \rho_0, \tag{2.41}$$

where ρ_0 is the average number density and $\delta\rho(\mathbf{r}, t)$ is the fluctuation of number density at point r and time t.

Since ρ_0 is independent of $\mathbf{r}$, it contributes only to forward scattering ($\mathbf{q} = 0$). Obviously, for $\mathbf{q} \neq 0$, Eq. (2.40) can be written in terms of the spatial Fourier transform of the density fluctuation $\delta\rho(\mathbf{r}, t)$ as

$$\sum_m \exp -i\mathbf{q}\cdot\mathbf{r}_m(t) = \int_V d^3r\, \delta\rho(\mathbf{r}, t) \exp -i\mathbf{q}\cdot\mathbf{r} = \delta\rho(\mathbf{q}, t). \tag{2.42}$$

Thus, instead of Eq. (2.35), we can write the dynamic-structure factor as

$$S(\mathbf{q}, \Omega) = \frac{1}{2\pi}\int_{-\infty}^{\infty} dt\, e^{-i\Omega t} \langle \delta\rho(\mathbf{q}, t)\, \delta\rho^*(\mathbf{q}, 0)\rangle. \tag{2.43}$$

Since the light-scattering spectral density from translational motion is proportional to the dynamic-structure factor, the treatment of the dynamic-structure factor of the density–density-correlation function holds the key to understanding the light-scattering spectrum resulting from translational motion.

Next we shall consider a system of macromolecules dissolved in a solvent consisting of small molecules. In such a solution, the scattering spectral density is reflected mainly by the dynamical state of the macromolecules, because the polarizability density of macromolecules is greater than that of small solvent molecules. Furthermore, the macromolecules move more slowly than the solvent molecules. Therefore, the scattering spectrum for this system is expected to consist of a sharp peak associated with the translational motion of macromolecules followed by a broad spectral background that results from the molecular motion of solvent molecules.

2.5 TRANSLATIONAL DIFFUSION OF MACROMOLECULES IN DILUTE SOLUTION[§]

The time-correlation function that enters into the dynamic-structure factor $S(\mathbf{q}, \Omega)$ as given in Eq. (2.35) can be divided into two parts, a self term and a pair term:

$$\begin{aligned}\langle \delta\rho(\mathbf{q}\cdot t)\delta\rho^*(q)\rangle &= \left\langle \sum_{m,n} \exp\{-i\mathbf{q}\cdot[\mathbf{r}_m(t) - \mathbf{r}_n]\} \right\rangle \\ &= \left\langle \sum_{n} \exp\{-i\mathbf{q}\cdot[\mathbf{r}_n(t) - \mathbf{r}_n]\} \right\rangle \\ &\quad + \left\langle {\sum_{m,n}}' \exp\{-i\mathbf{q}\cdot[\mathbf{r}_m(t) - \mathbf{r}_n]\} \right\rangle. \end{aligned} \tag{2.44}$$

The first term denotes the self-correlation and the second, the pair correlation. The prime sign in the second term indicates that m is not equal to n. We consider here only the narrow, sharp spectral component caused mainly by the translational motion of the macromolecules; thus, the position vectors in Eq. (2.44) refer only to those of macromolecules. Furthermore, we assume that the concentration of macromolecules is so small that the positions of different macromolecules are not correlated. As a result, the second term on the right-hand side of Eq. (2.44) can be neglected, and we need to consider only the self-correlation term given by

$$F(\mathbf{q}, t) = \frac{1}{N}\left\langle \sum_{n} \exp\{-i\mathbf{q}\cdot[\mathbf{r}_n(t) - \mathbf{r}_n]\} \right\rangle = \frac{1}{N}\left\langle \sum_{n} \exp[-i\mathbf{q}\cdot\Delta\mathbf{r}_n(t)] \right\rangle. \tag{2.45}$$

The quantity $\Delta\mathbf{r}_n(t)$ is the displacement of the macromolecule n in the time interval between 0 and t.

It is useful to introduce a correlation function

$$G(\mathbf{r}, t) = \frac{1}{N}\left\langle \sum_{n} \delta[r - \Delta\mathbf{r}_n(t)] \right\rangle. \tag{2.46}$$

This function is known as the van Hove self-space- time-correlation function (Van Hove, 1954), originally introduced to describe the incoherent neutron-scattering process from a molecular liquid. The correlation function $G(\mathbf{r}, t)$ can be considered as the probability that any particle in the system will suffer a displacement of approximately d^3r at the point $\mathbf{r}$ between time 0 and t.

[§] (Berne and Pecora, 1976).

Since

$$G(\mathbf{r}, 0) = \delta(\mathbf{r}), \tag{2.47}$$

it can be regarded as the conditional probability that some particle is located in d^3r at $\mathbf{r}$ and time t, given that the particle was at the origin at $t = 0$.

Clearly, $G(\mathbf{r}, t)$ is related to $F(\mathbf{q}, t)$ [Eq. (2.46)] by a spatial Fourier transform

$$\int_V d^3r\, G(\mathbf{r}, t) \exp -i\mathbf{q} \cdot \mathbf{r} = F(\mathbf{q}, t). \tag{2.48}$$

We shall show in Chapter 4 that for a system of particles that undergoes Brownian motion, the time-correlation function $G(\mathbf{r}, t)$ satisfies the diffusion equation

$$\partial G(\mathbf{r}, t)/\partial t = D\nabla^2 G(\mathbf{r}, t), \tag{2.49}$$

with the initial condition given by Eq. (2.47). Here D is the translational-diffusion coefficient.

Equation (2.49) can be simplified by converting the spatial dependence of $G(\mathbf{r}, t)$ into the q dependence through a Fourier transformation. This is carried out by first multiplying both sides of the equation by the function $\exp -i\mathbf{q} \cdot \mathbf{r}$ and then integrating over the scattering volume. Performing these two successive integrations by parts converts Eq. (2.49) into

$$\partial F(\mathbf{q}, t)/\partial t = -Dq^2 F(\mathbf{q}, t), \tag{2.50}$$

which immediately yields the solution

$$F(\mathbf{q}, t) = \exp - t/\tau_q, \tag{2.51}$$

where $\tau_q = (q^2D)^{-1}$. Note that τ_q becomes infinitely long as $q \to 0$. Any fluctuation in this property is called a hydrodynamic mode. Therefore, translational diffusion of a Brownian particle is a hydrodynamic mode.

The dynamic-structure factor for the dilute macromolecular solution is the temporal Fourier transform of $NF(\mathbf{q}, t)$:

$$S(q, \omega) = N \frac{q^2 D}{\omega^2 + (q^2 D)^2} \tag{2.52}$$

where we have replaced Ω by ω. Thus, the half-width at half-height of the scattering spectrum is equal to $q^2D = \tau_q^{-1}$, where τ_q is the relaxation time of $F(\mathbf{q}, t)$. Plotting τ_q^{-1} versus q^2 gives a straight line with the slope equal to the translational-diffusion coefficient. Hence, the light-scattering technique provides a very convenient method for measuring the translational-diffusion coefficient of macromolecules in dilute solution. Typically, the relaxation times τ_q for macromolecules (observed at $\theta = 90°$) are on the order of 10^{-3} to 10^{-5} sec. Light mixing techniques such as homodyne and heterodyne are used for this type of lightscattering study.

2.6 LIGHT SCATTERING FROM TRANSLATIONAL MOTION OF A PURE DENSE FLUID

The treatment of the density–density-correlation function in terms of the diffusion equation neglects the correlation of positions between different molecules. This assumption is applicable only for a system of particles in dilute concentration. As the concentration increases, the pair-correlation term cannot be neglected. Inclusion of both the self- and pair-correlation terms requires the full consideration of the density–density-correlation function. We shall next show that, in such a situation, the density fluctuation is manifested in the production of sound waves and entropy. The sound wave is associated with the adiabatic part of the density fluctuation and is oscillatory. Scattering from sound waves yields a doublet structure shifted symmetrically up and down about the excitation-laser frequency. The doublet is known as the Brillouin doublet, and the frequency shift of the Brillouin doublet is related to the velocity of the longitudinal sound wave in the gigahertz (10^9 Hz) frequency region. The width of the Brillouin doublet is determined by the attenuation mechanism of the gigahertz hypersonic wave.

The part owing to entropy production corresponds to the isobaric part of the density fluctuation and is not oscillatory. Scattering from this part does not result in a frequency shift but contributes to the intensity at the position of the incident-laser frequency (i.e., $\Omega = 0$) and is known as Rayleigh scattering. For an ordinary viscous fluid, the width of the central Rayleigh component is determined by the thermal diffusivity of the scattering medium.

Rayliegh–Brillouin light-scattering experiments have been performed on many viscous fluids. The spectrum of an ordinary viscous fluid has a three-

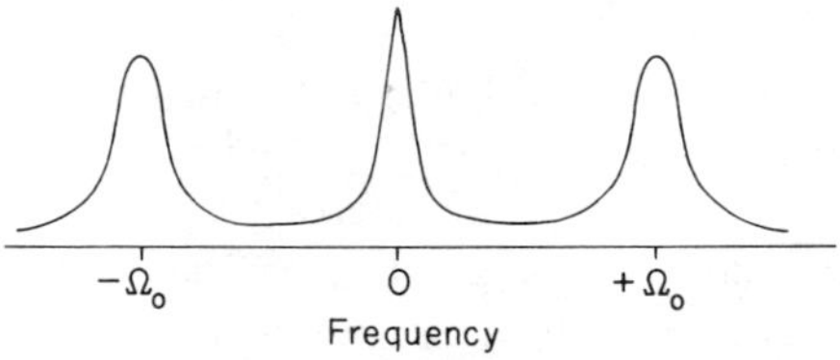

FIG. 2.2 Rayleigh–Brillouin spectrum of a viscous fluid.

peak structure, as shown in Fig. 2.2. In the next section we shall calculate the spectral structure of Rayleigh–Brillouin scattering in terms of the hydrodynamic theory (Mountain, 1966).

2.7 HYDRODYNAMIC EQUATIONS

The hydrodynamic theory rests on the assumption that the system can be treated as a continuous medium. This means that the macroscopic (observable) quantities such as density, pressure, and temperature are continuous functions of space and time. We can formulate a macroscopic theory by subdividing our system into a large number of subvolume elements. Each element must contain enough molecules to permit its macroscopic properties to be measured or calculated by taking averages of the molecular properties within the sub volume; and at the same time, each element must be sufficiently small so that the values obtained do not vary appreciably. In other words, the subvolumes must be microscopically large and macroscopically small.

The continuous theory is justified for calculating the Rayleigh–Brillouin spectrum for the following reasons. Using the convolution theorem, we can write the density–density-correlation function given in Eq. (2.43) as

$$\langle \delta\rho(\mathbf{q}, t)\delta\rho^*(\mathbf{q}, 0)\rangle = \int_V d^3r \exp(i\mathbf{q}\cdot\mathbf{r}) \int_V d^3r' \langle \delta\rho(\mathbf{r} - \mathbf{r}', t)\, \delta\rho(\mathbf{r}', 0)\rangle. \tag{2.53}$$

Obviously, Rayleigh–Brillouin scattering occurs as a result of the correlation of density fluctuations at different points and at different times. When the correlation disappears, there will be no Rayleigh–Brillouin scattering. In a dense fluid, the intermolecular distance is short compared with the wavelength of light used in a typical light-scattering experiment. Thus, within one wavelength of light many molecules are present, and we expect that a continuum theory will be sufficient to explain the Rayleigh–Brillouin spectrum. Indeed, we shall show that much information about the Rayleigh–Brillouin spectrum of an ordinary dense fluid can be obtained from the hydrodynamic equations. On the other hand, if the gas density is so low that the mean free path of molecules becomes long or comparable with the wavelength of the light, the continuum theory of hydrodynamics will become inadequate, and in this case the kinetic theory will be needed to predict the spectrum from the translational motion of molecules.

In a continuum theory, each hydrodynamic process is associated with a conserved local variable. There are five conserved local variables: the number density $\rho(\mathbf{r}, t)$, the momentum density $\mathbf{g}(\mathbf{r}, t)$, and the energy density $\varepsilon(\mathbf{r}, t)$. These density functions are defined as

$$\rho(\mathbf{r}, t) = \left\langle \sum_l \delta[\mathbf{r} - \mathbf{r}_l(t)] \right\rangle \tag{2.54}$$

$$\mathbf{g}(\mathbf{r}, t) = \left\langle \sum_l \mathbf{\Pi}_l(t)\delta[\mathbf{r} - \mathbf{r}_l(t)] \right\rangle \tag{2.55}$$

$$\varepsilon(\mathbf{r}, t) = \left\langle \sum_l \frac{\Pi_l^2}{2m_l} \delta[\mathbf{r} - \mathbf{r}_l(t)] + \frac{1}{2}\sum_{k,l}{}' \phi(r_{kl})\, \delta[\mathbf{r} - \mathbf{r}_l(t)] \right\rangle, \tag{2.56}$$

where $\mathbf{\Pi}_l$ is the linear momentum of molecule l and $\phi(\mathbf{r}_{kl})$ the interaction potential of molecules k and l, which is assumed to be short range. The prime sign associated with the second term in Eq. (2.56) denotes that k is not equal to l. For a system of identical particles, the momentum density can be written as $\mathbf{g}(\mathbf{r}, t) = \rho_m \mathbf{u}$, where $\rho_m = m\rho$.

In the hydrodynamic assumption, each subvolume is an open system in which the local densities can vary over time. We therefore begin with a discussion of the balance equations for number, momentum, and energy; these are conserved variables in hydrodynamics (deGroot and Mazur, 1962).

Consider an arbitrary volume element within the hydrodynamic system. We are interested in the time evolution of extensive quantities $F(t)$. We distinguish two mechanisms for changing $F(t)$:

$$dF(t)/dt = F_{\mathrm{p}}(t) + F_{\mathrm{f}}(t), \tag{2.57}$$

where $F_{\mathrm{p}}(t)$ and $F_{\mathrm{f}}(t)$ represent, respectively, the production of the quantity F within the volume element and the rate of flow of F into the volume through the boundary surface S. The corresponding intensive quantities (local densities) are defined by

$$F(t) = \int_V d^3r\, f(\mathbf{r}, t), \tag{2.58a}$$

$$F_{\mathrm{p}}(t) = \int_V d^3r \sum (\mathbf{r}, t) \tag{2.58b}$$

$$F_{\mathrm{f}}(t) = -\int_S \mathbf{J}(\mathbf{r}, t) \cdot d\mathbf{S} = -\int_V d^3r\, \nabla \cdot \mathbf{J}(\mathbf{r}, t), \tag{2.58c}$$

where $\mathbf{J}$ is the current density (or flux) corresponding to the rate of flow of F per unit area per unit time across the differential area $d\mathbf{S}$ into the volume V. The last term in Eq. (2.58c) is obtained through the use of the divergence theorem. In terms of the intensive quantities, the balance equation [Eq. 2.59] takes the local form

$$\partial f(\mathbf{r}, t)/\partial t = \sum(\mathbf{r}, t) - \nabla \cdot \mathbf{J}(\mathbf{r}, t). \tag{2.59}$$

The minus sign in the flow term results from the opposite sense in the definitions of the current density and the surface element. It should be noted that the flow term will, in general, consist of separate contributions owing to convection ($\mathbf{J}_{\text{conv}}$) and conduction ($\mathbf{J}_{\text{conv}}$). The convection contribution to the current density results from the local flow of matter with velocity $\mathbf{u}$. Thus, $\mathbf{J} = f\mathbf{u} + \mathbf{J}_{\text{cond}}$. The conduction flux occurs even in a system at rest.

We see that the balance equation (2.59) takes a particularly simple form for quantities that are conserved (i.e., $\sum = 0$):

$$\partial f/\partial t + \nabla \cdot \mathbf{J} = 0. \tag{2.60}$$

Thus, the conservation laws for mass (or the number of molecules in the absence of chemical reactions), momentum, and energy (with no external forces) are expressed by Eq. (2.60). Entropy, on the other hand, is not conserved and satisfies the inequality $\sum(s) \geq 0$.

These conserved local-density functions are described explicitly in the hydrodynamic theory by the following continuity equations:

$$\partial \rho(\mathbf{r}, t)/\partial t + \nabla \cdot \mathbf{J}(\mathbf{r}, t) = 0, \tag{2.61}$$

$$\partial \mathbf{g}(\mathbf{r}, t)/\partial t + \nabla \cdot \boldsymbol{\tau}'(\mathbf{r}, t) = 0, \tag{2.62}$$

$$\partial \varepsilon(\mathbf{r}, t)/\partial t + \nabla \cdot \mathbf{J}_{\text{h}}(\mathbf{r}, t) = 0. \tag{2.63}$$

Here we emphasize that the quantities ρ, ε, $\mathbf{J}$, $\boldsymbol{\tau}'$, and $\mathbf{J}_{\text{h}}$ are quantities at local equilibrium and not the true equilibrium values. Their amplitudes vary in time and location and relax only slowly to true equilibrium values. We shall not make a distinction between these local-equilibrium quantities and those that enter into the correlation functions. For example, in the density–density-correlation function $\langle \rho(\mathbf{r}, t)\, \rho(\mathbf{r}', t')\rangle$, $\rho(\mathbf{r}, t)$ in the rigorous sense should be a microscopic variable; in hydrodynamic theory, it is regarded as a preaveraged local-equilibrium quantity. This approximation is valid only in long-time and long-wavelength limits and can be justified by the linear-response theory.

The continuity equations given in Eqs. (2.61)–(2.63) are not useful unless they are supplemented with the constitutive equations. The number flux $\mathbf{J}(\mathbf{r}, t)$ is equal to $\rho\mathbf{u}$, because the conduction current does not contribute to the transport of mass (or number of molecules). The first hydrodynamic equation is then given by

$$\partial \rho/\partial t + \nabla \cdot (\rho\mathbf{u}) = 0. \tag{2.64}$$

The derivation of the momentum flux will now be considered. The momentum flux owing to convection (the molecular flow) is simply $\rho\mathbf{uu}$, and the part owing to conduction results from intermolecular interaction. This intermolecular interaction part of the momentum flux will be called the stress tensor and is designated as $\boldsymbol{\sigma}$. The element σ_{ij} can be considered the ith component of the force per unit area exerted on a plane perpendicular to direction j. We assume that the stress tensor is symmetric and thus has six components. The three diagonal elements σ_{xx}, σ_{yy}, and σ_{zz} are of normal stress. They represent forces per unit area acting normal to a given plane. The other three off-diagonal components are shear stresses; they represent forces per unit area acting parallel to the planes of reference.

To see how the stress tensor comes about, consider for simplicity the case of an ideal liquid in which no intermolecular interaction is present. The total force acting on the volume V because of the hydrostatic pressure P is given as (see Fig. 2.3)

$$F = \int_V d^3r\, \rho_m \frac{d\mathbf{u}}{dt} = -\int_A d\mathbf{S}\, P, \tag{2.65}$$

where ρ_m is the mass density and P the hydrostatic pressure. We can transform the surface integral, given in the last step, into the volume integral by Green's theorem. Equating the integrands, we obtain

$$\rho_m\, d\mathbf{u}/dt = \rho_m[\partial\mathbf{u}/\partial t + (\mathbf{u}\cdot\nabla)\mathbf{u}] = -\nabla P. \tag{2.66}$$

Thus, the ith component of velocity satisfies the equation

$$\partial u_i/\partial t = -(1/\rho_m)\, \partial P/\partial x_i - u_k(\partial u_i/\partial x_k), \tag{2.67}$$

where the repeated index k denotes summation over the three Cartesian components.

We want to incorporate Eq. (2.67) into an equation for the rate of change of the momentum density over time. The ith component of the momentum density is given by $\rho_m u_i$. Its rate of change at a fixed spatial point is given by

$$\partial(\rho_m u_i)/\partial t = \rho_m\, \partial u_i/\partial t + u_i\, (\partial\rho_m/\partial t). \tag{2.68}$$

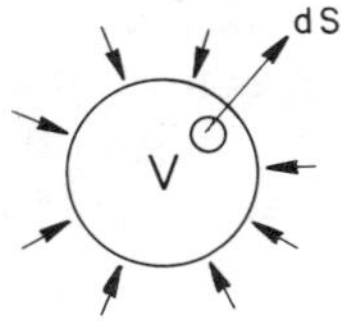

FIG. 2.3 The force exerted on a fluid due to hydrostatic pressure.

Substituting in the expressions of Eqs. (2.64) and (2.67) we obtain respectively, for the $\partial\rho/\partial t$ and $\partial u_i/\partial t$ terms,

$$\partial g_i/\partial t = \partial(\rho_m u_i)/\partial t = -\partial(\rho_m u_k u_i)/\partial x_k - \partial P/\partial x_i = \frac{\partial}{\partial x_k}(-\rho_m u_i u_k - P\delta_{ik}). \tag{2.69}$$

Comparing this equation with Eq. (2.62) gives an expression for the momentum-flux tensor of the ideal fluid as

$$\boldsymbol{\tau}' = \rho_m \mathbf{uu} + P\mathbf{I}, \tag{2.70}$$

where $\mathbf{I}$ is the unit tensor. Thus, in the absence of intermolecular forces, the conduction part of the stress tensor $\boldsymbol{\tau}'$ is given by the pressure term $P\mathbf{I}$. However, in a real system it is necessary to add the contribution from intermolecular interaction to the stress tensor.

In the case of real fluids, intermolecular interaction can cause momentum transport even in the absence of net molecular flow. It is this intermolecular-interaction contribution to the stress tensor that determines the viscosity of fluids. In the case of elastic solids, the stress tensor is related to strain. Here we shall consider only the fluid case, leaving the discussion of the elastic response of solids to another section in this chapter.

The stress tensor of a fluid consists of a hydrostatic pressure part and an intermolecular part σ_{ij}:

$$\tau'_{ij} = \rho u_i u_j + P\delta_{ij} + \sigma_{ij}. \tag{2.71}$$

Since a fluid cannot sustain a steady strain when subjected to external stress, σ_{ij} must be related to the velocity gradient that determines the flow of the fluid. For simplicity, we consider the case when σ_{ij} is proportional to the rate of strain (i.e., the velocity gradient $\partial u_i/\partial x_j$.) For a fluid that is rotationally isotropic, two invariants can be constructed from the velocity gradient tensors $\partial u_i/\partial x_j$. The two invariants are the scalar component given by $(\nabla \cdot \mathbf{u})\delta_{ij}$ and the traceless component given by $(\partial u_i/\partial x_j + \partial u_j/\partial x_i - \frac{2}{3}(\nabla \cdot \mathbf{u})\delta_{ij})$. Thus, the intermolecular part of the stress tensor can be written as

$$\sigma_{ij} = -\eta_S[\partial u_i/\partial x_j + \partial u_j/\partial x_i - \tfrac{2}{3}(\nabla \cdot \mathbf{u})\delta_{ij}] - \eta_V(\nabla \cdot \mathbf{u})\delta_{ij}, \tag{2.72}$$

where the proportionality coefficient η_s and η_v are known as the shear and bulk viscosities, respectively. The minus signs in front of η_s and η_v in Eq. (2.72) are warranted because η_S and η_V are positive. A fluid with σ_{ij} given by

Eq. (2.72) is called a Newtonian fluid. The conservation-of-momentum equation for the Newtonian fluid is thus given by

$$\frac{\partial}{\partial t}(\rho_{\rm m} u_i) = \frac{\partial}{\partial x_k}\left\{-\rho_{\rm m} u_i u_k - P\delta_{ik} + \eta_{\rm S}\left[\frac{\partial u_i}{\partial x_k} + \frac{\partial u_k}{\partial x_i} - \frac{2}{3}(\nabla\cdot\mathbf{u})\delta_{ik}\right] + \eta_{\rm V}(\nabla\cdot\mathbf{u})\delta_{ik}\right\}, \tag{2.73}$$

which, after carrying out some algebraic manipulation, can be written as

$$\rho_{\rm m}\left(\frac{\partial u_i}{\partial t} + u_k\frac{\partial u_i}{\partial x_k}\right) = -\frac{\partial P}{\partial x_i} + \eta_{\rm S}\frac{\partial^2 u_i}{\partial x_k \partial x_k} + \left(\eta_{\rm V} + \frac{1}{3}\eta_{\rm S}\right)\frac{\partial}{\partial x_i}(\nabla\cdot\mathbf{u}). \tag{2.74}$$

In terms of vector notation, this equation may be written as

$$\rho_{\rm m}[\partial\mathbf{u}/\partial t + (\mathbf{u}\cdot\nabla)\,\mathbf{u}] = -\nabla P + \eta_{\rm S}\nabla^2\mathbf{u} + (\eta_{\rm V} + \tfrac{1}{3}\eta_{\rm S})\,\nabla(\nabla\cdot\mathbf{u}). \tag{2.75}$$

Equation (2.74) [or equivalently Eq. (2.75)] is known as the Navier–Stokes equation. This is the second basic hydrodynamic equation.

The local-energy flux consists of three contributions. The first represents the energy transport caused by material flow; this term is given by $\varepsilon\mathbf{u}$. The second contribution represents the work done by stress from fluids outside the volume element d^3r on fluids inside the volume element; this is given by $P\mathbf{u} + \boldsymbol{\sigma}\cdot\mathbf{u}$. The third results from heat conduction, which, according to Fourier's law, is given by

$$\mathbf{Q} = -\lambda\nabla T, \tag{2.76}$$

where λ is the thermal-conductivity coefficient.

Combining these three contributions gives the energy equation

$$\partial\varepsilon/\partial t = -\nabla\cdot(\varepsilon\mathbf{u} + P\mathbf{u} + \mathbf{u}\cdot\boldsymbol{\sigma} - \lambda\nabla T), \tag{2.77}$$

where again for brevity of notation we have used ε for $\varepsilon(\mathbf{r}, t)$. We also have not written out the argument in $\mathbf{u}$, $\boldsymbol{\sigma}$, and T; however, it should be understood that these quantities are functions of $\mathbf{r}$ and t.

The local-energy density consists of the contributions owing to the kinetic-energy density $\frac{1}{2}m\rho u^2$ and the internal-energy density $e(\mathbf{r}, t)$:

$$\varepsilon = \tfrac{1}{2}m\rho u^2 + e. \tag{2.78}$$

Substituting Eq. (2.78) into Eq. (2.77) and eliminating the kinetic-energy term with the help of the equation of continuity [Eq. (2.64)] and the momentum conservation equation [Eq. (2.73)], we obtain, after some algebraic manipulation, an equation for internal energy:

$$\partial e/\partial t = -\nabla \cdot (e\mathbf{u} - \lambda \nabla T) + P(\nabla \cdot \mathbf{u}) + \boldsymbol{\sigma}\colon\nabla\mathbf{u}, \tag{2.79}$$

where the quantity $\boldsymbol{\sigma}\colon\nabla\mathbf{u} = \sigma_{ij}\,\partial u_i/\partial x_j$. Again, the repeated indices denote summation. Equation (2.79) is the third basic hydrodynamic equation.

Equations (2.64), (2.75), and (2.79) represent the fundamental hydrodynamic equations for a Newtonian liquid. These equations contain a number of coefficients that remain undetermined. They are thermodynamic derivatives and transport coefficients, and they are the parameters of the hydrodynamic theory.

Following are a set of five nonlinear, coupled, partial-differential equations. The application of these equations is the subject of fluid dynamics. To calculate the dynamic-structure factor of a fluid away from its critical point it suffices to linearize these equations by assuming small fluctuations in the values for local-equilibrium density $\delta\rho$, local-equilibrium internal-energy density δe, local velocity $\delta\mathbf{u}$, local temperature δT, and local pressure δP around their respective true equilibrium values ρ_0, ε_0, $\mathbf{u}_0$ $(=0)$, T_0, and P_0.

In this case, Mountain (1966) assumed

$$\begin{aligned} \rho &= \rho_0 + \delta\rho, \\ e &= e_0 + \delta e, \\ \mathbf{u} &= \mathbf{u}_0 + \delta\mathbf{u} = \delta\mathbf{u}, \\ T &= T_0 + \delta T, \\ P &= P_0 + \delta P, \end{aligned} \tag{2.80}$$

and he then substituted these into Eqs. (2.64), (2.75), and (2.79) and rejected all terms higher than the first order in the fluctuation variables. After doing so, he obtained a set of linearized hydrodynamics equations:

$$\partial\delta\rho/\partial t + \rho_0 \nabla \cdot \mathbf{u} = 0, \tag{2.81}$$

$$m\rho\,\partial\mathbf{u}/\partial t = -\nabla(\delta P) + \eta_S \nabla^2\mathbf{u} + (\eta_V + \tfrac{1}{3}\eta_S)\,\nabla(\nabla\cdot\mathbf{u}), \tag{2.82}$$

$$\partial\delta e/\partial t + (P_0 + e_0)\,\nabla\cdot\mathbf{u} = \lambda\nabla^2(\delta T). \tag{2.83}$$

These linearized equations can now be used to determine the dynamic-structure factor and, hence, the Rayleigh–Brillouin spectrum of an ordinary viscous fluid.

2.8 SOLUTION OF HYDRODYNAMIC EQUATIONS AND INTERPRETATION OF THE RAYLEIGH–BRILLOUIN SPECTRUM OF A VISCOUS FLUID

To solve the set of linearized hydrodynamic equations given in Eqs. (2.81)–(2.83), it is convenient to first separate the velocity vector **u** into a longitudinal and a transverse part:

$$\mathbf{u}(\mathbf{r}, t) = \mathbf{u}_L(\mathbf{r}, t) + \mathbf{u}_T(\mathbf{r}, t), \tag{2.84}$$

where $\mathbf{u}_L$ and $\mathbf{u}_T$ are the longitudinal and transverse components, respectively. They satisfy

$$\nabla \times \mathbf{u}_L = 0, \qquad \text{and} \qquad \nabla \cdot \mathbf{u}_T = 0. \tag{2.85}$$

Equations (2.81) and (2.83) contain only the longitudinal velocity component. The linearized Navier–Stokes equation [Eq. (2.82)] has both the longitudinal and transverse components. The transverse component is given by

$$m\rho_0\, \partial \mathbf{u}_T/\partial t - \eta_S \nabla^2 \mathbf{u}_T = 0. \tag{2.86}$$

This equation can be solved by the Fourier–Laplace transform technique:

$$\hat{\mathbf{u}}_T(\mathbf{q}, z) = \int_0^\infty dt\, e^{-zt} \int d^3r\, e^{+i\mathbf{q}\cdot\mathbf{r}} \mathbf{u}_T(\mathbf{r}, t). \tag{2.87}$$

The transformation converts Eq. (2.86) into an initial-value problem,

$$\hat{u}_T(\mathbf{q}, z) = \frac{u_T(\mathbf{q}, 0)}{z + \eta_S q^2/m\rho_0}, \tag{2.88}$$

which gives a pole at $z = -\eta_S q^2/m\rho_0$ corresponding to the lifetime of the non oscillating hydrodynamic transverse wave at $\tau(q) = m\rho_0/\eta_s q^2$, which becomes extremely long as $q \to 0$. This result is very similar to the hydrodynamic translational-diffusion mode of a system of Brownian particles described previously.

The longitudinal part of Eq. (2.82) is a little more complicated because it couples to Eqs. (2.81) and (2.83). To facilitate solving these coupled equations, it is convenient to eliminate the terms containing $\nabla \cdot \mathbf{u}$ in both Eqs. (2.81) and (2.83) to obtain

$$(\partial/\partial t)\, \{\delta e - [(e_0 + P_0)/\rho_0]\, \delta\rho\} = \lambda \nabla^2(\delta T). \tag{2.89}$$

From the first law of thermodynamics, we have

$$T\delta S = \delta E + P\delta V = \delta(eV) + P\delta V = V\delta e + (e + P)\,\delta V,$$

or

$$T\delta s \equiv T\delta S/V = \delta e + [(e + P)/V]\,\delta V \simeq \delta e - [(e_0 + P_0)/\rho_0]\,\delta\rho. \tag{2.90}$$

In terms of the entropy-density fluctuation $\delta s = \delta S/V$, Eq. (2.89) can be written as

$$T_0(\partial(\delta s)/\partial t = \lambda\nabla^2(\delta T). \tag{2.91}$$

Note that in Eq. (2.91) we have replaced T by T_0, to be consistent with the linear approximation.

We next express δs and δP in terms of $\delta\rho$ and δT, because $\delta\rho$ and δT are statistically independent; i.e., $\langle\delta\rho\;\delta T\rangle = 0$.

Using the usual thermodynamic manipulations we have

$$\delta P = (\partial P/\partial\rho)_{\mathrm{T}}\,\delta\rho + (\partial P/\partial T)_\rho\,\delta T, \tag{2.92}$$

$$\delta s = (1/V)\,\delta S = (1/V)[(\partial S/\partial\rho)_{\mathrm{T}}\,\delta\rho + (\partial S/\partial T)_\rho\,\delta T]. \tag{2.93}$$

Substituting Eq. (2.92) into the longitudinal part of (2.82) and substituting Eq. (2.93) into (2.91), we obtain

$$(\partial/\partial t - D_1\nabla^2)\mathbf{u}_{\mathrm{L}} + (1/m\rho_0)(\partial P/\partial\rho)_{\mathrm{T}}\nabla(\delta\rho) + (1/m\rho_0)(\partial P/\partial T)_\rho\,\nabla(\delta T) = 0, \tag{2.94}$$

$$-(\partial T/\partial\rho)_{\mathrm{S}}\,(\partial(\delta\rho)/\partial t) + [\partial/\partial t - (\lambda V/T_0)\,(\partial T/\partial S)_\rho\nabla^2]\,(\delta T) = 0, \tag{2.95}$$

where $D_1 = (\eta_{\mathrm{v}} + \frac{4}{3}\eta_{\mathrm{s}})/m\rho_0$ is the longitudinal kinematic viscosity.

Equations (2.94) and (2.95) together with Eq. (2.81) form a set of coupled equations. These can be solved by converting them to the initial-value problem by using the Fourier–Laplace transformation technique. As was done previously for the transverse velocity, we define

$$\delta\hat{\rho}\,(\mathbf{q}, z) = \int_0^\infty dt\; e^{-zt}\int d^3r\; e^{i\mathbf{q}\cdot\mathbf{r}}\;\rho(\mathbf{r}, t), \tag{2.96}$$

$$\hat{\mathbf{u}}_{\mathrm{L}}\,(\mathbf{q}, z) = \int_0^\infty dt\; e^{-zt}\int d^3r\; e^{i\mathbf{q}\cdot\mathbf{r}}\;\mathbf{u}_{\mathrm{L}}(\mathbf{r}, t), \tag{2.97}$$

$$\delta\hat{T}(\mathbf{q}, z) = \int_0^\infty dt\; e^{-zt}\int d^3r\; e^{i\mathbf{q}\cdot\mathbf{r}}\;\delta T(\mathbf{r}, t). \tag{2.98}$$

The Fourier–Laplace transformations convert Eqs. (2.81), (2.94), and (2.95) to algebraic equations, which can be expressed in matrix form as

$$\begin{bmatrix} z & -iq\rho_0 & 0 \\ \dfrac{-iq}{m\rho_0^2 x_T} & z + D_l q^2 & \dfrac{-iq\alpha}{m\rho_0 x_T} \\ \dfrac{-\alpha T_0}{C_V x_T \rho_0} z & 0 & z + \dfrac{\lambda}{C_V} q^2 \end{bmatrix} \begin{bmatrix} \delta\hat{\rho}(\mathbf{q}, z) \\ \hat{u}_L(\mathbf{q}, z) \\ \delta T(\mathbf{q}, z) \end{bmatrix} = \begin{bmatrix} 1 & 0 & 0 \\ 0 & 1 & 0 \\ \dfrac{-T_0\alpha}{\rho_0 C_V x_T} & 0 & 1 \end{bmatrix} \begin{bmatrix} \delta\rho(\mathbf{q}) \\ u_L(\mathbf{q}) \\ \delta T(\mathbf{q}) \end{bmatrix} \tag{2.99}$$

where x_T is the isothermal compressibility, α the isobaric expansion coefficient, and C_V the specific heat at constant volume per unit volume, i.e.,

$$\begin{aligned} x_T &= -(1/V)\,(\partial V/\partial p)_T, \\ \alpha &= (1/V)\,(\partial V/\partial T)_p, \\ C_V &= (1/V)\,(\partial E/\partial T)_V. \end{aligned} \tag{2.100}$$

Equation (2.99) can be inverted to obtain a solution for $\delta\rho(\mathbf{q}, z)$, $\mathbf{u}_L(\mathbf{q}, z)$, and $\delta T(\mathbf{q}, z)$, as given by

$$\begin{aligned} \delta\hat{\rho}(\mathbf{q}, z) = [1/\Delta(z)]\,\{&(z + q^2\gamma D_t)(z + D_1 q^2)\,\delta\rho(\mathbf{q}) - iq(z + q^2\gamma D_t)u_L(\mathbf{q}) \\ &+ q^2\alpha\rho_0 v_T^2[(\alpha T_0 m/C_v)\,v_T^2\,\delta\rho(\mathbf{q}) - \delta T(\mathbf{q})]\}, \end{aligned} \tag{2.101}$$

$$\hat{u}_L(\mathbf{q}, z) = [1/i\Delta(z)q\rho_0][\delta\rho(0) - z\,\delta\rho(\mathbf{q}, z)], \tag{2.102}$$

$$\begin{aligned} \delta\hat{T}(\mathbf{q}, z) = [1/\Delta(z)]\{&(z\alpha T_0 m/C_V)\,v_T^2[(z + D_1 q^2)\delta\rho(\mathbf{q}) - iqu_L(\mathbf{q}) \\ &+ [z^2 + q^2(v_T^2 + D_1 z)]\,[(-\alpha T_0 m/C_V)v_T^2\,\delta\rho(\mathbf{q}) + \delta T(\mathbf{q})]\}, \end{aligned} \tag{2.103}$$

where $v_T = (\rho_0 m\chi_T)^{-1/2}$ is the isothermal sound velocity, $D_t = \lambda/c_p$ the thermal diffusivity, and $\gamma = C_p/C_V$. The quantity $\Delta(z)$ is the determinant of the square matrix on the left-hand side of Eq. (2.99), given by

$$\begin{aligned} \Delta(z) &= z(z + D_1 q^2)(z + \gamma D_t q^2) + zq^2\rho_0\alpha^2 v_T^4 T_0 m/C_V + q^2 v_T^2(z + \gamma D_t q^2) \\ &= z^3 + z^2(D_1 + \gamma D_t)q^2 + z(v_s^2 + \gamma D_1 D_t q^2)q^2 + q^4 v_T^2\gamma D_t, \end{aligned} \tag{2.104}$$

where v_s is the adiabatic sound velocity and is given by

$$v_s^2 = \gamma v_T^2 = v_T^2[1 + (\alpha^2 T_0/c_v\chi_T)], \tag{2.105}$$

where c_v is the specific heat per unit mass.

We now multiply both sides of Eq. (2.153) by $\delta\rho^*(\mathbf{q})$ and take an ensemble average to obtain the correlation function for the density fluctuations,

$$\langle\delta\hat{\rho}(\mathbf{q}, z)\,\delta\rho(\mathbf{q})^*\rangle = [B(z)/\Delta(z)]\,\langle|\delta\rho(\mathbf{q})|^2\rangle, \tag{2.106}$$

where

$$B(z) = (z + \gamma D_t q^2)(z + D_l q^2) + (\gamma - 1)v_T^2 q^2. \tag{2.107}$$

Note that Eq. (2.106) does not contain terms such as $\langle u_L(\mathbf{q})\, \delta\rho^*(\mathbf{q})\rangle$ and $\langle \delta T(\mathbf{q})\, \delta\rho^*(\mathbf{q})\rangle$. This is because $u_L(\mathbf{q})$ and $\delta\rho(\mathbf{q})$ have opposite time-reversal symmetry and $\delta T(\mathbf{q})$ and $\delta\rho(\mathbf{q})$ are statistically independent, as was previously mentioned.

The correlation function for velocity u_L and temperature δT can also be obtained from Eqs. (2.102) and (2.103), but these are of no interest for calculating the dynamic-structure factor.

The time-correlation function $\langle \delta\rho(\mathbf{q}, t)\, \delta\rho^*(\mathbf{q})\rangle$ can be calculated by taking the inverse Laplace transform of Eq. (2.106). Unfortunately, the exact result is rather complicated. However, we can get pertinent physics out of what we have by finding the zeros of the function $\Delta(z)$.

Recognizing that for practically all viscous liquids $q^2 D_l$ or $q^2 D_t$ is small compared with v_s, we can find the roots of $\Delta(z)$ in terms of a power series of $q^2 D_l/v_s$ or $q^2 D_t/v_s$. The procedure is carried out by first setting $\Delta(z)$ equal to zero,

$$\Delta(z) = z^3 + z^2(D_l + \gamma D_t)q^2 + z(\omega_0^2 + D_l D_t q^4) + q^2\omega_0^2 D_t = 0, \tag{2.108}$$

where $\omega_0 = qv_s^2$ is the adiabatic sound frequency.

We then let

$$z = z_0 + z_1 + z_2 + \cdots.$$

The zero-order equation is

$$z_0^3 + z_0\omega_0^2 = 0. \tag{2.109a}$$

The first-order equation is

$$3z_0^2 z_1 + z_0^2(D_l + \gamma D_t)q^2 + z_1\omega_0^2 + \omega_0^2 q^2 D_t = 0. \tag{2.109b}$$

Equation (2.109a) gives three roots for z_0 as $z_0 = 0,\ i\omega_0,\ -i\omega_0$. We substitute the zero-order solution into Eq. (2.109b) to obtain first-order solutions. Thus the three roots of $\Delta(z)$ up to first order in $D_t q^2$ (or $D_1 q^2$) are

$$z_1^{(1)} = -D_t q^2, \tag{2.110}$$

$$z_1^{(2)} = i\omega_0 - \Gamma q^2, \tag{2.111}$$

$$z_1^{(3)} = -i\omega_0 - \Gamma q^2, \tag{2.112}$$

where

$$\begin{aligned}\Gamma &= \tfrac{1}{2}[D_l + (\gamma - 1)D_t] \\ &= \tfrac{1}{2}[(\eta_V + \tfrac{4}{3}\eta_S)/m\rho_0 + (\gamma - 1)\lambda/C_p]\end{aligned} \tag{2.113}$$

The dynamic-structure factor is obtained by setting $z = i\Omega$ in Eq. (2.106) and then taking its real part

$$
\begin{aligned}
S(\mathbf{q}, \Omega) &= \frac{1}{\pi} \operatorname{Re}\langle \delta\rho(\mathbf{q}, z)\delta\rho^*(\mathbf{q})\rangle|_{z=i\Omega} \\
&= \frac{1}{\pi}\langle|\delta\rho(\mathbf{q})|^2\rangle \operatorname{Re}\left\{\left[\frac{B(z_1)}{(z_1 - z_2)(z_1 - z_3)}\right]\frac{1}{z - z_1}\right. \\
&\quad + \left[\frac{B(z_2)}{(z_2 - z_1)(z_2 - z_3)}\right]\frac{1}{z - z_2} \\
&\quad \left.+ \left[\frac{B(z_3)}{(z_3 - z_1)(z_3 - z_2)}\right]\frac{1}{z - z_3}\right\}\Bigg|_{z=i\Omega},
\end{aligned}
$$

or

$$
\begin{aligned}
S(\mathbf{q}, \Omega) = \frac{\langle|\delta\rho(\mathbf{q})|^2\rangle}{\pi}&\left\{\left(1 - \frac{1}{\gamma}\right)\frac{q^2 D_t}{\Omega^2 + (q^2 D_t)^2}\right. \\
&\left.+ \frac{1}{\gamma}\left[\frac{\Gamma q^2}{(\Omega - \omega_0)^2 + (\Gamma q^2)^2} + \frac{\Gamma q^2}{(\Omega + \omega_0)^2 + (\Gamma q^2)^2}\right]\right\}.
\end{aligned}
\tag{2.114}
$$

Equation (2.114) indicates that in the hydrodynamic limit, $S(\mathbf{q}, \Omega)$ consists of three terms. The first term represents an unshifted central Rayleigh component having a Lorentzian shape with half-width at half-maximum intensity, given by

$$
\Delta\Omega_{\text{Ray}} = D_t q^2 = (\lambda/C_p)q^2. \tag{2.115}
$$

This mode is diffusive and corresponds to the temperature-fluctuation part in δs. The fluctuation increases local entropy with a time constant given by $(D_t q^2)^{-1}$ through a random-walk process. At low temperature, $(\partial T/\partial\rho)_s$ becomes small, and according to Eq. (2.95), the temperature-fluctuation part becomes the main process responsible for the central Rayleigh component.

The other two components represented by the second and third terms on the right-hand side of Eq. (2.114) are the Brillouin doublet. These are two Lorentzian lines, symmetrically shifted from the central Rayleigh component by $\omega_0 = qv_s$. The half-width at half-maximum intensity of each component is

$$
\Delta\Omega_B = \Gamma q^2 = \tfrac{1}{2}q^2(\eta_V + \tfrac{4}{3}\eta_S)/m\rho_0 + (\gamma - 1)\lambda/C_p. \tag{2.116}
$$

Thus, the measurement of the linewidth of a Brillouin doublet can provide the values of various transport coefficients, especially the quantity η_V, which can rarely be obtained by other techniques.

It should be emphasized that the speed of sound involved in the Brillouin peaks is adiabatic rather than isothermal. This is because of the coupling of density to energy fluctuations. The coupling provides a mechanism for the transfer of internal energy to translational energy, which causes the longitudinal sound wave to propagate with the adiabatic rather than the isothermal velocity.

From Eq. (2.114), we find that the integrated intensity of the central Rayleigh component is

$$I_c = \langle|\delta\rho(\mathbf{q})|^2\rangle(1 - 1/\gamma), \tag{2.117}$$

and the intensity of the Brillouin doublet is

$$2I_B = \langle|\delta\rho(\mathbf{q})|^2\rangle/\gamma. \tag{2.118}$$

The ratio of I_c to $2I_B$, known as the Landau–Placzek ratio, is

$$\text{LPR} = I_c/2I_B = \gamma - 1. \tag{2.119}$$

For liquids at low temperature, C_p and C_V are nearly equal and the Landau–Placzek ratio becomes small.

This consideration applies to a viscous fluid when the viscosity and thermal conductivity are independent of frequency. However, when the fluid becomes dispersive, i.e., when the dispersive mechanisms vary with frequency, the sound wave will no longer propogate with a unique velocity but will vary with frequency (Mountain, 1968, 1970). When the dispersion is large, as is the case with high-polymer liquids and glass-forming molecular liquids, the classical hydrodynamic approach presented here is no longer valid. We now consider the dispersive viscoelastic system.

2.9 VISCOELASTIC LIQUIDS

In its most general form, the constitutive equation for the linear stress–strain relation is given by (Ferry, 1980)

$$\sigma_{ij}(t) = \int_0^t dt'\, \Gamma_{ijkl}(t - t')\dot{e}_{kl}(t'). \tag{2.120}$$

In Eq. (2.120), the Einstein summation convention, that repeated indices indicate summation, is implied. The term $\dot{e}_{kl}(t)$ is the strain rate tensor and can be represented in terms of the velocity gradient tensor as

$$\dot{e}_{ij} = \left(\frac{\partial u_i}{\partial x_j} + \frac{\partial u_j}{\partial x_i}\right)\Big/2,$$

which gives $\dot{e}_{ij} = \nabla \cdot \mathbf{u}$ when $i = j$. The term $\Gamma_{ijkl}(t)$ is known as the stress-relaxation modulus tensor. For a homogeneous, amorphous viscoelastic system, the stress-relaxation modulus does not vary with x_j, the position of the sample. Furthermore, in such a system, there are only two independent components for $\Gamma_{ijkl}(t)$, i.e., the bulk modulus $K(t)$ and the shear modulus $G(t)$. The bulk and shear components combine in such a manner that Eq. (2.120) becomes

$$\sigma_{ij}(t) = \int_0^t dt' \, \{K(t - t')\dot{e}_{kk}(t') \, \delta_{ij} + 2G(t - t')[\dot{e}_{ij}(t') - \tfrac{1}{3}\dot{e}_{kk}(t') \, \delta_{ij}]\}. \tag{2.121}$$

Substituting Eq. (2.121) into the conservation of momentum equation [Eq. (2.62)] gives

$$\rho_m\left(\frac{\partial u_i}{\partial t} + u_k \frac{\partial u_i}{\partial x_k}\right) = -\frac{\partial P}{\partial x_i} + \int_0^t dt' \, \{G(t - t') \, \nabla^2 u_i(t') + [K(t - t') + \frac{1}{3} G(t - t')] \frac{\partial}{\partial x_i} \nabla \cdot \mathbf{u}(t')\}. \tag{2.122}$$

Equation (2.122) is a generalization of the Novier–Stokes equation for viscoelastic liquids [Eq. (2.74)]. A linearization of this equation corresponds to dropping the second term of the left-hand side of Eq. (2.122) and replacing ρ_m by $m\rho_0 = \rho_m^0$.

Equation (2.122) can be separated into a longitudinal and a transverse part in the same manner as described in the last section. The transverse component is given by

$$\rho_m^0 \frac{\partial u_{\mathrm{T}}}{\partial t} = \int_0^t dt \, G(t - t') \, \nabla^2 u_{\mathrm{T}}(t'). \tag{2.123}$$

Taking the Fourier–Laplace transform of this equation, we obtain the solution of $\hat{u}_T(\mathbf{q}, z)$ as

$$\hat{u}_T(\mathbf{q}, z) = \frac{u_T(\mathbf{q})}{z + q^2 G(z)/\rho_m^0}, \tag{2.124}$$

where $\hat{u}_T(\mathbf{q}, z)$ is defined in Eq. (2.87) and $G(z)$ is the Laplace transform of $G(t)$ given by

$$G(z) = \int_0^\infty dt\, e^{-zt} G(t). \tag{2.125}$$

To obtain Eq. (2.124), we employed the convolution theorem associated with the Laplace transform. The correlation function of $\hat{u}_T(\mathbf{q}, z)$ is obtained by multiplying $\hat{u}_T(\mathbf{q}, z)$ by $u_T^*(\mathbf{q})$ and then taking the ensemble average. The result is

$$C_T(\mathbf{q}, z) \equiv \langle \hat{u}_T(\mathbf{q}, z) u_T^*(\mathbf{q}) \rangle = \frac{\langle |u_T(\mathbf{q})|^2 \rangle}{z + q^2 G(z)/\rho_m^0}. \tag{2.126}$$

This equation is the generalization of Eq. (2.88) and is appropriate for the viscoelastic system. The term $G(z)$ plays the role of the dynamic-shear viscosity. At $q = 0$, $C_T(\mathbf{q}, z)$ has a pole at $z = 0$. As q becomes finite, the pole moves away from the origin. For short time or high frequency, we may approximate $G(z)$ as

$$\begin{aligned} G(z) &= \int_0^\infty dt\, e^{-zt} \left[G(0) + \frac{1}{2} \ddot{G}(0) t^2 + \cdots \right] \\ &= \frac{1}{z} G(0) + O\left(\frac{1}{z^3} \right). \end{aligned} \tag{2.127}$$

Putting this into Eq. (2.126) we obtain

$$C_T(\mathbf{q}, z) = \frac{\langle |u_T(\mathbf{q})|^2 \rangle}{z^2 + q^2 G(0)/\rho_m^0}. \tag{2.128}$$

This gives poles at $z = \pm iq\sqrt{G(0)/\rho_m^0}$. Thus, at high frequency where the approximation of Eq. (2.127) is valid, we find oscillatory, rather than diffusive, behavior. This is a transverse sound mode, with the propogation velocity V_T given by $\omega_T = qV_T$, where $V_T = \sqrt{G(0)/\rho_m^0}$.

To discuss the general case, it is useful to define a quantity $G^*(\omega)$, known as the dynamic shear modulus as

$$G^*(\omega) = i\omega \int_0^\infty dt\, e^{-i\omega t} G(t) = G'(\omega) + iG''(\omega), \tag{2.129}$$

where the real and imaginary parts of the dynamic-shear modulus are given by

$$G'(\omega) = \omega \int_0^\infty dt\, G(t) \sin \omega t \tag{2.130}$$

and

$$G''(\omega) = \omega \int_0^\infty dt\, G(t) \cos \omega t. \tag{2.131}$$

The real $[\eta'(\omega)]$ and imaginary $[\eta''(\omega)]$ parts of the dynamic viscosity are related to $G'(\omega)$ and $G''(\omega)$ by $\eta'(\omega) = G''(\omega)/\omega$ and $\eta''(\omega) = G'(\omega)/\omega$.

Clearly, in the limit of vanishing ω, $\eta'(\omega) \to \eta_s$ and $\eta''(\omega) \to 0$. At finite ω, one can express $G''(\omega)$ [or $\eta'(\omega)$] in terms of the time-correlation function of the transverse velocity. This is obtained by first rewriting Eq. (2.126) as

$$z + q^2 G(z)/\rho_m^0 = \frac{\langle |u_{\mathrm{T}}(\mathbf{q})|^2 \rangle}{\langle u_{\mathrm{T}}(\mathbf{q}, z) u_{\mathrm{T}}^*(\mathbf{q}) \rangle}. \tag{2.132}$$

We next set $z = i\omega$ and then equate the imaginary parts on both sides of the equation to obtain

$$\frac{q^2}{\rho_m^0} G''(\omega) = \frac{\omega' A(\mathbf{q}, \omega)}{[A(\mathbf{q}, \omega)]^2 + [B(\mathbf{q}, \omega)]^2}, \tag{2.133}$$

where $A(\mathbf{q}, \omega)$ and $B(\mathbf{q}, \omega)$ are the cosine and sine transforms, respectively, of the time-correlation function of the transverse velocity given by

$$A(\mathbf{q}, \omega) = \int_0^\infty dt \cos \omega t \frac{\langle u_{\mathrm{T}}(\mathbf{q}, t) u_{\mathrm{T}}^*(\mathbf{q}) \rangle}{\langle |u_{\mathrm{T}}(\mathbf{q})|^2 \rangle} \tag{2.134}$$

and

$$B(\mathbf{q}, \omega) = \int_0^\infty dt \sin \omega t \frac{\langle u_{\mathrm{T}}(\mathbf{q}, t) u_{\mathrm{T}}^*(\mathbf{q}) \rangle}{\langle |u_{\mathrm{T}}(\mathbf{q})|^2 \rangle} \tag{2.135}$$

In the limit of vanishing ω we have

$$\frac{q^2}{\rho_m^0} \eta_s = \lim_{\omega \to 0} \left(\frac{q^2}{\rho_m^0} \right) \frac{G''(\omega)}{\omega} = \frac{1}{\hat{A}(\mathbf{q}, 0)}, \tag{2.136}$$

which is what we would have obtained from Eq. (2.88) by taking the $z \to 0$ limit.

Equation (2.133) provides a method of calculating $G''(\omega)$ if the time-correlation function of the transverse velocity can be measured by some experimental technique. While the longitudinal modulus can be obtained by measurement from the dynamic light-scattering technique (to be discussed below), the mechanical relaxation method is the one that is currently used to obtain $G''(\omega)$ (Ferry, 1980). However, the mechanical method is rather limited in the range of frequency.

The longitudinal part of Eq. (2.122) is given by

$$\rho_m^0 \frac{\partial \mathbf{u}_L}{\partial t} = -\nabla P + \int_0^t dt'\, M(t - t')\, \nabla^2 u_L(t'), \tag{2.137}$$

where $M(t)$ is the longitudinal modulus given by

$$M(t) = K(t) + \tfrac{4}{3}G(t). \tag{2.138}$$

Although it is possible to provide a complete analysis of longitudinal velocity for the viscoelastic liquid similar to that given for the transverse velocity, to demonstrate the effect of viscoelasticity we shall only provide an approximate analysis by neglecting the temperature fluctuation contribution to the ∇P term. This approximation amounts to neglecting the effect of thermal diffusion. Thermal diffusion contributes to the density fluctuation spectrum in the 10 to 100 MHz ranges (Herzfeld and Litovitz, 1959). A complete analysis including the effect of thermal diffusion is given in Chapter 4.

In this approximation, we can write

$$\nabla P = \left(\frac{\partial P}{\partial \rho}\right)_T \nabla(\delta\rho) = \frac{1}{\rho_0 \chi_T} \nabla(\delta\rho). \tag{2.139}$$

Substituting this equation into Eq. (2.137) gives

$$\rho_m^0 \frac{\partial \mathbf{u}_L}{\partial t} = \frac{-1}{\rho_0 \chi_T} \nabla(\delta\rho) + \int_0^t dt'\, M(t - t')\, \nabla^2 u_L(t'). \tag{2.140}$$

Equations (2.140) and 2.81) can now be solved by transforming them into two coupled algebraic equations using the Fourier–Laplace transform techniques as

$$\begin{bmatrix} z & iq\rho_0 \\ \dfrac{iq}{m\rho_0^2 \chi_T} & z + \dfrac{q^2}{\rho_m} \hat{M} \end{bmatrix} \begin{bmatrix} \delta\hat{\rho}(\mathbf{q}, z) \\ u_L(\mathbf{q}, z) \end{bmatrix} = \begin{bmatrix} \delta\rho(\mathbf{q}) \\ u_L(\mathbf{q}) \end{bmatrix}, \tag{2.141}$$

where

$$\hat{M} = \int_0^\infty dt\, e^{-zt} M(t) = \int_0^\infty dt\, e^{-zt} \{K(t) + \tfrac{4}{3}G(t)\}. \tag{2.142}$$

The matrix equation can be readily inverted to provide the solution of the $\delta\hat{\rho}(\mathbf{q}, z)$ and $\hat{u}_L(\mathbf{q}, z)$ terms of $\delta\rho(\mathbf{q})$ and $u_L(\mathbf{q})$. The autocorrelation function of $\delta\hat{\rho}(\mathbf{q}, z)$ [or $\hat{u}_L(\mathbf{q}, z)$] can then be constructed by multiplying them by $\delta\rho^*(\mathbf{q})$ [or $u_L^*(\mathbf{q})$] and then taking the ensemble average. Using the fact that $\langle u_L(\mathbf{q})\, \delta\rho^*(\mathbf{q})\rangle = \langle\delta\rho(\mathbf{q})\, u_L^*(\mathbf{q})\rangle = 0$, we obtain

$$\hat{C}_\rho(\mathbf{q}, t) = \langle\delta\hat{\rho}(\mathbf{q}, z)\, \delta\rho^*(\mathbf{q})\rangle = \frac{z + q^2\hat{M}/\rho_m^0}{z(z + q^2\hat{M}/\rho_m^0) + \omega_T^2}\langle|\delta\rho(\mathbf{q})|^2\rangle \tag{2.143}$$

$$\hat{C}_L(\mathbf{q}, t) = \langle\hat{u}_L(\mathbf{q}, z)u_L^*(\mathbf{q})\rangle = \frac{z}{z(z + q^2\hat{M}/\rho_m^0) + \omega_T^0}\langle|u_L(q)|^2\rangle, \tag{2.144}$$

where $\omega_T^2 = q^2/(\rho_m^0\chi_T)$ is the square of the isothermal sound frequency.

Except for the presence of the ω_T^2 term, the form of the longitudinal-velocity correlation function is similar to the correlation function of the transverse velocity. The ω_T^2 term comes from the pressure fluctuation that results in the propagation of the longitudinal sound wave, as discussed above. We are particularly interested in the density-correlation function because it is related to the Rayleigh–Brillouin spectrum.

It is important to note that Eq. (2.143) provides a method of determining the longitudinal modulus by measuring the Rayleigh–Brillouin spectrum. To describe this technique, we use Eq. (2.143) and express $\hat{M}$ in terms of $\hat{S}(= \hat{C}/\langle|\delta\rho(\mathbf{q})|^2\rangle)$ as

$$q^2\hat{M}/\rho_m^0 = \frac{(\omega_T^2 + z^2)\hat{S} - z}{1 - z\hat{S}}. \tag{2.145}$$

The real and imaginary parts of the dynamic longitudinal modulus, $M'(\omega)$ and $M''(\omega)$, can now be obtained by setting $z = i\omega$ in Eq. (2.145) and then identifying the real and imaginary parts on both sides of the equation.

We first note that the dynamic longitudinal modulus is given by

$$M^*(\omega) = i\omega M(i\omega) = i\omega\int_0^\infty dt\, e^{-i\omega t}\, M(t) = M'(\omega) + iM''(\omega), \tag{2.146}$$

where

$$M'(\omega) = \omega\int_0^\infty dt\, M(t)\sin\omega t \tag{2.147}$$

and

$$M''(\omega) = \omega\int_0^\infty dt\, M(t)\cos\omega t. \tag{2.148}$$

Further, we obtain

$$\hat{S}(\mathbf{q}, \omega) = \frac{\hat{C}(\mathbf{q}, i\omega)}{\langle|\delta\rho(\mathbf{q})|^2\rangle} = \int_0^\infty dt\, e^{-i\omega t} \frac{\langle\delta\rho(\mathbf{q}, t)\, \delta\rho^*(\mathbf{q})\rangle}{\langle|\delta\rho(\mathbf{q})|^2\rangle}$$
$$= \hat{I}(\mathbf{q}, \omega) - i\hat{H}(\mathbf{q}, \omega), \tag{2.149}$$

where $\hat{I}(\mathbf{q}, \omega)$ and $\hat{H}(\mathbf{q}, \omega)$ are the cosine and sine transforms of the normalized density–time correlation function. The function $\hat{I}(\mathbf{q}, \omega)$ is proportional to the measured spectrum.

Multiplying both sides of Eq. (2.145) by z, setting $z = i\omega$, and then taking the imaginary part of the equation, we obtain

$$M''(\omega) = \left(\frac{1}{\chi_T}\right) \frac{\omega\hat{I}(\mathbf{q}, \omega)}{[1 - \omega\hat{H}(\mathbf{q}, \omega)]^2 + [\omega\hat{I}(\mathbf{q}, \omega)]^2}. \tag{2.150}$$

Having obtained $M''(\omega)$, one can calculate $M'(\omega)$ from the Kramers–Kronig relation.

Equation (2.150) provides the method of determining $M''(\omega)$ from the dynamic light-scattering data (Wang *et al.*, 1984). This result is especially useful when the density–time correlation function can be directly measured, as in the case of photon-correlation spectroscopy. In this case, $\hat{I}(\mathbf{q}, \omega)$ and $\hat{H}(\mathbf{q}, \omega)$ can be readily calculated.

2.10 STATIC-STRUCTURE FACTOR AND LONGITUDINAL COMPLIANCE

The total integrated intensity of the Rayleigh–Brillouin spectrum is the sum of Eqs. (2.117) and (2.118) and is equal to $\langle|\delta\rho(\mathbf{q})|^2\rangle$. The quantity $\langle|\delta\rho(\mathbf{q})|^2\rangle$ is known as the static-structure factor. In the next section we shall show that in the limit of $\mathbf{q} \to 0$, the static-structure factor is equal to $kTV\rho_0^2\chi_T$. However, this result becomes invalid when the fluid is near its critical point.

The static-structure factor $S(\mathbf{q})$ is obtained by integrating $S(\mathbf{q}, \Omega)$ over frequency:

$$S(\mathbf{q}) = \int_{-\infty}^{\infty} d\Omega\, S(\mathbf{q}, \Omega) = \langle|\delta\rho(\mathbf{q})|^2\rangle. \tag{2.151}$$

Since

$$\delta\rho(\mathbf{q}) = \int_V d^3r \exp(i\mathbf{q} \cdot \mathbf{r})\, \delta\rho(\mathbf{r}), \tag{2.152}$$

the static-structure factor can be written as

$$S(\mathbf{q}) = \int_V d^3r \int_V d^3r' \exp[i\mathbf{q} \cdot (\mathbf{r} - \mathbf{r}')] \langle \delta\rho(\mathbf{r})\, \delta\rho(\mathbf{r}') \rangle. \tag{2.153}$$

We now define the static density–density-correlation function as

$$G(\mathbf{r}, \mathbf{r}') = \langle \delta\rho(\mathbf{r})\, \delta\rho(\mathbf{r}') \rangle, \tag{2.154}$$

where $G(\mathbf{r}, \mathbf{r}')$ measures the correlation of density fluctuations at two different points of the fluid, $\mathbf{r}$ and $\mathbf{r}'$. Since a fluid is spatially uniform, $G(\mathbf{r}, \mathbf{r}')$ depends only on the relative distance, i.e.,

$$G(\mathbf{r}, \mathbf{r}') = G(\mathbf{r} - \mathbf{r}') = G(\mathbf{R}) = G(R). \tag{2.155}$$

The last step follows from the isotropic symmetry of the fluid.

As $R \to \infty$, density fluctuations at the two end points become uncorrelated so that

$$G(\mathbf{r} - \mathbf{r}') \to \langle \delta\rho(\mathbf{r}) \rangle \langle \delta\rho(\mathbf{r}') \rangle = 0. \tag{2.156}$$

The dependence of $G(R)$ on R is determined by the intermolecular potential.

Using Eq. (2.155) and transforming the integration variables $\mathbf{r}$ and $\mathbf{r}'$ to $\mathbf{R}$ and $\mathbf{r}'$, and after integrating out the variable $\mathbf{r}'$ we obtain from Eq. (2.153)

$$S(\mathbf{q}) = V \int d^3R \exp(i\mathbf{q} \cdot \mathbf{R}) G(R). \tag{2.157}$$

Equation (2.157) is an important result. It indicates that the static-structure factor is proportional to the spatial Fourier transform of the pair-correlation function of density fluctuations.

The static density–density-correlation function $G(R)$ is related to the fluctuations of the total number of particles within the scattering volume. Since

$$\delta N = \int_V d^3r\, \delta\rho(\mathbf{r}), \tag{2.158}$$

we have

$$\begin{aligned} \langle (\delta N)^2 \rangle &= \int d^3r \int d^3r' \langle \delta\rho(\mathbf{r})\, \delta\rho(\mathbf{r}') \rangle = \int d^3r \int d^3r'\, G(\mathbf{r} - \mathbf{r}') \\ &= V \int d^3R\, G(R). \end{aligned} \tag{2.159}$$

Comparing Eq. (2.157) with Eq. (2.159), we find that the mean-square fluctuation in the total number is related to $S(\mathbf{q})$ by

$$\langle(\delta N)^2\rangle = \lim_{\mathbf{q}\to 0} S(\mathbf{q}). \tag{2.160}$$

The quantity $\langle(\delta N)^2\rangle$ can be calculated according to the method of statistical thermodynamics.

The probability of observing N particles of a system in a grand-canonical ensemble is given by

$$P_N = e^{\beta N\mu} Q_N(T, V)/\Xi(\mu, T, V), \tag{2.161}$$

where $Q_N(T, V)$ is the canonical-partition function for a system consisting of N particles, $\Xi(\mu, T, V)$ the grand-canonical-partition function, $\beta = 1/kT$, and μ the chemical potential.

Since

$$\langle N\rangle = \sum_{N=0}^{\infty} NP_N = \sum_{N=0}^{\infty} Ne^{\beta N\mu}\frac{Q_N}{\Xi} = \beta^{-1}\left(\frac{\partial \ln\Xi}{\partial\mu}\right)_{\mathrm{T,V}}, \tag{2.162}$$

and

$$\begin{aligned}\langle N^2\rangle &= \sum_{N=0}^{\infty} N^2 P_N = \sum_{N=0}^{\infty} N^2 e^{\beta N\mu} Q_N/\Xi \\ &= \beta^{-2}(1/\Xi)(\partial/\partial\mu)(\partial\Xi/\partial\mu)_{\mathrm{T,V}} = (1/\beta\Xi)[(\partial/\partial\mu)(\langle N\rangle\Xi)]_{\mathrm{T,V}} \\ &= \langle N\rangle^2 + \beta^{-1}(\partial\langle N\rangle/\partial\mu)_{\mathrm{T,V}},\end{aligned} \tag{2.163}$$

we have

$$\begin{aligned}\langle(\delta N)^2\rangle &= \langle N^2\rangle - \langle N\rangle^2 = \beta^{-1}(\partial\langle N\rangle/\partial\mu)_{\mathrm{T,V}} \\ &= -kT\rho_0\,(\partial V/\partial\mu)_{\mathrm{T},\langle N\rangle} = kT\rho_0^2 V\chi_{\mathrm{T}}.\end{aligned} \tag{2.164}$$

Therefore

$$\lim_{\mathbf{q}\to 0} S(\mathbf{q}) = \lim_{\mathbf{q}\to 0}\langle|\delta\rho(\mathbf{q})|^2\rangle = kTV\rho_0^2\chi_{\mathrm{T}}. \tag{2.165}$$

It is useful to make a connection between the light-scattering spectrum and the dynamic longitudinal compliance that one measures in the mechanical relaxation experiment. Using Eq. (2.165), we define the longitudinal compliance at a finite q to be

$$D(\mathbf{q}) = \frac{1}{kT\rho_0^2 V} S(\mathbf{q}). \tag{2.166}$$

The fluctuation–dissipation theorem as given in Eq. (1.167) and in the Appendix can be written as (in the limit of $kT \gg h\omega$)

$$\chi''(\mathbf{q}, \omega)/\omega = (\pi/kT)\, S(\mathbf{q}, \omega), \tag{2.167}$$

where we have included the q dependence and used $S(\mathbf{q}, \omega)$ to represent the spectral power density.

The integration of Eq. (2.167) over frequency gives

$$\int_{-\infty}^{\infty} d\omega \frac{\chi''(\mathbf{q}, \omega)}{\omega} = \frac{\pi}{kT} S(\mathbf{q}). \tag{2.168}$$

Comparing Eq. (2.168) with Eq. (2.166), we obtain

$$D(\mathbf{q}) = \frac{1}{\pi \rho_0^2 V} \int_{-\infty}^{\infty} d\omega \frac{\chi''(\mathbf{q}, \omega)}{\omega}. \tag{2.169}$$

The dynamic longitudinal compliance $D^* = D' - iD''$ is related to $D(\mathbf{q})$ as

$$D(\mathbf{q}) = \int_{-\infty}^{\infty} d\omega \frac{D''(\mathbf{q}, \omega)}{\omega}, \tag{2.170}$$

where $D''(\mathbf{q}, \omega)$ is the imaginary part of the dynamic longitudinal compliance. From Eqs. (2.167), (2.169), and (2.170) we obtain

$$D''(\mathbf{q}, \omega) = \frac{\omega}{\rho_0^2 VkT} S(\mathbf{q}, \omega) = \frac{\omega}{\pi \rho_0^2 VkT} \int_0^{\infty} dt\, \langle \delta\rho(\mathbf{q}, t)\, \delta\rho^*(\mathbf{q}) \rangle \cos \omega t. \tag{2.171}$$

This equation shows that the imaginary part of the dynamic longitudinal compliance is proportional to the product of ω and $S(\mathbf{q}, \omega)$.

The real part of the dynamic compliance can be calculated from the Kramers–Kronig relation [Eq. (1.125)] as

$$\begin{aligned} D'(\mathbf{q}, \omega) - D_e(\mathbf{q}) &= \frac{1}{\pi} p \int_{-\infty}^{\infty} d\omega' \frac{D''(\mathbf{q}, \omega')}{\omega' - \omega} \\ &= \frac{1}{\pi \rho_0^2 VkT} p \int_{-\infty}^{\infty} d\omega \left[\frac{\omega' S(q, \omega')}{\omega' - \omega} \right], \end{aligned} \tag{2.172}$$

where $D_e(\mathbf{q})$ is equal to χ_T at vanishing $\mathbf{q}$.

It is simple matter to show that Eq. (2.172) reduces to

$$\begin{aligned} &D_e(\mathbf{q}) - D'(\mathbf{q}, \omega) \\ &= \frac{1}{\pi \rho_0^2 VkT} \left[\langle |\delta\rho(\mathbf{q})|^2 \rangle - \omega \int_0^{\infty} dt\, \langle \delta\rho(\mathbf{q}, t)\, \delta\rho^*(\mathbf{q}) \rangle \sin \omega t \right]. \end{aligned} \tag{2.173}$$

The real part of the dynamic compliance is thus given by

$$D'(\mathbf{q}, \omega) = \frac{\omega}{\pi \rho_0^2 VkT} \int_0^\infty dt \, \langle \delta\rho(\mathbf{q}, t) \, \delta\rho^*(\mathbf{q}) \rangle \sin \omega t. \tag{2.174}$$

Since, the dynamic compliance $D''(\mathbf{q}, \omega)$ is defined in terms of the creep compliance $D(\mathbf{q}, t)$ by

$$D''(\mathbf{q}, \omega) = \omega \int_0^\infty dt \, [D_e(\mathbf{q}) - D(\mathbf{q}, t)] \cos \omega t. \tag{2.175}$$

Comparing Eq. (2.171) with Eq. (2.175), we obtain

$$D_e(\mathbf{q}) - D(\mathbf{q}, t) = \frac{1}{\pi \rho_0^2 VkT} \langle \delta\rho(\mathbf{q}, t) \, \delta\rho^*(\mathbf{q}) \rangle. \tag{2.176}$$

Equation (2.176) is an important result. It relates the longitudinal creep compliance to the time-correlation function of density fluctuations. This relation provides a basis for measuring the creep compliance using photon-correlation spectroscopy (Wang and Fischer, 1984).

2.11 CRITICAL OPALESCENCE

Except in the vicinity of its critical point, χ_T of a fluid is in general a well-behaved function of temperature and density. Near the critical point, $(\partial V/\partial P)_T$ becomes very large and hence χ_T becomes divergent. Equation (2.165), or the total light-scattering intensity, becomes enormous. This is the phenomenon of critical opalescence.

Near the critical point, particles are correlated at a long distance. If the range of correlation becomes the order of the wavelength of light, the $q \to 0$ limit given in Equation (2.165) is no longer valid, and we need a more refined theory to describe the phenomenon of critical opalescence. An improved theory was first given by Ornstein and Zernike in the discussion of phase transitions and critical phenomena (Stanley, 1971).

To introduce the Ornstein–Zernike theory, we return to Eq. (2.154) and write the static density–density-correlation function $G(\mathbf{r} - \mathbf{r}')$ as (Stanley, 1971)

$$G(\mathbf{r} - \mathbf{r}') = \langle \rho(\mathbf{r})\rho(\mathbf{r}') \rangle - \rho_0^2 = \left\langle \sum_i \sum_j \delta(\mathbf{r} - \mathbf{r}_i) \, \delta(\mathbf{r}' - \mathbf{r}_j) \right\rangle - \rho_0^2. \tag{2.177}$$

We see that the density–density-correlation function contains two contributions. There are terms with $i = j$ (self-correlation) and terms with $i \neq j$ (pair correlation). Obviously, the self-correlation part is proportional to $\delta(\mathbf{r} - \mathbf{r}')$, and we represent the pair-correlation part by $\rho_0^2\,\Gamma(\mathbf{r} - \mathbf{r}')$. Thus,

$$G(\mathbf{r} - \mathbf{r}') = \rho_0 \delta(\mathbf{r} - \mathbf{r}') + \rho_0^2\,\Gamma(\mathbf{r} - \mathbf{r}'). \tag{2.178}$$

Notice that $\Gamma(\mathbf{r} - \mathbf{r}')$ is dimensionless, owing to the ρ_0^2 term. The calculation of $G(\mathbf{r} - \mathbf{r}')$ lies in our knowledge of $\Gamma(\mathbf{r} - \mathbf{r}')$. For this, Ornstein and Zernike argue that there are two contributions to $\Gamma(\mathbf{r} - \mathbf{r}')$: a direct correlation between points $\mathbf{r}$ and $\mathbf{r}'$ given by $C(\mathbf{r} - \mathbf{r}')$ and an indirect correlation through an intermediate point $\mathbf{r}''$ owing to the propagation of this direct correlation (see, for example, Fig. 2.4).

The direct-correlation function $C(\mathbf{r} - \mathbf{r}')$ is short range, similar to the intermolecular potential, but $\Gamma(\mathbf{r} - \mathbf{r}')$ can be long range through propagation of the direction correlation.

This intuitive argument leads to an equation for $\Gamma(\mathbf{r} - \mathbf{r}')$ as

$$\Gamma(\mathbf{r} - \mathbf{r}') = C(\mathbf{r} - \mathbf{r}') + \rho_0 \int d^3r''\, C(\mathbf{r} - \mathbf{r}'')\,\Gamma(\mathbf{r}'' - \mathbf{r}'),$$

or

$$\Gamma(R) = C(R) + \rho_0 \int d^3r'\, C(R - R')\,\Gamma(R'). \tag{2.179}$$

This equation is known as the Ornstein–Zernike integral equation, which can equally well be chosen as the defining equation for $C(R)$.

We now multiply by $\exp i\mathbf{q} \cdot \mathbf{R}$ on both sides of Eq. (2.179), integrate over all volume, and obtain by the convolution theorem of Fourier transform

$$\hat{\Gamma}(\mathbf{q}) = \hat{C}(\mathbf{q}) + \rho_0 \hat{C}(\mathbf{q})\,\hat{\Gamma}(\mathbf{q}), \tag{2.180}$$

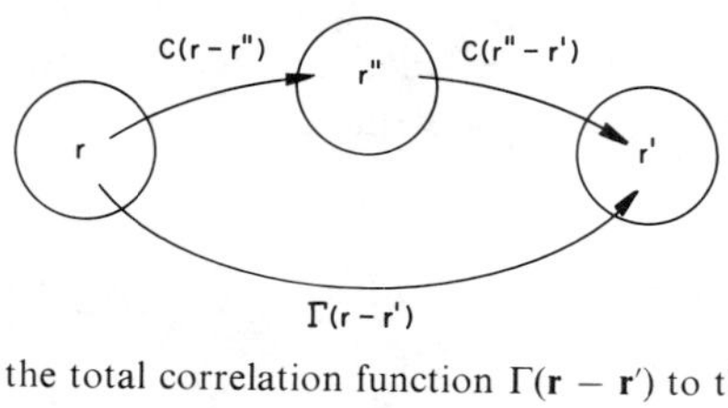

FIG. 2.4 The relation of the total correlation function $\Gamma(\mathbf{r} - \mathbf{r}')$ to the direct pair-correlation function $C(\mathbf{r} - \mathbf{r}')$.

where $\hat{\Gamma}(\mathbf{q})$ and $\hat{C}(\mathbf{q})$ are spatial Fourier transforms of $\Gamma(\mathbf{R})$ and $C(\mathbf{R})$, respectively.

Solving for $\hat{\Gamma}(\mathbf{q})$ in terms of $\hat{C}(\mathbf{q})$, we obtain

$$\Gamma(\mathbf{q}) = \frac{\hat{C}(\mathbf{q})}{1 - \rho_0 C(\mathbf{q})}. \tag{2.181}$$

Note that $C(\mathbf{R})$ is short range and does not depend strongly on temperature. However, at the temperature when $\hat{C}(\mathbf{q}) \to \rho_0^{-1}$, $\Gamma(\mathbf{q})$ becomes divergent; Thus, the determination of $\hat{C}(\mathbf{q})$ is essential to the discussion of the phenomenon of critical opalescence.

According to Eqs. (2.157) and (2.178), we have

$$S(\mathbf{q}) = V \int_V d^3R \exp(i\mathbf{q}\cdot\mathbf{R})\, G(\mathbf{R}) = V\left[\rho_0 + \rho_0^2 \frac{\hat{C}(\mathbf{q})}{1 - \rho_0 \hat{C}(\mathbf{q})}\right] = \frac{V\rho_0}{1 - \rho_0 \hat{C}(\mathbf{q})}. \tag{2.182}$$

Ornstein and Zernike assumed that $C(q)$ can be expanded in a Taylor series around $q = 0$ for all temperatures up to T_c. This is permissible because $C(R)$ is short range. Thus, for an isotropic system

$$\begin{aligned} C(\mathbf{q}) &= \int d^3R\, C(\mathbf{R})\, e^{i\mathbf{q}\cdot\mathbf{R}} = \int_0^{2\pi} d\phi \int_0^{\pi} \sin\theta\, d\theta \int_0^{\infty} dR\, R^2 C(R) e^{iqR\cos\theta} \\ &= 4\pi \int_0^{\infty} dR\, R^2 C(R) \frac{\sin qR}{qR} = C_0 - q^2 C_2 + \cdots, \end{aligned} \tag{2.183}$$

where

$$C_0 = 4\pi \int_0^{\infty} dR\, R^2 C(R), \tag{2.184}$$

$$C_2 = \frac{2\pi}{3} \int_0^{\infty} dR\, R^4 C(R). \tag{2.185}$$

Substituting Eq. (2.183) into (2.182), we obtain

$$S(\mathbf{q}) = \langle N \rangle \frac{1}{(1 - \rho_0 C_0) + \rho_0 q^2 C_2} = \frac{\langle N \rangle}{R_0^2 (q_0^2 + q^2)}, \tag{2.186}$$

where $q_0 = (1 - \rho_0 C_0)^{1/2}/R_0$ and $R_0 = (\rho_0 C_2)^{1/2}$. Since $C(R)$ is short range, C_0 and C_2 are finite; consequently, q_0 and R_0^2 are expected to be finite for all temperatures. Furthermore, according to Eq. (2.165), we have

$$\lim_{q \to 0} S(\mathbf{q}) = kTV\rho_0^2\chi_T = \frac{\langle N \rangle}{R_0^2 q_0^2}, \tag{2.187}$$

and

$$(R_0 q_0)^{-2} = kT\rho_0\chi_T; \tag{2.188}$$

hence,

$$S(\mathbf{q}) = \frac{kT\rho_0\chi_T\langle N \rangle}{1 + (q\xi)^2}, \tag{2.189}$$

where $\xi = 1/q_0$ has a dimension of length called the correlation length. At high temperature $C(\mathbf{q}) \to 0$, and ξ is small. Near T_c, $C(\mathbf{q}) \to \rho_0^{-1}$ and $q_0 \to 0$, i.e., ξ becomes very long.

Equation (2.189) predicts that the scattered intensity is a Lorentzian function of $|\mathbf{q}|$. For a normal fluid away from T_c, ξ is small and $S(\mathbf{q})$ is independent of q; therefore, there is no preferred direction for light to be scattered. However, near T_c when ξ increases to a length comparable with q^{-1}, the scattered intensity becomes q dependent. The scattered-light intensity distributes anisotropically, with more intensity in the forward direction. Equation (2.189) predicts that a plot of $S(\mathbf{q})^{-1}$ versus q^2 should yield a straight line with a slope given by $\xi^2/kT\rho_0\chi_T\langle N \rangle$ and an intercept by $(kT\rho_0\chi_T\langle N \rangle)$. This result has been tested experimentally and found to be in qualitative agreement.

Finally, let us mention that while in the condensed phase, the Brillouin peaks are much more prominent than the central peak because $C_p \simeq C_V$. Near the critical point, χ_T becomes very large and $\gamma - 1$ diverges because

$$C_p - C_V = VT\,(\partial P/\partial T)_V, \qquad \text{and} \qquad \chi_T \to \infty, \quad \text{as} \quad T \to T_c. \tag{2.190}$$

Thus, according to Eq. (2.117), most of the intensity near the critical point is in the central peak. The linewidth of the central peak is proportional to λ/C_p, which becomes very narrow as $C_p \to \infty$ near T_c. A narrow linewidth means a long relaxation time. Thus, near the critical point, the entropy fluctuation relaxes very slowly. This is the phenomenon of critical slowdown.

With regard to Brillouin scattering, the frequency of the Brillouin sidebands will become small near T_c, but there is nothing anomalous about the intensity and the linewidth. This is expected, since the critical phenomenon near T_c is caused mainly by entropy fluctuation and not by sound-wave propagation. However, when the damping mechanism of the sound wave is not caused by viscosity and conductivity but rather by another mechanism that slows down near the critical region, the linewidth anomaly will also occur.

2.12 LIGHT SCATTERING FROM FLUIDS OF OPTICALLY ANISOTROPIC MOLECULES

We have considered in detail the scattering from a fluid consisting of optically isotropic molecules. In such a fluid, the scattered light has the same polarization characteristics as the incident light, because the induced polarization vector lies in the same direction as the incident electric field. On the other hand, in a system of nonspherical molecules, the index of refraction is a function of molecular orientation. For example, the index of refraction would be a maximum if all the molecules were oriented with their long axes parallel to the direction of the externally applied electric field; the index is minimum if they are oriented perpendicular to it. Since the orientations of the molecular axes fluctuate within the scattering volume, the index of refraction of the medium in the scattering volume is time dependent. This results in frequency modulation of the scattered light. Consequently, one can probe the time-dependent orientation fluctuations through analysis of the spectrum of the scattered light. In the liquid state, the scattered light resulting from orientation fluctuations can account for a small percentage of the total scattered intensity for slightly anisotropic molecules and for over 50% for highly anisotropic macromolecules.

To analyze the light-scattering spectrum from orientation fluctuations, we return to Eq. (2.33). First we choose a laboratory coordinate system by directing the incident light and the scattered light beams in the xz plane in such a way that the scattering vector $\mathbf{q}$ is along the z axis of the chosen laboratory coordinate system. The incident light is assumed to be plane polarized in the vertical y axis. Figure 2.5 gives the scattering geometry (Pecora, 1968). The xz plane is the scattering plane.

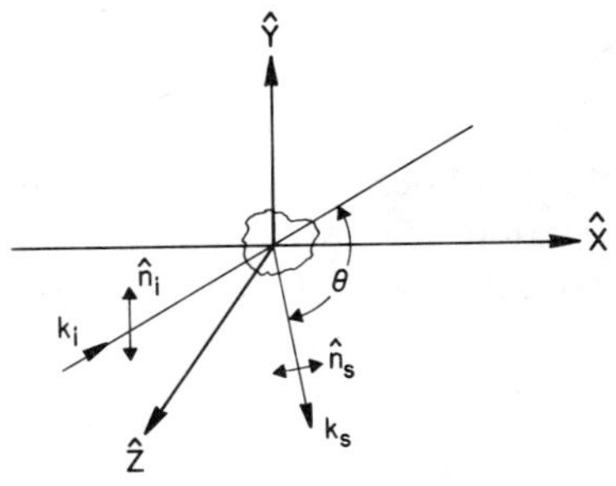

FIG. 2.5 Scattering geometry related to orientation fluctuations. The vector $\mathbf{k}_i$ indicates the incident propagation direction, $\hat{n}_i$ is the polarization of the incident light in the y direction, $\mathbf{k}_s$ the propagation of the scattered light, and $\mathbf{q}$ the scattering vector in the z direction.

In accordance with this scattering geometry, time-correlation functions of the polarizability density for the polarized (VV) and depolarized (VH) scattered light are given, respectively, by

$$C_{\mathrm{VV}}(\mathbf{q}, t) = \sum_{m,n} \langle \alpha_{yy}^{(m)}(t)\, \alpha_{yy}^{(n)} \exp\{-i\mathbf{q}\cdot[\mathbf{r}_m(t) - \mathbf{r}_n]\}\rangle \tag{2.191}$$

$$C_{\mathrm{VH}}(\mathbf{q}, t) = \sum_{m,n} (\langle \alpha_{yz}^{(m)}(t)\, \alpha_{yz}^{(n)} \exp\{-i\mathbf{q}\cdot[\mathbf{r}_m(t) - \mathbf{r}_n]\}\rangle \cos^2 \theta/2$$

$$+ \langle \alpha_{yx}^{(m)}(t)\, \alpha_{yx}^{(n)} \exp\{-i\mathbf{q}\cdot[\mathbf{r}_m(t) - \mathbf{r}_n]\}\rangle \sin 2\, \theta/2), \tag{2.192}$$

where θ is the scattering angle, as defined previously.

The various molecular-polarizability tensor elements present in these equations are expressed with respect to the laboratory plane. However, as the molecules reorient, these tensor elements will change with time. For the following discussion, we shall assume that the molecule is not optically active, so that the molecular-polarizability tensor $\boldsymbol{\alpha}^{(m)}$ is symmetric. The coordinate frame that diagonalizes the molecular polarizability tensor will be called the molecular frame. In the molecular frame, the polarizability tensor has three real principal components: $\alpha_1^{(m)}$, $\alpha_2^{(m)}$, and $\alpha_3^{(m)}$. For simplicity, we shall also assume here that $\alpha_1^{(m)}$ and $\alpha_2^{(m)}$ are equal and $\alpha_3^{(m)}$ is unique. Thus, we denote $\alpha_1^{(m)} = \alpha_2^{(m)} = \alpha_\perp^{(m)}$ and $\alpha_3^{(m)} = \alpha_\parallel^{(m)}$, leaving the general case to the next chapter, where a sophisticated technique for treating the molecular reorientation problem will be developed. Here, we consider a system consisting of molecules that are cylindrically symmetric.

Thus, in the molecular frame, the molecular polarization tensor can be expressed as

$$\boldsymbol{\alpha}^{(m)} = \begin{bmatrix} \alpha_\perp^{(m)} & 0 & 0 \\ 0 & \alpha_\perp^{(m)} & 0 \\ 0 & 0 & \alpha_\parallel^{(m)} \end{bmatrix}. \tag{2.193}$$

The projection of $\boldsymbol{\alpha}^{(m)}$ on the laboratory coordinate system as defined by the polarization vectors of the incident and scattered light can now be found by specifying these polarization vectors with respect to the molecular frame. If we let the direction cosines of the incident polarization vector with respect to the molecular frame be given by (n_{i1}, n_{i2}, n_{i3}) and the direction cosines of the polarization vector of the scattered light by (n_{j1}, n_{j2}, n_{j3}), i.e.,

$$\hat{n}_\alpha = \begin{bmatrix} n_{\alpha 1} \\ n_{\alpha 2} \\ n_{\alpha 3} \end{bmatrix}, \tag{2.194}$$

where $\alpha = i$ indicates the incident light and $\alpha = j$ indicates the scattered light, then the projected element α_{ij} (for brevity of notation, we suppress superscript m) can be written as

$$\alpha_{ij} = (n_{i1} n_{i2} n_{i3}) \begin{bmatrix} \alpha_\perp & 0 & 0 \\ 0 & \alpha_\perp & 0 \\ 0 & 0 & \alpha_\parallel \end{bmatrix} \begin{bmatrix} n_{j1} \\ n_{j2} \\ n_{j3} \end{bmatrix}$$

$$= \alpha_\perp (n_{i1} n_{j1} + n_{i2} n_{j2}) + \alpha_\parallel (n_{i3} n_{j3}),$$

or

$$\alpha_{ij} = \bar{\alpha}(\hat{n}_i \cdot \hat{n}_j) + \beta[n_{i3} n_{j3} - \tfrac{1}{3}(\hat{n}_i \cdot \hat{n}_j)], \tag{2.195}$$

where $\bar{\alpha}$ is the average polarizability given by

$$\bar{\alpha} = \tfrac{1}{3}(\alpha_\parallel + 2\alpha_\perp), \tag{2.196}$$

and β is the coefficient of optical anistropy given by

$$\beta = \alpha_\parallel - \alpha_\perp. \tag{2.197}$$

As the molecule rotates, the direction cosines will change with time. The direction cosines thus express the part of α_{ij} that depends on the molecular orientation.

Using Eq. (2.195), we rewrite $C_{VV}(\mathbf{q}, t)$ and $C_{VH}(\mathbf{q}, t)$ in Eqs. (2.191) and (2.192) as

$$C_{VV}(\mathbf{q}, t) = (\bar{\alpha})^2 \sum_m \sum_n \langle \exp\{-i\mathbf{q} \cdot [\mathbf{r}_m(t) - \mathbf{r}_n]\}\rangle$$
$$+ \beta^2 \sum_m \sum_n \langle P_2^{(m)}(yy, t) P_2^{(n)}(yy, 0) \exp\{-i\mathbf{q} \cdot [\mathbf{r}_m(t) - \mathbf{r}_n]\}\rangle, \tag{2.198}$$

and

$$C_{VH}(\mathbf{q}, t) = \beta^2 \sum_m \sum_n \{\langle P_2^{(m)}(yz, t)\, P_2^{(n)}(yz, 0)$$
$$\times \exp\{-i\mathbf{q} \cdot [\mathbf{r}_m(t) - \mathbf{r}_n]\}\rangle \cos^2 \theta/2$$
$$+ \langle P_2^{(m)}(yx, t)\, P_2^{(n)}(yx, 0) \exp\{-i\mathbf{q} \cdot [\mathbf{r}_m(t) - \mathbf{r}_n]\}\rangle \sin^2 \theta/2\}, \tag{2.199}$$

where

$$P_2(ij, t) = [n_{i3}(t) n_{j3}(t) - \tfrac{1}{3}\hat{n}_i(t) \cdot \hat{n}_j(t)]. \tag{2.200}$$

The subscript 2 on P indicates that the P_2 functions are related to the Legendre polynomial of order 2. In obtaining Eqs. (2.198) and (2.199), we have assumed that $\bar{\alpha} = \bar{\alpha}^{(m)}$ and $\beta = \beta^{(m)}$ for all m. In the $C_{VV}(\mathbf{q}, t)$ expression, there are also cross terms containing the factor $\bar{\alpha}\beta$. These terms vanish in an isotropic liquid because of the lack of preferential molecular orientation.

Note that the first term in Eq. (2.198) does not contain orientation dependence. This term is identical to the density-fluctuation term

$$\langle \delta\rho(\mathbf{q}, t)\, \delta\rho^*(\mathbf{q}, 0)\rangle,$$

previously considered in detail for the case of spherical particles. This density-fluctuation term is responsible for the three-peak structure in the Rayleigh–Brillouin spectrum of a molecular liquid. The second term in Eq. (2.198) has both a position and an orientation dependence. Similarly, in the $C_{VH}(\mathbf{q}, t)$ expression both orientation and position dependence are present. These orientation-dependent terms give rise to so-called anisotropic scattering.

At high temperatures, orientation fluctuations of small nonspherical molecules are very rapid. The position-dependent function

$$\exp[-i\mathbf{q} \cdot (\mathbf{r}_m(t) - \mathbf{r}_n)],$$

which is related to translational motion, contributes insignificantly in liquids consisting of small molecules, for many reasons. For instance, if we assume that there is an orientation correlation present, the correlation must be short range involving only near neighbors. As pointed out previously, in light scattering q is typically $10^5\ \mathrm{cm}^{-1}$, and the distance between near neighbors in a liquid is about $10^{-7}\ \mathrm{cm}^{-1}$ at most. Thus, at $t = 0$, the position-dependent term $\exp\{i\mathbf{q} \cdot [\mathbf{r}_m(t) - \mathbf{r}_n]\}$ is approximately equal to one. Hence, for this term to vary significantly in time, the molecules must either self-diffuse ($m = n$) or must move relatively ($m \neq n$) over a distance approximately equal to q^{-1}. Since the time for this to occur is generally longer than the average reorientation times of small molecules, unless the effect of translational motion on the rotational spectrum is examined under a high-resolution spectrometer so that the influence of the translational motion can be detected, the translational function can be neglected in the anisotropic scattering expressions, Eqs. (2.198) and (2.199). As a result of this approximation, these equations become

$$C_{VV}(\mathbf{q}, t) = (\bar{\alpha})^2 \langle \delta\rho(\mathbf{q}, t)\, \delta\rho^*(\mathbf{q}, 0)\rangle + \beta^2 \langle Q(yy, t)\, Q(yy, 0)\rangle, \quad (2.201)$$

and

$$C_{VH}(\mathbf{q}, t) = \beta^2 \langle Q(yz, t) Q(yz, 0)\rangle, \quad (2.202)$$

where the orientation function $Q(ij, t)$ is given by

$$Q(ij, t) = \sum_m P_2^{(m)}(ij, t). \quad (2.203)$$

To obtain Eq. (2.202), we used the fact that the fluid is isotropic and the orientational time-correlation function $\langle Q(yz,t)Q(yz,0)\rangle$ is equal to $\langle Q(yx,t)Q(yx,0)\rangle$. Since the orientational function $Q(ij,t)$ is a collective variable involving all molecules in the scattering volume, the time dependence of the orientational time-correlation function will be determined by the dynamics of a single reorienting molecule as well as by the pair correlation between near neighbors. When a single molecule is assumed to undergo rotational Brownian motion (to be discussed in more detail in Chapter 3), the single-particle reorientational dynamics can be described by a rotational-diffusion equation. In Chapter 3, we shall show that the rotational-diffusion equation for all molecules m gives

$$\begin{aligned}\langle P_2^{(m)}(ij,t)P_2^{(m)}(ij)\rangle &= \langle P_2^{(m)}(ij)^2\rangle \exp(-6D_R t)\\ &= \tfrac{1}{5}\exp(-6D_R t),\end{aligned} \tag{2.204}$$

where D_R is the rotational diffusion coefficient, which is controlled by the medium's viscosity and temperature, and the shape of the rotating molecule. Equation (2.204) is given by only a single exponential because of the axial symmetry of the molecular polarizability tensor. In this case, only one coefficient of optical anisotropy is nonzero.

The Fourier transform of Eq. (2.204) gives the orientational spectrum

$$I_{VH}(\omega) = \frac{1}{5\pi}\,\mathrm{Re}\int_0^\infty \exp(i\omega t)\exp(-6D_R t) = \frac{\beta^2}{5\pi}\frac{\tau_S}{1+\omega^2\tau_S^2}. \tag{2.205}$$

This represents a Lorentzian line of half-width at half-height equal to

$$\tau_S^{-1} = 6D_R. \tag{2.206}$$

Equation (2.205) gives the depolarized spectrum in terms of the single-particle reorientational time. On the basis of the Stokes–Einstein–Debye equation, the rotational-diffusion coefficient D_R is related to the friction coefficient ξ_R by $D_R = kT/\xi_R$, where $\xi_R = 6\eta_S V_h$, η_S is the shear viscosity, and V_h the effective hydrodynamic volume determined by the molecular volume V_m and boundary conditions that enter the derivation of the Stokes–Einstein–Debye equation (Hu and Zwanzig, 1974). Writing $V_h = aV_m$, it is found experimentally that for small molecules not interacting strongly with their neighbors (e.g., not forming hydrogen bonds), a is less than one. This indicates that for most small molecules, slip-boundary conditions are more appropriate for describing rotation of small molecules in the liquid state, as stick-boundary conditions would yield $a = 1$ (Alms *et al.*, 1973, 1974; Kivelson and Madden, 1980).

Experimental results also show that when τ_S is plotted as a function of η_S/T, a nonvanishing intercept τ_0 is found for a number of molecules. The finite intercept τ_0 has been interpreted as owing to free rotation; however, this is inconsistent with the derivation of the Stokes–Einstein–Debye equation. At present, the physical origin of τ_0 is not clear. To interpret the presence of the finite intercept, hypotheses have been made about the nonlinearity of the τ_S versus η_S/T plot at small η_S/T; however, very few experimental measurements are carried out at small η_S/T that can definitively support these hypotheses (Kivelson and Madden, 1980).

Equation (2.205) is appropriate for a dilute solution of anisotropic scatterers in a solvent consisting of optically isotropic molecules, because in this case the pair-orientation correlational effects can be neglected. However, for a concentrated solution or neat liquid, the pair-correlation effect may be important. It will be shown in Chapter 4 that, in this case, the time-correlation function that describes the relaxation of $Q(ij, t)$ for a symmetric top given in Eq. (2.203) can still be written as

$$\langle Q(ij, t)Q(ij, 0)\rangle = \langle Q(ij)^2\rangle \, e^{-\Gamma t}, \tag{2.207}$$

provided that Eq. (2.204) is valid, where Γ^{-1} is the time constant describing the relaxation of the collective orientational variable $Q(ij, t)$ and is related to the single-particle orientational-relaxation time by

$$\Gamma^{-1} = \tau_S(1 + fN)/(1 + gN), \tag{2.208}$$

where N is the number density of the anisotropic scatterers and f and g, which are defined in Chapter 4, represent, respectively, the static and dynamic orientational pair-correlation factors. Experimentally, it is found that τ_{Ray} is a linear function of n, indicating that $g \ll f$. Furthermore, it is found in most cases that τ_{Ray} is a linear function of η_S/T, suggesting that the pair orientational factor is not a function of η_S/T.

To sum up this discussion, for the small wave vector q we find a simple exponential relaxation equation for $Q(ij|t)$ involving one relaxation rate:

$$\partial Q(ij|t)/\partial t = -\Gamma Q(ij|t), \tag{2.209}$$

where Γ is independent of q. Instead of by Eq. (2.205), the VH spectrum for a system of orientationally correlated isotropic liquid is given by

$$I_{VH}(\omega) = (\beta^2/\pi)\, \Gamma/(\Gamma^2 + \omega^2)\, \langle |Q(ij)|^2\rangle, \tag{2.210}$$

where Γ is proportional to T/η_S.

2.13 COUPLING OF POLARIZABILITY ANISOTROPY TO TRANSVERSE WAVES

Equation (2.209) is not valid when reorientational motion is coupled to transverse stress. From Eq. (2.88), we note that in the absence of coupling, the shear (transverse) velocity $u_T(\mathbf{q}, t)$ decays with a time constant $\tau_T(\mathbf{q})$ given by $m\rho_0/\eta_S q^2$. Taking the direction specified by T as y and the direction of $\mathbf{q}$ to be along the z axis of the laboratory coordinate system, we note that $u_T(\mathbf{q}, t)$ [or $u_y(\mathbf{q}, t)$] has the same translation and reflection symmetry as $\alpha_{yz}(\mathbf{q}, t)$. Thus, in the presence of translation–rotation coupling, $u_y(\mathbf{q}, t)$ and $\alpha_{yz}(\mathbf{q}, t)$ are not independent variables (Rytov, 1957). Allowing for such a coupling, we write the strain tensor as

$$\partial\alpha_{yz}/\partial t = -\Gamma\alpha_{yz} + \mu'\dot{e}_{yz}, \tag{2.211}$$

where μ' is the coupling coefficient and $\dot{e}_{yz}$ the strain rate. In Eq. (2.211), instead of using the orientation-density function $Q(yz\,|\,t)$, we use $\alpha_{yz}(\mathbf{r}, t)$ to represent the orientation density.

In the absence of coupling, the stress tensor σ_{yz} is related to the strain-rate tensor by the usual relation: $\sigma_{yz} = \eta_S\dot{e}_{yz}$, where η_S is the shear viscosity. However, when translation–rotation coupling is incorporated, we have

$$\sigma_{yz} = \eta_S\dot{e}_{yz} + \mu\dot{\alpha}_{yz}, \tag{2.212a}$$

or

$$\sigma_{yz} = \eta\dot{e}_{yz} - \mu\Gamma\alpha_{yz}. \tag{2.212b}$$

To obtain Eq. (2.212b), we have substituted Eq. (2.211) into (2.212a): η is a renormalized viscosity equal to $(\eta_S + \mu\mu')$, where μ is a coupling coefficient proportional to the negative value of μ' (because α_{yz} and u_y have opposite parity).

Inserting Eq. (2.211) into the Newton equation of motion for continuum mechanics, we obtain

$$\rho_m\,\partial u_y/\partial t = \partial(\sigma_{yz})/\partial z = \eta\,\partial^2 u_y/\partial z^2 - \mu\Gamma\,\partial\alpha_{yz}/\partial z. \tag{2.213}$$

Equations (2.212) and (2.213) constitute a set of coupled differential equations. They can be solved using the Fourier–Laplace transform technique previously employed. Spatial Fourier transforms of these equations give (note that $\mathbf{q}$ is along the z axis)

$$\partial u_y(\mathbf{q}, t)/\partial t = -(\eta q^2/\rho_m)\,u_y(\mathbf{q}, t) - (iq\mu\Gamma/\rho_m)\,\alpha_{yz}(\mathbf{q}, t), \tag{2.214}$$

and

$$\partial\alpha_{yz}(\mathbf{q}, t)/\partial t = iq\mu' u_y(\mathbf{q}, t) - \Gamma\alpha_{yz}(\mathbf{q}, t). \tag{2.215}$$

The foregoing time-dependent equations can now be solved by the usual Laplace transform technique. Defining

$$\hat{u}_y(\mathbf{q}, s) = \int_0^\infty dt\, e^{-st} u_y(\mathbf{q}, t) \tag{2.216}$$

and

$$\hat{\alpha}_{yz}(\mathbf{q}, s) = \int_0^\infty dt\, e^{-st} a_{yz}(\mathbf{q}, t), \tag{2.217}$$

we obtain from Eqs. (2.214) and (2.215) two coupled algebraic equations. The matrix representation of the coupled equations is

$$\begin{bmatrix} s + q^2\eta/\rho_m & iq\mu\Gamma/\rho_m \\ -i\mu' q & s + \Gamma \end{bmatrix} \begin{bmatrix} \hat{u}_y(\mathbf{q}, s) \\ \hat{\alpha}_{yz}(\mathbf{q}, s) \end{bmatrix} = \begin{bmatrix} u_y(\mathbf{q}, 0) \\ \alpha_{yz}(\mathbf{q}, 0) \end{bmatrix}. \tag{2.218}$$

The solutions for $\hat{u}_y(\mathbf{q}, s)$ and $\hat{\alpha}_{yz}(\mathbf{q}, s)$ are obtained by inverting this matrix to obtain

$$\begin{bmatrix} \hat{u}_y(\mathbf{q}, s) \\ \hat{\alpha}_{yz}(\mathbf{q}, s) \end{bmatrix} = \Delta^{-1} \begin{bmatrix} s + \Gamma & -iq\mu\Gamma/\rho_m \\ i\mu' q & s + q^2\eta/\rho_m \end{bmatrix} \begin{bmatrix} u_y(\mathbf{q}, 0) \\ \alpha_{yz}(\mathbf{q}, 0) \end{bmatrix},$$

or

$$\hat{u}_y(\mathbf{q}, s) = (1/\Delta)[(s + \Gamma)u_y(\mathbf{q}, 0) - (iq\mu\Gamma/\rho_m)\, \alpha_{yz}(\mathbf{q}, 0)] \tag{2.219}$$

and

$$\hat{\alpha}_{yz}(\mathbf{q}, s) = (1/\Delta)[iq\mu' u_y(\mathbf{q}, 0) + (s + q^2\eta/\rho_m)\alpha_{yz}(\mathbf{q}, 0)], \tag{2.220}$$

where

$$\Delta = (s + \Gamma)(s + \eta q^2/\rho_m) - \Gamma\mu\mu' q^2/\rho_m. \tag{2.221}$$

The correlation functions can now be obtained. For example,

$$\langle \hat{u}_y(\mathbf{q}, s) u_y^*(\mathbf{q}, 0) \rangle$$

is obtained from Eq. (2.219) if both sides of the equation are multiplied by $u_y^*(\mathbf{q}, 0)$ and an ensemble average is taken [note that $\langle \alpha_{yz}(q) u_y^*(q) \rangle = 0$ owing to opposite time-reversal symmetry]. Thus,

$$\langle \hat{u}_y(\mathbf{q}, s)\, u_y^+(\mathbf{q}) \rangle = [(s + \Gamma)/\Delta]\, \langle |u_y(\mathbf{q}, 0)|^2 \rangle. \tag{2.222}$$

Similarly, from Eq. (2.220) we obtain

$$\langle \hat{\alpha}_{yz}(\mathbf{q}, s)\, \alpha_{yz}^+(\mathbf{q}) \rangle = [\langle |\alpha_{yz}(\mathbf{q})|^2 \rangle / \Delta]\, (s + q^2\eta/\rho_m). \tag{2.223}$$

We can rewrite Eq. (2.222) as

$$\langle u_y(\mathbf{q}, s)u_y^+(\mathbf{q})\rangle = \frac{\langle |u_y(\mathbf{q}, 0)|^2\rangle}{s + \eta(s)q^2/\rho_m}, \tag{2.224}$$

where

$$\eta(s) = \eta + \mu\mu' s/(s + \Gamma) = \eta_S - \xi s/(s + \Gamma). \tag{2.225}$$

Equation (2.224) should be contrasted with Eq. (2.88) which is obtained without translation–rotation coupling. In the presence of rotation–translation coupling, the shear viscosity η_S is replaced by an effective viscosity $\eta(s)$, which is frequency dependent $\xi = -\mu\mu'$.

The frequency dependence is clearly seen if we set $s = i\omega$ and write

$$\eta(\omega) = \eta(s)|_{s=i\omega} = [\eta_S - \xi\omega^2/(\omega^2 + \Gamma^2)] - [i\xi\omega\Gamma/(\omega^2 + \Gamma^2)]. \tag{2.226}$$

Thus, the real part of $\eta(\omega)$ goes from η_S (for $\omega \ll \Gamma$) to $\eta_S - \xi/\Gamma$ (for $\omega \gg \Gamma$). The imaginary part is negligible for $\omega \ll \Gamma$ and equal to $\xi(\Gamma/\omega)$ for $\omega \gg \Gamma$, which is also negligible.

The correlation function $\langle \hat{\alpha}_{yz}(\mathbf{q}, s)\, \alpha_{yz}^+(\mathbf{q})\rangle$ is related to the VH spectrum. Setting $s = i\omega$, we obtain

$$\begin{aligned} I_{yz}(\mathbf{q}, \omega) &= \frac{1}{\pi}\,\mathrm{Re}\langle \hat{\alpha}_{yz}(\mathbf{q}, s)\alpha_{yz}^+(\mathbf{q})\rangle|_{s=i\omega} \\ &= \frac{\langle |\alpha_y\,(\mathbf{q})|^2\rangle}{\pi}\,\frac{b\omega_S^2 + \Gamma\omega^2}{(-\omega_S^2 + \omega^2)^2 + \Gamma_S^2\omega^2}, \end{aligned} \tag{2.227}$$

where $b = \eta q^2/\rho_m$, $\Gamma_S = \Gamma + b$, $\omega_S^2 = b\Gamma + \xi q^2\Gamma/\rho_m$. Since η is related to shear viscosity η_S by $\eta = \eta_S(1 - R)$, R being the fraction of viscosity originated from translation–rotation coupling defined as

$$R = \xi/\eta_S, \tag{2.228}$$

it follows that $\omega_S^2 = \eta_S q^2\Gamma/\rho_m$, and Eq. (2.227) becomes

$$I_{yz}(\mathbf{q}, \omega) = \frac{\langle |\alpha_{yz}(\mathbf{q})|^2\rangle}{\pi}\left\{\frac{\Gamma[\omega^2 + (\nu_S q^2)^2(1 - R)]}{(\omega^2 - q^2\nu_S\Gamma)^2 + \omega^2[\Gamma + (1 - R)q^2\nu_S]^2}\right\}, \tag{2.229}$$

where ν_S is the kinematic viscosity equal to η_S/ρ_m.

The complete VH spectrum is the Fourier transform of Eq. (2.192). We have just calculated the part associated with the $\cos^2\theta/2$ term, as given in Eq. (2.229). The part associated with the $\sin^2\theta/2$ term will not be coupled to the transverse velocity because $\alpha_{yx}(\mathbf{q}, t)$ does not have the same symmetry as $u_y(\mathbf{q}, t)$. Thus, we have simply

$$I_{yx}(\mathbf{q}, \omega) = [\langle|\alpha_{yx}(\mathbf{q})|^2\rangle/\pi]\,\Gamma/(\omega^2 + \Gamma^2). \tag{2.230}$$

The complete VH spectrum for an isotropic viscous fluid in the presence of rotation–translation coupling is given by (Andersen and Pecora, 1971a; Kivelson and Keyes, 1972)

$$I_{\mathrm{VH}}(\mathbf{q}, \omega) = \frac{kT\chi_\alpha}{\pi}\left\{\frac{\Gamma}{\omega^2 + \Gamma^2}\sin^2\frac{\theta}{2} + \frac{\Gamma[\omega^2 + (v_{\mathrm{S}}q^2)^2(1 - R)]}{(\omega^2 - q^2v_{\mathrm{S}}\Gamma)^2 + \omega^2[\Gamma + (1 - R)q^2v_{\mathrm{S}}]^2}\cos^2\frac{\theta}{2}\right\}. \tag{2.231}$$

This equation shows that the VH spectrum is not composed of a single lorentzian. Using an appropriate set of parameters, Eq. (2.231) shows fine structure corresponding to a dip at the center of the incident-laser frequency ($\omega = 0$) in the VH spectrum (Stegeman and Stoicheff, 1968), as shown in Fig. (2.6). In Eq. (2.231), $kT\chi_\alpha = \langle|\alpha_{yx}(\mathbf{q})|^2\rangle = \langle|\alpha_{yz}(\mathbf{q})|^2\rangle$, because of the rotational isotropy of the fluid.

In the high-temperature limit we have $q^2v_{\mathrm{S}}/\Gamma \ll 1$. Equation (2.231) reduces to the sum of two lorentzians. (Kivelson and Keys, 1972).

$$I_{\mathrm{VH}}(\mathbf{q}, \omega) = \frac{kT\chi_\alpha}{\pi}\left[\frac{\Gamma}{\omega^2 + \Gamma^2} - R\left(\frac{q^2v_{\mathrm{S}}}{\Gamma}\right)\frac{q^2v_{\mathrm{S}}}{\Gamma^2 + (q^2v_{\mathrm{S}})^2}\cos^2\frac{\theta}{2}\right]. \tag{2.232}$$

The spectrum predicted by Eq. (2.232) consists of a broad lorentzian of HWHH (half width at half height) Γ and a narrow negative lorentzian of

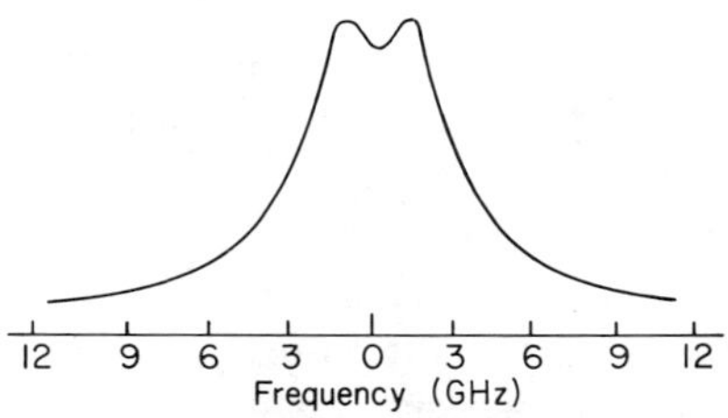

FIG. 2.6 Schematic of depolarized spectrum with fine structure.

HWHH $= q^2 \nu_S$. Obviously, this narrow lorentzian leads to the dip observed in the VH spectrum of many liquids. Note that the intensity of the dip is directly proportional to R. The fine structure is a direct consequence of "negative" coupling between orientation and transverse wave (or rotation–translation coupling).

In Chapter 4, a microscopic theory will be used to derive an expression for the coupling parameter R. Intuitively, we expect R to depend on some molecular parameters such as geometric shape, molecular size, and the aspects of steric or repulsive forces. Experimental results so far accumulated, however, indicate that R is independent of temperature and does not depend significantly on steric forces or the molecular dipole moment. Recent studies of a group of mono-substituted benzene and pyridine derivatives suggest that the details of molecular charge distribution are very important in determining the magnitude of the coupling parameter R as measured by depolarized light scattering (O'Steen, 1981).

As the molecular liquid is supercooled and becomes viscoelastic, the VH spectrum shows a triplet structure associated with the oscillation of the shear wave (Fablinski *et al.*, 1969; Starunov and Fablinski, 1974). In the viscoelastic state, the zero-frequency shear viscosity η_S is large, so that the value of $\nu_S q^2$ becomes greater than the orientational relaxation rate Γ, and the calculation of the VH spectrum given in Eq. (2.231) is inappropriate because the dynamic response of the viscosity is not included in the analysis. To include the dynamic effect, rather than using the $\sigma_{yz}(t) = \eta_S \dot{e}_{yz}$ relation, on the basis of linear viscoelasticity the constitutive equation is given by

$$\sigma_{yz}(t) = \int_0^\infty dt'\, G(t - t')\, \dot{e}_{yb}(t'), \tag{2.233}$$

where $G(t)$ is the shear-relaxation modulus and is assumed to be a function of time only. The equation incorporates the concept of classical elasticity theory for solids in which the stress is always directly proportional to the instantaneous strain (Hooke's law) and the concept of classical hydrodynamic theory for viscous liquids in which the stress is directly proportional to the instantaneous rate of strain (Newton's law). Substances that exhibit both solidlike and liquidlike behavior are said to be viscoelastic (Ferry, 1980). If the stress and strain rates are kept sufficiently small so that the ratio of stress to strain is a function of time (or frequency) only and is independent of the stress level, then the material is said to show linear viscoelastic behavior. The linear constitutive equation given in Eq. (2.233) reduces to Hooke's law for an amorphous solid if the shear (relaxation) modulus relaxes infinitely slowly, i.e., $G(t)$ is time independent. On the other hand, it reduces to Newton's law for a viscous liquid if the shear modulus relaxes infinitely fast, i.e., $G(t) = \eta_S \delta(t)$. With the time dependence of $G(t)$ situated between these two extremes, Eq. (2.233) represents a realistic phenomenological equation for the stress–strain relation of viscoelastic liquids. Consequently, the shear–stress relation given in Eq. (2.212a) for a translation–rotation-coupled viscoelastic fluid is replaced by (Wang, 1980; Wang *et al.*, 1984).

$$\sigma_{yz}(t) = \int_0^t dt'\, G(t - t')\, \dot{e}_{yz}(t') + \mu\dot{\alpha}_{yz}(t), \tag{2.234}$$

which, according to Eq. (2.211), becomes

$$\sigma_{yz}(t) = \int_0^t dt'\, \bar{G}(t - t')\, \dot{e}_{yz}(t) - \mu\Gamma\alpha_{yz}(t), \tag{2.235}$$

where $\bar{G}(t)$ is the renormalized shear-relaxation modulus given by

$$\bar{G}(t) = G(t) + \mu\mu'\, \delta(t). \tag{2.236}$$

Using Eq. (2.235), we can write Newton's equation of motion as

$$\rho_m \frac{\partial u_y}{\partial t} = \int_0^\infty dt'\, \bar{G}(t - t') \frac{\partial^2 u_y(t')}{\partial z^2} - \mu\Gamma \frac{\partial \alpha_{yz}}{\partial z}. \tag{2.237}$$

Using Eqs. (2.211) and (2.237), we can calculate the VH spectrum in accordance with the procedure shown above. The result for $\langle \alpha_{yz}(\mathbf{q}, s)\alpha_{yz}^{+}(\mathbf{q})\rangle$ is

$$\langle \alpha_{yz}(\mathbf{q}, s)\alpha_{yz}^{+}(\mathbf{q})\rangle = \frac{kT\chi_\alpha[s + q^2\, \bar{G}(s)/\rho_m]}{(s + \Gamma)(s + q^2\bar{G}(s)/\rho_m) - \Gamma\mu\mu' q^2/\rho_m}, \tag{2.238}$$

where $\bar{G}(s)$ is given by

$$\bar{G}(s) = \int_0^\infty dt\, e^{-st}\, [G(t) + \mu\mu'\delta(t)] = G(s) - \xi \tag{2.239}$$

with $\xi = -\mu\mu'$ (as previously defined).

To carry the calculation further it is necessary to have some knowledge about the time-dependent behavior of the shear-relaxation modulus. Consider first the Maxwell model (Fig. 2.7a), wherein the constitutive equation for the stress tensor σ_{yz} is assumed to be given by

$$\dot{e}_{yz} = \frac{1}{G_M}\dot{\sigma}_{yz} + \frac{1}{\eta_M}\sigma_{yz}, \tag{2.240}$$

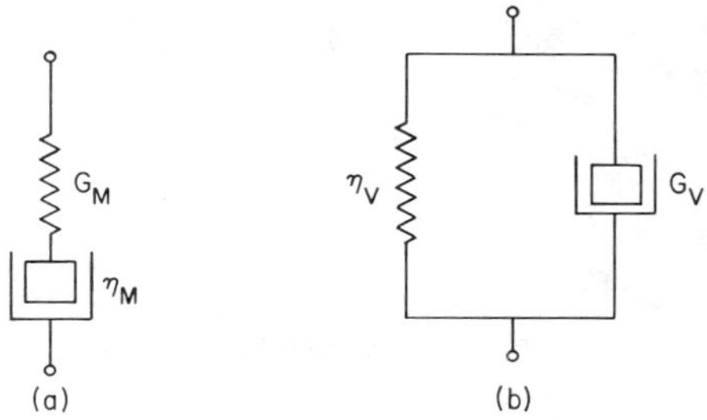

FIG. 2.7 (a) The Maxwell element. The elastic spring is in series with the viscous element (dashpot); thus, the strain rate is equal to the sum of contributions from the spring and the dashpot. (b) The Voigt element. The elastic element is in parallel with the dashpot; thus, the stress is equal to the sum of contributions from the spring and the dashpot.

which, when converted to the form of Eq. (2.233), becomes

$$\sigma_{yz}(t) = G_{\mathrm{M}} \int_0^t \exp[-(t - t')/\tau_{\mathrm{M}}]\, \dot{e}_{yz}(t')\, dt', \tag{2.241}$$

where τ_{M} is the Maxwell relaxation time and is equal to $\eta_{\mathrm{M}}/G_{\mathrm{M}}$. Comparing Eq. (2.241) with (2.233) shows that the shear modulus in a Maxwell liquid is given by

$$G(t) = G_{\mathrm{M}} \exp(-t/\tau_{\mathrm{M}}). \tag{2.242}$$

Since $G(t = 0) = G_\infty$, we can regard G_{M} as the shear modulus at infinite frequency. The term η_{M} is the Maxwell viscosity.

Thus, for a Maxwell liquid, we have

$$G(s) = G_\infty/(s + \tau_{\mathrm{M}}^{-1}) = \eta_{\mathrm{M}}/(1 + s\tau_{\mathrm{M}}). \tag{2.243}$$

Substituting Eq. (2.243) into (2.238) yields

$$\langle \alpha_{yz}(\mathbf{q}, s)\, \alpha_{yz}^{+}(\mathbf{q})\rangle = \frac{kT\chi_\alpha}{\pi}\left\{\frac{s(1 + s\tau_{\mathrm{M}}) + b_{\mathrm{M}} - s\tau_{\mathrm{M}}\delta_{\mathrm{M}}}{s^2(1 + s\tau_{\mathrm{M}}) + s[\Gamma_{\mathrm{M}} + s\tau_{\mathrm{M}}(\Gamma - \delta_{\mathrm{M}})] + \omega_{\mathrm{M}}^2}\right\}, \tag{2.244}$$

where

$$b_M = q^2(\eta_M - \xi)/\rho_m,\ \delta_M = q^2\xi/\rho_m,\ \Gamma_M = \Gamma + b_M,$$

and

$$\omega_M^2 = \Gamma q^2 \eta_M/\rho_m = q^2(\Gamma\tau_M)G_\infty/\rho_m = \Omega_\infty^2(\Gamma\tau_M),$$

where

$$\Omega_\infty^2 = q^2 G_\infty/\rho_M.$$

Setting $s = i\omega$ and taking the real part of Eq. (2.244), we obtain the spectral power density $I_{yz}(\mathbf{q}, \omega)$ as

$$I_{yz}(\mathbf{q}, \omega) = \frac{kT\chi_\alpha}{\pi} \times \frac{b_M[\omega_M^2 - \omega^2\tau_M(\Gamma - \delta_M)] + \omega^2[\Gamma - \tau_M(\omega_M^2 - \omega^2\Gamma\tau_M + \delta_M\Gamma_M)]}{\{\omega_M^2 - \omega^2[1 + \tau_M(\Gamma - \delta_M)]\}^2 + \omega^2(\Gamma_M - \omega^2\tau_M)^2}. \tag{2.245}$$

This expression contains four parameters: the molecular orientation rate Γ, the limiting high-frequency shear modulus G_∞ (or Ω_∞), the shear viscosity η_M, and the dynamic-coupling coefficient between the shear wave and the orientation density δ_m(or ξ). In the limit $\tau_M \to 0$, the expression is reduced to Eq. (2.227), derived for the ordinary viscous liquid. Equation (2.243) together with Eq. (2.230) can qualitatively account for the triplet structure of the VH spectrum observed in a viscoelastic liquid.

Another model that can also be used to illustrate the shear-relaxation modulus is the Voigt viscoelastic element (Fig. 2.7b). Using this model for the viscoelastic liquid, we write the constitutive equation as

$$\sigma_{yz}(t) = G_V e_{yz}(t) + \eta_V \dot{e}_{yz}(t), \tag{2.246}$$

where G_V is the Voigt elastic constant and η_V the Voigt shear viscosity. When this constitutive equation is expressed in the form of Eq. (2.233), we obtain the shear-relaxation modulus $G(t)$ as

$$G(t) = G_V + \eta_V\, \delta(t); \tag{2.247}$$

consequently,

$$G(s) = G_V/s + \eta_V. \tag{2.248}$$

Substituting Eq. (2.248) into (2.238), setting $s = i\omega$, and then taking the real part of the resulting expression, we obtain the spectral power density for the Voigt liquid as

$$I_{yz}(\mathbf{q}, \omega) = \frac{kT\chi_\alpha}{\pi}\left[\frac{q^2 v_V \omega^2 \omega_V^2 + \Gamma(-\omega^2 + \Omega_V^2)^2}{(-\omega^2\Gamma_V + \Gamma\Omega_V^2)^2 + \omega^2(-\omega^2 + \Omega_V^2 + q^2\Gamma v_V)^2}\right], \tag{2.249}$$

where $v_V = (\eta_V - \xi)/\rho_m$, $\omega_V^2 = \Gamma q^2 \eta_V/\rho_m$, and $\Omega_V^2 = q^2 G_V/\rho_m$. This equation, together with Eq. (2.230), can also be used to fit the VH spectrum for the viscoelastic liquid.

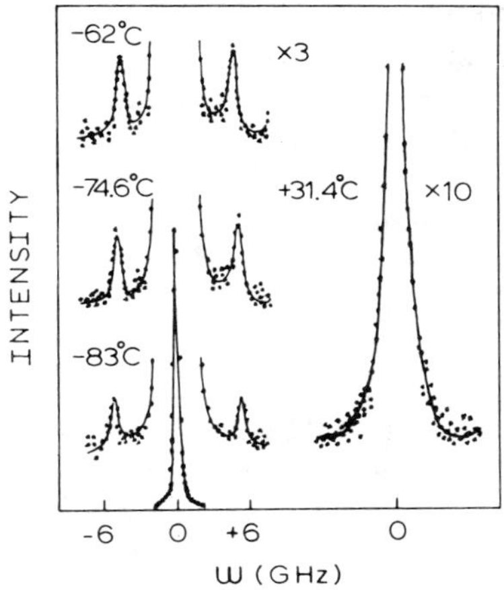

FIG. 2.8 Comparison of the experimental data of α-phenyl *o*-cresol with the theoretical spectra calculated using the Voigt model.

Both the Maxwell and Voigt models have been used to fit the experimental spectra of α-phenyl *o*-cresol at temperatures where the viscoelastic behavior is most pronounced (Wang and Liu, 1983). The two models give similar values for Ω and Γ, although the value for η predicted by the Maxwell fit is smaller than that from the Voigt fit. However, the Maxwell fit gives a significantly higher root means squares error. While the difficulty associated with the single Maxwell fit or the Voigt model for the description of the viscoelastic relaxation behavior in supercooled liquids is well known, the Voigt model has been shown to yield correct angular dependent values for Ω, η and δ (Wang *et al.*, 1984). The comparison of the experimental spectra with the theoretical spectra calculated in accordance with the Voigt equation [Eq. (2.249)] together with Eq. (2.230) is shown in Fig. 2.8.

2.14 PHOTOELASTIC EFFECTS

The modern era of photoelasticity studies began with Brillouin's pioneering work on light scattering from thermal fluctuations in solids. Because of the coupling of fluctuations in the dielectric tensor to acoustic waves, the frequency of the monochromatic laser light scattered from an anisotropic elastic solid is shifted by an amount determined by the elastic constants of the solid as well as by the scattering angle. The intensity of the scattered light is determined by the elastic and photoelastic constants. Thus Brillouin scattering provides a useful method for determining the elastic and photoelastic constants of an anisotropic solid.

The interpretation of the photoelastic effect in an anisotropic solid was originally based on the Pockels formulation, which states that the change in the inverse dielectric tensor $\delta\boldsymbol{\varepsilon}^{-1}$ was proportional to the strain tensor $\mathbf{e}$ [according to Born and Huang (1962)]:

$$(\delta\varepsilon^{-1})_{ij} = \sum_{k,l} p_{ijkl} e_{kl}, \tag{2.250}$$

where p_{ijkl} is the Pockels photoelastic tensor. Because $(\delta\varepsilon^{-1})_{ij}$ and e_{kl} are symmetric tensors, p_{ijkl} are symmetric on interchange of i and j and on interchange of k and l. The existence of the various photoelastic tensor components is determined by the crystal symmetry of an ordinary fourth-rank tensor.

In Eq. (2.26), the spectral power density of the scattered light is expressed in terms of the time-correlation function of the polarizability density. To incorporate the effect of acoustooptic interaction in the formulation, we use the relation $\delta\boldsymbol{\varepsilon}(\mathbf{q}, t) = 4\pi\boldsymbol{\alpha}(\mathbf{q}, t)$ and rewrite Eq. (2.26) in terms of the spectral power density of the fluctuations of the dielectric tensor as

$$S_{ij}(\mathbf{q}, \omega) = \frac{1}{2\pi}\int_0^\infty dt\, e^{-i\omega t} \langle \delta\varepsilon_{ij}(\mathbf{q}, t)\, \delta\varepsilon_{ij}(\mathbf{q})\rangle, \tag{2.251}$$

where $S_{ij}(\mathbf{q}, \omega) = (1/8\pi)\, I_{ij}(\mathbf{q}, \omega)$, and ω, which plays the role of Ω, is the shift from the incident-laser frequency. In Eq. (2.251), the first index i denotes the polarization direction of the incident light, and the second index j denotes the polarization direction of the scattered light; $\delta\varepsilon_{ij}$ is the projection of $\delta\boldsymbol{\varepsilon}$ in the i and j directions.

Since the Pockels law expresses the photoelastic effect in terms of $\delta\varepsilon_{-ij}$, to use this law, it is necessary to relate $\delta\boldsymbol{\varepsilon}$ to $\delta\boldsymbol{\varepsilon}^{-1}$. This is accomplished by using the identity

$$\boldsymbol{\varepsilon}\boldsymbol{\varepsilon}^{-1} = \mathbf{I} \tag{2.252}$$

where $\mathbf{I}$ is a unit tensor. Thus

$$(\delta\boldsymbol{\varepsilon})\boldsymbol{\varepsilon}^{-1} + \boldsymbol{\varepsilon}(\delta\boldsymbol{\varepsilon}^{-1}) = 0,$$

or

$$\delta\boldsymbol{\varepsilon} = -\boldsymbol{\varepsilon}(\delta\boldsymbol{\varepsilon}^{-1})\,\boldsymbol{\varepsilon}. \tag{2.253}$$

Thus, according to Eq. (2.250), we obtain (to first order in e_{kl})

$$\begin{aligned}\delta\varepsilon_{ij} &\simeq \sum_{ij} \varepsilon^0_{ii'} (\delta\varepsilon^{-1})_{i'j'} \varepsilon^0_{j'j} \\ &= -\varepsilon^0_{ii'} \varepsilon^0_{jj'} p_{i'j'kl} e_{kl},\end{aligned} \tag{2.254}$$

where the Einstein summation convention is used, i.e., a repeated summation index is understood to imply a summation over this index.

If the i and j directions are along the optical axes of the crystal, Eq. (2.254) is simplified to

$$\delta\varepsilon_{ij}(\mathbf{q}, t) = -n_i^2 n_j^2 p_{ijkl} e_{kl}(\mathbf{q}, t), \tag{2.255}$$

where $n_i = \sqrt{\varepsilon^0_{ii}}$ and $n_j = \sqrt{\varepsilon^0_{jj}}$ are the refractive indices along the optical axes i and j, respectively.

Using Eq. (2.255), the spectral density of the scattering light given in Eq. (2.251) becomes

$$S_{ij}(\mathbf{q}, \omega) = \frac{1}{2\pi} n_i^4 n_j^4 p_{ijkl} p_{ijk'l'} \int_0^\infty dt\, e^{-i\omega t} \langle e_{kl}(\mathbf{q}, t)\, e^*_{k'l'}(\mathbf{q})\rangle. \tag{2.256}$$

The general equation for $S_{ij}(\mathbf{q}, \omega)$ is quite complicated (Vacher and Boyer, 1972). To illustrate the application, we consider a specific example in which the photoelastic tensor has hexagonal symmetry. The photoelastic tensor in this case is given by (Nye, 1960)

$$p_{ijkl} = p_{\alpha\beta} = \begin{bmatrix} p_{11} & p_{12} & p_{13} & 0 & 0 & 0 \\ p_{12} & p_{11} & p_{13} & 0 & 0 & 0 \\ p_{31} & p_{31} & p_{33} & 0 & 0 & 0 \\ 0 & 0 & 0 & p_{44} & 0 & 0 \\ 0 & 0 & 0 & 0 & p_{44} & 0 \\ 0 & 0 & 0 & 0 & 0 & p_{66} \end{bmatrix}, \tag{2.257}$$

where $p_{66} = \frac{1}{2}(p_{11} - p_{12})$.

In Eq. (2.256), the contracted subscript notation of Voigt (with the index 1 for xx, 2 for yy, 3 for zz, 4 for yz, 5 for zx, and 6 for xy) is used for p_{ijkl}. For example, p_{xxzz} is p_{13}, p_{yzyz} is p_{44}, etc. These photoelastic constants can be determined by measuring the integrated intensity of the Brillouin lines.

Using Eqs. (2.215) and (2.213), we find that the six components of $\delta\varepsilon_{ij}$ are given by

$$\delta\varepsilon_{xx} = -\eta_x^4(p_{11}e_{xx} + p_{12}e_{yy} + p_{13}e_{zz}), \tag{2.258a}$$

$$\delta\varepsilon_{yy} = -n_y^4(p_{12}e_{xx} + p_{11}e_{yy} + p_{13}e_{zz}), \tag{2.258b}$$

$$\delta\varepsilon_{zz} = -n_z^4(p_{31}e_{xx} + p_{31}e_{yy} + p_{33}e_{zz}), \tag{2.258c}$$

$$\delta\varepsilon_{yz} = -n_y^2 n_z^2 p_{44} e_{yz}, \tag{2.258d}$$

$$\delta\varepsilon_{zx} = -n_z^2 n_x^2 p_{44} e_{zx}, \tag{2.258e}$$

$$\delta\varepsilon_{xy} = -n_x^2 n_y^2 p_{66} e_{xy}. \tag{2.258f}$$

Since $e_{ij}(\mathbf{r}) = \frac{1}{2}(\partial U_i/\partial x_j + \partial U_i/\partial x_i)$, $e_{ij}(\mathbf{q})$ vanishes unless the scattering vector $\mathbf{q}$ has a component along the i or j axis. Furthermore, since the x and y axes are equivalent, $\delta\varepsilon_{xx} = \delta\varepsilon_{yy}$ and $\delta\varepsilon_{yz} = \delta\varepsilon_{zx}$. According to Eqs. (2.258), two distinct depolarized scattering spectra obtained at the 90° scattering angle are given by

$$S_{yz}(\mathbf{q}, \omega) = n_y^4 n_z^4 p_{44}^2 \frac{1}{2\pi} \int_{-\infty}^{\infty} dt\, e^{-i\omega t} \langle \varepsilon_{yz}(\mathbf{q}, t)\, e_{yz}^*(\mathbf{q})\rangle, \tag{2.259}$$

$$S_{xy}(\mathbf{q}, \omega) = n_x^8 p_{66}^2 \frac{1}{2\pi} \int_{-\infty}^{\infty} dt\, e^{-i\omega t} \langle e_{xy}(\mathbf{q}, t)\, e_{xy}^*(\mathbf{q})\rangle \tag{2.260}$$

The depolarized spectra are associated with the transverse (shear) acoustic phonons. Thus p_{44} and p_{66} are also known as transverse photoelastic constants.

The Porto notation $a(ij)b$ is useful to describe the propagation and polarization directions of the incident and scattered light in a scattering experiment. The letters a and b outside the parentheses indicate the propagation directions of the incident and scattered light, respectively. The polarization direction of the incident light is indicated by the letter i and that of the scattered light by j. According to this notation, it is clear that the spectrum

$S_{yz}(\mathbf{q}, \omega)$ can be measured using the scattering geometry $z(yz)y$, and that the $y(xy)x$ geometry gives the $S_{xy}(\mathbf{q}, \omega)$ spectrum.

The polarized spectra associated with $\delta\varepsilon_{xx}$, $\delta\varepsilon_{yy}$, and $\delta\varepsilon_{zz}$ yield results on the other longitudinal photoelastic constants, p_{11}, p_{12}, p_{13}, p_{31}, and p_{33}. They are called longitudinal because the scattering spectra are associated with the longitudinal acoustic phonons through the strain-tensor components e_{xx}, e_{yy}, and e_{zz}. Note that because the electromagnetic radiation propagating along the optical axis of the crystal is always transverse, p_{11} and p_{33} cannot be measured using the scattering geometry with the polarization vectors of the incident and scattered radiation directed along the optical axes. For example, according to Eq. (2.258a), to observe the contribution from the p_{11} term, the scattering vector $\mathbf{q}$ must have a component along the optical x axis so that $e_{xx}(\mathbf{q})$ will not vanish. This is impossible, because light polarized along x cannot propagate along x. Similarly, p_{33} cannot be measured using the scattering geometry that yields Eq. (2.258b). To measure p_{11} and p_{33}, the propagation of the incident and scattered light is intentionally directed away from the optical axes. In this case, a more general and thus more complex expression such as that given in Eq. (2.212) is needed to carry out the spectral analysis.

Backscattering geometries are useful for obtaining p_{12}, p_{13}, and p_{32}. For example, if the scattering geometry $y(xx)\bar{y}$ is used, then, according to Eq. (2.258a), $\delta\varepsilon_{xx} = -n_x^4 p_{12} e_{yy}$. On the other hand, the scattering geometry $z(xx)\bar{z}$ gives $\delta\varepsilon_{xx} = -n_x^4 p_{13} e_{zz}$.

Calculation of the time-correlation function $\langle e_{kl}(\mathbf{q}, t)\, e^*_{k'l'}(\mathbf{q})\rangle$, which enters Eq. (2.256), can be carried out by using a constitutive equation that relates the stress tensor $\boldsymbol{\sigma}$ to the strain tensor. The ideal Hooke's law relation for an anisotropic body is

$$\sigma_{ij} = C_{ijkl} e_{kl}, \tag{2.261}$$

where C_{ijkl} are elements of the elastic (stiffness) constant tensor. However, the expression does not include damping. Wavelike vibrations calculated according to Hooke's law will propagate without any attenuation of amplitude. The time-correlation function of the strain tensor given in Eq. (2.256) will cause resonant oscillations and will yield the δ function type of spectral lines for $S_{ij}(\mathbf{q}, \omega)$. Real materials generally do not obey a stress–strain relation of this type, although the constitutive equation of weakly damped materials can sometimes be approximated.

Elastic damping depends in a rather complex manner on temperature as well as frequency and type of vibration. An impressive number of physical mechanisms contribute to the damping phenomena, and it is impossible to represent them all by a single constitutive equation. The Maxwell and Voigt equations, presented previously for viscoelastic liquids, are but two types of model equations for the constitutive equation. However, at room temperature, the constitutive equation including attenuation in the acoustic response in many solids can be adequately represented by a Voigt equation, [compare with Eq. (2.246)] given by

$$\sigma_{ij} = C_{ijkl}e_{kl} + \eta_{ijkl}\dot{e}_{kl}. \tag{2.262}$$

The viscosity tensor η_{ijkl}, relating the strain-rate tensor to the stress tensor, is assumed to have the same symmetry as the elastic-constant tensor C_{ijkl}. For an anisotropic solid with cylindrical symmetry, the general forms of the elastic-constant tensor and the viscosity tensor are given by

$$C_{ijkl} \to C_{\alpha\beta} = \begin{bmatrix} C_{11} & C_{12} & C_{13} & 0 & 0 & 0 \\ C_{12} & C_{11} & C_{13} & 0 & 0 & 0 \\ C_{13} & C_{13} & C_{33} & 0 & 0 & 0 \\ 0 & 0 & 0 & C_{44} & 0 & 0 \\ 0 & 0 & 0 & 0 & C_{44} & 0 \\ 0 & 0 & 0 & 0 & 0 & C_{66} \end{bmatrix}, \tag{2.263}$$

$$\eta_{ijkl} \to \eta_{\alpha\beta} = \begin{bmatrix} \eta_{11} & \eta_{12} & \eta_{13} & 0 & 0 & 0 \\ \eta_{12} & \eta_{11} & \eta_{13} & 0 & 0 & 0 \\ \eta_{13} & \eta_{13} & \eta_{33} & 0 & 0 & 0 \\ 0 & 0 & 0 & \eta_{44} & 0 & 0 \\ 0 & 0 & 0 & 0 & \eta_{44} & 0 \\ 0 & 0 & 0 & 0 & 0 & \eta_{66} \end{bmatrix}, \tag{2.264}$$

where the elastic-constant and the viscosity tensors are represented in accordance with the Voigt notation in a manner similar to the photoelastic-constant tensor. Moreover, $C_{66} = \frac{1}{2}(C_{11} - C_{12})$ and $\eta_{66} = \frac{1}{2}(\eta_{11} - \eta_{12})$. With the constitutive equation given by Eq. (2.262), we can now calculate the time-correlation functions of the strain tensor present in Eq. (2.256) by using Newton's equation for continuum mechanics:

$$\begin{aligned} \rho\, \partial^2 U_i/\partial t^2 &= \partial\sigma_{ij}/\partial x_j \\ &= \tfrac{1}{2}(\partial/\partial x_j)[C_{ijkl}(\partial U_k/\partial x_l + \partial U_l/\partial x_k) + \eta_{ijkl}(\partial \dot{U}_k/\partial x_l + \dot{U}_l/\partial x_k)]. \end{aligned} \tag{2.265}$$

This gives three equations for the three components of the vector $\mathbf{U}$:

$$\begin{aligned}\rho\, \partial^2 U_x/\partial t^2 &= (C_{11}\, \partial^2 U_x/\partial x^2 + \eta_{11}\, \partial^2 \dot{U}_x/\partial x^2) + (C_{12}\, \partial^2 U_y/\partial x\, \partial y \\ &\quad + \eta_{12}\, \partial^2 \dot{U}_y/\partial x\, \partial y) + (C_{13}\, \partial^2 U_z/\partial x\, \partial z + \eta_{13}\, \partial^2 \dot{U}_z/\partial x\, \partial z) \\ &\quad + [C_{66}\, (\partial^2 U_x/\partial y^2 + \partial^2 U_y/\partial x\, \partial y) + \eta_{66}\, (\partial^2 \dot{U}_x/\partial y^2 \\ &\quad + \partial^2 U_y/\partial x\, \partial y)] + [C_{44}\, (\partial^2 U_x/\partial z^2 + \partial^2 U_z/\partial x^2) \\ &\quad + \eta_{44}\, (\partial^2 \dot{U}_x/\partial z^2 + \partial^2 \dot{U}_z/\partial x^2)]; \end{aligned} \tag{2.266a}$$

$$\rho\, \partial^2 U_y/\partial t^2 = [\text{interchange } x \text{ and } y \text{ in Eq. (2.224a)}]; \tag{2.266b}$$

$$\begin{aligned}\rho\, \partial^2 U_z/\partial t^2 &= (C_{33}\, \partial^2 U_z/\partial z^2 + \eta_{33}\, \partial^2 \dot{U}_z/\partial z^2) + [C_{13}\, (\partial^2 U_x/\partial x\, \partial z \\ &\quad + \partial^2 U_y/\partial y\, \partial z) + \eta_{13}\, (\partial^2 \dot{U}_x/\partial x\, \partial z + \partial^2 \dot{U}_y/\partial y\, \partial z)] \\ &\quad + [C_{44}\, (\partial^2 U_z/\partial x^2 + \partial^2 U_z/\partial y^2 + \partial^2 U_x/\partial x\, \partial z + \partial^2 U_y/\partial y\, \partial z) \\ &\quad + \eta_{44}\, (\partial^2 \dot{U}_z/\partial x^2 + \partial^2 \dot{U}_z/\partial y^2 + \partial^2 \dot{U}_x/\partial x\, \partial z + \partial^2 \dot{U}_y/\partial y\, \partial z)]; \end{aligned} \tag{2.266c}$$

where $\mathbf{U}$ is the particle displacement and $\dot{\mathbf{U}}$ the particle velocity.

As an example, we consider specifically the time-correlation function $\langle e_{xy}(q, t) e^*_{xy}(q)\rangle$ associated with the $S_{xy}(\mathbf{q}, \omega)$ spectrum given in Eq. (2.260). The time-correlation function depends on the direction of $\mathbf{q}$. If we use the $y(xy)x$ scattering geometry, the scattering vector $\mathbf{q}$ is given by $q(l_y \hat{y} + l_x \hat{x})$, l_x and l_y being the direction cosines of the scattering vector with respect to the $\hat{x}$ and $\hat{y}$ axes. In this case, we write

$$\begin{aligned}\langle e_{xy}(\mathbf{q}, t)\, e^*_{xy}(\mathbf{q})\rangle &= q^2/4\{l_y^2\, \langle U_x(\mathbf{q}, t) U^*_x(\mathbf{q})\rangle \\ &\quad + l_x^2\, \langle U_y(\mathbf{q}, t) U^*_x(\mathbf{q})\rangle + l_x l_y [\langle U_x(\mathbf{q}, t) U^*_y(\mathbf{q})\rangle \\ &\quad + \langle U_y(\mathbf{q}, t)\, U^*_x(\mathbf{q})\rangle]\}. \end{aligned} \tag{2.267}$$

From Eqs. (2.266), we obtain for the cylindrical solid equation for U_x and U_y as

$$\begin{aligned}\rho\, \partial^2 U_x(\mathbf{q}, t)/\partial t^2 &= -q^2[(C_{11} l_x^2 + C_{66}\, l_y^2)\, U_x(\mathbf{q}, t) + l_x l_y (C_{12} + C_{66})\, U_y(\mathbf{q}, t) \\ &\quad + (\eta_{11} l_x^2 + \eta_{66}\, l_y^2)\, U_x(\mathbf{q}, t) + l_x l_y (C_{12} + C_{66})\, \dot{U}_y(\mathbf{q}, t)], \end{aligned} \tag{2.268a}$$

$$\begin{aligned}\rho\, \partial^2 U_y(\mathbf{q}, t)/\partial t^2 &= -q^2[(C_{11} l_y^2 + C_{66}\, l_x^2)\, U_y(\mathbf{q}, t) + l_x l_y (C_{12} + C_{66}) U_x(\mathbf{q}, t) \\ &\quad + \eta_{11} l_y^2 + C_{66}\, l_x^2) U_y(\mathbf{q}, t) + l_x l_y (C_{12} + C_{66}) U_x(\mathbf{q}, t)]. \end{aligned} \tag{2.268b}$$

Equations (2.268a) and (2.268b) can be solved by the usual Laplace transform technique. The results for the various time-correlation functions in Eqs. (2.268) are

$$\langle U_x(\mathbf{q}, s)U_x^*(\mathbf{q})\rangle = \frac{1}{\Delta}\left[s^2 + \frac{q^2}{2\rho}(C_{11} + C_{66} + s\eta_{11} + s\eta_{66})\right] \times \left[s + \frac{q^2}{2\rho}(\eta_{11} + \eta_{66})\right]\langle |U_x|^2\rangle \qquad (2.269a)$$

$$\langle U_y(\mathbf{q}, s)U_y^*(\mathbf{q})\rangle = \frac{1}{\Delta}\left[s^2 + \frac{q^2}{2\rho}(C_{11} + C_{66} + s\eta_{11} + s\eta_{66})\right] \times \left[s + \frac{q^2}{2\rho}(\eta_{11} + \eta_{66})\right]\langle |U_y|^2\rangle \qquad (2.269b)$$

$$\langle U_x(\mathbf{q}, s)U_y^*(\mathbf{q})\rangle = \frac{-1}{\Delta}\left[\frac{q^2}{2\rho}(C_{12} + C_{66} + s\eta_{11} + s\eta_{66})\right] \times \left[s + \frac{q^2}{2\rho}(\eta_{11} + \eta_{66})\right]\langle |U_y|^2\rangle \qquad (2.269c)$$

$$\langle U_y(\mathbf{q}, s)U_x^*(\mathbf{q})\rangle = \frac{-1}{\Delta}\left[\frac{q^2}{2\rho}(C_{12} + C_{66} + s\eta_{11} + s\eta_{66})\right] \times \left[s + \frac{q^2}{2\rho}(\eta_{11} + \eta_{66})\right]\langle |U_x|^2\rangle, \qquad (2.269d)$$

where we have used $l_x = l_y = 1/\sqrt{2}$ and

$$\Delta = [s^2 + (q^2/\rho)(C_{11} + s\eta_{11})][s^2 + (q^2/\rho)(C_{66} + s\eta_{66})]. \qquad (2.270)$$

Since the solid is cylindrically symmetric, we have $\langle |U_x|^2\rangle = \langle |U_y|^2\rangle$. Substituting Eqs. (2.269) into Eqs. (2.268) we obtain

$$\langle e_{xy}(\mathbf{q}, 2)e_{xy}^*(\mathbf{q})\rangle = \frac{q^2[s + (q^2/2\rho)(\eta_{11} + \eta_{66})]\langle |U_x|^2\rangle}{2[s^2 + q^2(C_{11} + s\eta_{11})/\rho]}, \qquad (2.271)$$

which gives two poles at

$$s_\pm = \pm iq(C_{11}/\rho)^{1/2}[1 - \eta_{11}^2/(4\rho C_{11})]^{1/2} - q^2\eta_{11}/2\rho. \qquad (2.272)$$

The real part of $s_\pm$ corresponds to damping and the imaginary part to oscillation. The result indicates that viscous damping induces a softening in frequency by the amount $\eta_{11}^2(4\rho C_{44}$. Since for most solids the viscosity coefficient is sufficiently small that $\eta_{11} \ll (\rho C_{11})^{1/2}$, the resonance frequency is still given by $q(C_{11}/\rho)^{1/2}$. However, the linewidth equal to $q^2\eta_{11}/\rho$ induced by viscous damping is not negligible. Note that the linewidth is proportional to q^2.

If, on the other hand, the $z(xy)\bar{z}$ scattering geometry is used, then the scattering vector becomes $\mathbf{q} = 2q\hat{z}$. In this case, $e_{xy}(\mathbf{q}, t)$ vanishes, and no scattering spectrum is expected. Similarly, if the scattering geometry $y(xy)z$ is used, the scattering vector $\mathbf{q}$ is equal to $q(l_y y + l_z z)$ and $e_{xy}(\mathbf{r}, t)$ reduces to $\frac{1}{2}\,\partial U_x(\mathbf{r}, t)/\partial y$ or $e_{xy}(\mathbf{q}, t) = \frac{1}{2}\,iql_y U_x(\mathbf{q}, t)$. However, examining Eq. (2.266a) shows that for this $\mathbf{q}$ vector, $\partial^2 U_x(\mathbf{q}, t)/\partial t^2$ vanishes. As a result, no spectrum is expected to appear in the $y(xy)z$ scattering geometry.

In short, to observe the $S_{xy}(\mathbf{q}, \omega)$ spectral density, only the $y(xy)x$ scattering geometry can result in a spectral line, with the peak position of this spectral line given by $q(C_{11}/\rho)^{1/2}$ and the width by $q^2\eta_{11}/2\rho$. All other scattering geometric configurations yield negative results. This example illustrates the method needed for describing the activity of photoelastic effects in anisotropic solids. To predict the activity, the polarizability tensor (or dielectric tensor) must first be expressed in terms of the strain tensor. The components of the strain tensor involved in the scattering process are projected by the polarization vectors of the incident and scattered light. Next the selected scattering geometry is used to determine the direction of the scattering vector $\mathbf{q}$. The direction of the scattering vector and the Newton equation for continuum mechanics finally determine the light-scattering activity of the projected strain-tensor component.

Shown in Fig. 2.9 are the Brillouin spectra of extruded polypropylene rods (having cylindrical symmetry) obtained using $x(zz)y$ and $z(yy)x$ scattering geometries (Cavanaugh and Wang, 1982). Using the procedure outlined here, elastic constants were obtained as a function of extrusion ratio (the ratio of cross-section area of the unextruded sample to that of the extruded sample). The Brillouin light-scattering technique provides an effective method for determining elastic constants of a solid that is subject to deformation.

Finally, it should be pointed out that experimental evidence has shown that the Pockels equation as given by Eq. (2.250) is not completely correct,

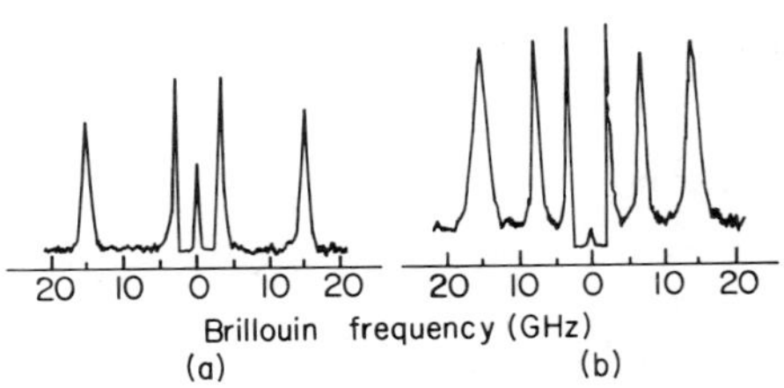

FIG. 2.9 Brillouin scattering spectra of polypropylene extruded to R (extrusion ratio) = 6.3 at room temperature, using scattering geometries $x(zz)y$ (a) and $z(yy)x$ (b).

because the contribution owing to the rotation of the volume element within an acoustic wave is not included. The scattering of light from an acoustic shear wave in an optically anisotropic medium is closely associated with the rotation of the volume element, since the rotational effect can cause a change in the dielectric tensor. To include the rotational contribution, Eq. (2.250) is modified to give

$$(\delta\varepsilon^{-1})_{ij} = p_{ijkl}\,(\partial U_k/\partial x_l. \tag{2.273}$$

Since $\partial U_k/\partial x_l$ can be written as a summation of the symmetric $[\frac{1}{2}(\partial U_k/\partial x_l + \partial U_l/\partial x_k) = e_{kl}]$ and the antisymmetric parts $[\frac{1}{2}(\partial U_k/\partial x_l - \partial U_l/\partial x_k)]$, the Pockels law is recovered from the symmetric component if the antisymmetric component is negligible. [The antisymmetric component is related to body rotation. Analysis of the contribution of the rotational component was given by Nelson and Lax (1970, 1971).]

2.15 RAMAN SCATTERING

To conclude this chapter, it should be mentioned that optical-lattice vibration, or molecular vibration, provides another mechanism for modulating the polarizability density. Light scattering resulting from modulation of the polarizability density through this mechanism is called Raman scattering. Raman scattering is a fruitful technique for the study of molecular motion and of the mechanism of phase transition in condensed media, because the Raman polarizability tensor is sensitive to molecular-reorientation processes and vibrational relaxation as well as to the structural environment in which molecules are located. A detailed description of Raman-scattering spectroscopy will be given in Chapter 6.

3
Angular Momentum and Molecular Rotation

3.1 INTRODUCTION

Molecules in the condensed phase, especially in the fluid state, undergo frequent reorientation. The observed spectra, such as those in nuclear magnetic resonance (NMR), in infrared absorption, or in Raman spectroscopy, correspond to the average value of the spectral lines of many molecules in different orientations. To systematically describe the reorientational motion of molecules in the condensed phase, it is useful to explore the general transformational properties of the spherical-tensor operator functions associated with molecules that undergo rotation. Moreover, in NMR spectroscopy we are concerned with the spin angular-momentum operator and its transformation properties when the spin system is subject to the action of an externally applied magnetic field or when it is subject to the internal local magnetic field, such as that associated with the dipole–dipole interaction, the spin–rotation coupling, or the chemical-shift shielding effects, etc.; knowledge of the theory of angular momentum should facilitate the analysis of these problems.

In this chapter, we shall first study the transformation property of the irreducible spherical-tensorial functions. The result will then be used to manipulate the time-correlation functions and to obtain expressions that deal with the effects of reorientational motion of anisotropic molecules in a dense molecular fluid. Specifically, we shall study the anisotropic rotational-diffusion equation and work out expressions useful in NMR relaxation and light-scattering spectroscopy. Different models prepared for describing the rotational motion will also be discussed.

3.2 ROTATION MATRIX

For simplicity, we shall begin by considering a point fixed in a rigid body. The position of this point with respect to the origin of a reference-coordinate system is given by a vector **r**. The three Cartesian components with respect to the given reference-coordinate frame are

$$\mathbf{r} = r\begin{bmatrix} \sin\theta\cos\phi \\ \sin\theta\sin\phi \\ \cos\theta \end{bmatrix}. \tag{3.1}$$

The rigid body is now rotated about the z axis by an angle γ. Since the vector **r** is fixed in the rigid body, the rotation transforms the vector **r** to a position specified as $\mathbf{r}'$ with respect to the reference frame (Fig. 3.1) according to the following equation:

$$\mathbf{r}' = r\begin{bmatrix} \sin\theta\cos(\phi+\gamma) \\ \sin\theta\sin(\phi+\gamma) \\ \cos\theta \end{bmatrix}. \tag{3.2}$$

Apparently vector $\mathbf{r}'$ is related to **r** by a rotation operator $R_z(\gamma)$ according to

$$\mathbf{r}' = R_z(\gamma)\mathbf{r}. \tag{3.3}$$

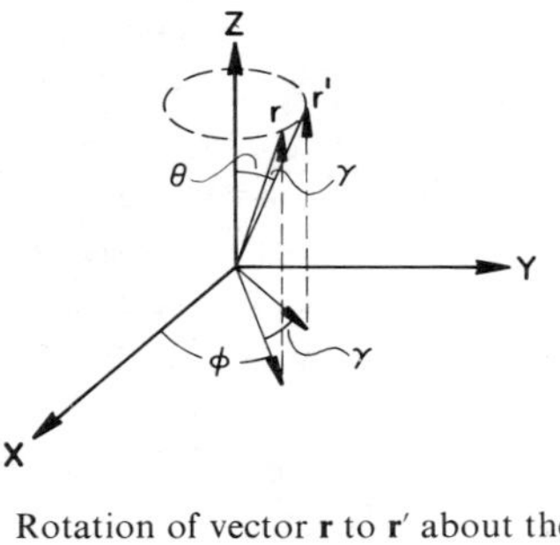

FIG. 3.1 Rotation of vector **r** to $\mathbf{r}'$ about the z axis.

We want to obtain an expression for $R_z(\gamma)$ that accomplishes the rotational transformation. According to the Taylor series expansion formula, we can write

$$\cos(\phi + \gamma) = \cos\phi + \gamma\frac{\partial}{\partial\phi}\cos\phi + \frac{\gamma^2}{2!}\frac{\partial^2}{\partial\phi^2}\cos\phi + \cdots + \frac{\gamma^n}{n!}\frac{\partial^n}{\partial\phi^n}\cos\phi + \cdots$$

$$= \sum_{n=0}^{\infty}\frac{\gamma^n}{n!}\frac{\partial^n}{\partial\phi^n}\cos\phi = e^{\gamma\,\partial/\partial\phi}\cos\phi \tag{3.4}$$

$$\sin(\phi + \gamma) = \sin\phi + \gamma\frac{\partial}{\partial\phi}\sin\phi + \cdots + \frac{\partial^n}{n!}\frac{\partial^n}{\partial\phi^n}\sin\phi + \cdots$$

$$= \sum_{n=0}^{\infty}\frac{\gamma^n}{n!}\frac{\partial^n}{\partial\phi^n}\sin\phi = e^{\gamma\,\partial/\partial\phi}\sin\phi. \tag{3.5}$$

Using Eqs. (3.4) and (3.5), we can rewrite Eq. (3.2) as

$$\mathbf{r}' = r\,e^{\gamma\,\partial/\partial\phi}\begin{bmatrix}\sin\theta\cos\phi\\ \sin\theta\sin\phi\\ \cos\theta\end{bmatrix} = e^{\gamma\,\partial/\partial\phi}\mathbf{r} \tag{3.6}$$

Comparison of Eq. (3.6) with (3.3) gives

$$R_z(\gamma) = e^{\gamma\,\partial/\partial\phi}. \tag{3.7}$$

Since the zth component of the dimensionless angular momentum is given by $J_z = i^{-1}\partial/\partial\phi$, the rotation operator defined by Eq. (3.7) can be written as

$$R_z(\gamma) = e^{i\gamma J_z} = e^{i\gamma(\hat{z}\cdot\mathbf{J})}. \tag{3.8}$$

Instead of rotating about the z axis, we can generalize this result by considering the rotation of vector $\mathbf{r}$ through an angle γ about an arbitrary direction defined by a unit vector $\hat{n}$. The operator that accomplishes this rotation is given by

$$R_n(\gamma) = e^{i\gamma(\hat{n}\cdot\mathbf{J})}. \tag{3.9}$$

The angular momentum operator $\mathbf{J}$ thus determines the rotational-transformation property of a vector attached to a rigid body. The operator $R_n(\gamma)$ is known as the rotation operator.

Consider next a state vector given by $|jm\rangle$ that describes the orientational state of the total angular momentum associated with a rigid body. Here j determines the amplitude and m the direction of the state vector. With reference to the coordinate system given in Fig. 3.1, the state vector is a function of the angles θ and ϕ. If the state vector $|jm\rangle$ is rotated through an angle

γ about a direction specified by $\hat{n}$, then the new state vector after rotation is given by $R_n(\gamma)|jm\rangle$ [c.f. Eq. (3.3)]. The new state vector can be written as

$$R_n(\gamma)|jm\rangle = \sum_{m'} |jm'\rangle\langle jm'|R_n(\gamma)|jm\rangle \tag{3.10}$$

because the set of state vectors $\{|jm'\rangle\}$ is complete and normalized. In Eq. (3.10), summation over m' goes from $-j$ to $+j$. Since rotation does not change the amplitude of a vector, the rotation operator $R_n(\gamma)$ will not connect states with a different j. In Eq. (3.10), the matrix element $\langle jm'|R_n(\gamma)|jm\rangle$ is known as the rotation matrix (Rose, 1957). The rotation matrix plays a central role in the discussion of rotational-transformation properties of angle-dependent functions (Edmonds, 1960). Therefore, we shall now give a more detailed account of its properties.

To begin with, we choose a coordinate system to specify the orientation of the rigid body. The coordinate system fixed to the rigid body is called the body-fixed coordinate system. The z axis of the body-fixed coordinate system is chosen to coincide with the direction $\hat{n}$; thus rotation of the rigid body about $\hat{n}$ by an angle γ is identical to rotation through an angle γ about the body-fixed z axis. It is of interest to specify the final orientation of the rigid body with reference to an external coordinate system known as the space-fixed system. For this purpose, it is necessary to specify only the direction of $\hat{n}$, because the orientation of the rigid body associated with the rotation about $\hat{n}$ is already specified by the angle γ. It takes two angles to define the direction of $\hat{n}$, and these two angles are α and β, which correspond respectively to the azimuthal and polar angles of $\hat{n}$ with respect to the space-fixed coordinate system. The three angles α, β, γ, defined in this way, are known as the Euler angles.

A rotation through an arbitrary angle γ of a rigid body about the general direction $\hat{n}$ is clearly specified by the three Euler angles α, β, γ. Thus, instead of using $R_n(\gamma)$, we specify the rotation operator as $R(\alpha, \beta, \gamma)$, which can accomplish a rotation of any angle-dependent function through an angle γ about $\hat{n}$ with reference to the space-fixed coordinate system.

The explicit form of $R(\alpha, \beta, \gamma)$ can be found as follows:

(1) Rotation about the space-fixed z axis through an angle α is accomplished by the rotation operator

$$R_z(\alpha) = \exp(i\alpha J_z). \tag{3.11}$$

As a result of this rotation, the new coordinate system is now given by (x', y', z'), the z' axis being coincident with the z axis.

(2) A second rotation, about the y' axis through an angle β, is carried out. This rotation brings the coordinate system from (x', y', z') to (x'', y'', z'')

with the z'' axis coinciding with $\hat{n}$. The rotation operator that accomplishes this is

$$R_{y'}(\beta) = \exp(i\beta J_{y'}). \tag{3.12}$$

(3) The final rotation, of angle γ about z'' (or $\hat{n}$), is accomplished by the operator

$$R_{z''}(\gamma) = \exp(i\gamma J_{z''}). \tag{3.13}$$

The three operations $R_z(\alpha)$, $R_{y'}(\beta)$, and $R_{z''}(\gamma)$, applied successively, give the rotation operator

$$\mathscr{R}(\alpha, \beta, \gamma) = R_{z''}(\gamma)\, R_{y'}(\beta)\, R_z(\alpha) = \exp(i\gamma J_{z''}) \exp(i\beta J_{y'}) \exp(i\alpha J_z). \tag{3.14}$$

Equation (3.14) is inconvenient, however, because it involves the use of three different coordinate systems. We now express this equation in terms of one coordinate system, the space-fixed system. Since $J_{y'}$ is related to J_y by a transform about the z axis by an angle α, we have

$$J_{y'} = \exp(i\alpha J_z)\, J_y \exp(-i\alpha J_z). \tag{3.15}$$

It then follows that

$$\exp(i\beta J_{y'}) = \exp(i\alpha J_z) \exp(i\beta J_y) \exp(-i\alpha J_z). \tag{3.16}$$

Similarly, $J_{z''}$ is related to J_z by

$$\begin{aligned} J_{z''} &= \exp(i\beta J_{y'})\, J_{z'} \exp(-i\beta J_{y'}) \\ &= \exp(i\alpha J_z) \exp(i\beta J_y)\, J_z \exp(-i\beta J_y) \exp(-i\alpha J_z). \end{aligned} \tag{3.17}$$

Thus,

$$\exp(i\gamma J_{z''}) = \exp(i\alpha J_z) \exp(i\beta J_y) \exp(i\gamma J_z) \exp(-i\beta J_y) \exp(-i\alpha J_z). \tag{3.18}$$

Substituting Eq. (3.16) and (3.18) into (3.14), we obtain an expression for $D(\alpha, \beta, \gamma)$ in terms of the angular-momentum operators defined in the space-fixed coordinate system as

$$D(\alpha, \beta, \gamma) = \exp(i\alpha J_z) \exp(i\beta J_y) \exp(i\gamma J_z). \tag{3.19}$$

Note that $D(\alpha, \beta, \gamma)$, when expressed in terms of the space-fixed coordinate system, has the order of three successive inverted rotations.

With the help of Eq. (3.19), we can also rewrite Eq. (3.10) as

$$D(\alpha, \beta, \gamma)|jm\rangle = \sum_{m'} |jm'\rangle\langle jm'|D(\alpha, \beta, \gamma)|jm\rangle = \sum_{m'} D^{(j)}_{m'm}(\Omega)|jm'\rangle, \tag{3.20}$$

where Ω collectively represents (α, β, γ), and $D^{(j)}_{m'm}(\Omega)$ is the rotation matrix given by

$$\begin{aligned} D^{(j)}_{m'm}(\Omega) &= \langle jm'|\exp(i\alpha J_z)\exp(i\beta J_y)\exp(i\gamma J_z)|jm\rangle \\ &= \exp[i(\alpha m' + \gamma m)]\langle jm'|\exp(i\beta J_y)|jm\rangle \\ &= \exp[i(\alpha m' + \gamma m)]\, d^{(j)}_{m'm}(\beta). \end{aligned} \tag{3.21}$$

It is easy to verify that the rotation matrices possess the following orthonormal properties:

$$\sum_m D^{(j)*}_{m'm}(\Omega)\, D^{(j)}_{m''m}(\Omega) = \sum_m D^{(j)*}_{mm'}(\Omega)\, D^{(j)}_{mm''}(\Omega) = \delta_{m'm''}. \tag{3.22}$$

We state without proof that

$$D^{(j)*}_{m'm}(\Omega) = (-1)^{m'-m} D^{(j)}_{-m',-m}(\Omega), \tag{3.23}$$

and

$$\int d\Omega\, D^{(j_1)*}_{m_1n_1}(\Omega)\, D^{(j_2)}_{m_2n_2}(\Omega) = \frac{8\pi^2}{2j_1 + 1}\delta_{j_1j_2}\delta_{m_1m_2}\delta_{n_1n_2}, \tag{3.24}$$

where the symbol $\int d\Omega$ stands for integration over three Euler angles,

$$\int d\Omega = \int_0^{2\pi} d\alpha \int_0^{\pi} d\beta \sin\beta \int_0^{2\pi} d\gamma. \tag{3.25}$$

Other useful expressions are

$$D^{(j)}_{m0}(\alpha, \beta, \gamma) = \left(\frac{4\pi}{2j + 1}\right)^{1/2} Y^*_{jm}(\beta, \alpha) = (-1)^m\left(\frac{4\pi}{2j + 1}\right)^{1/2} Y_{j-m}(\beta, \alpha), \tag{3.26a}$$

$$D^{(j)}_{0m}(\alpha, \beta, \gamma) = \left(\frac{4\pi}{2j + 1}\right)^{1/2} Y_{jm}(\beta, \gamma), \tag{3.26b}$$

and in particular

$$D^{(j)}_{00}(\alpha, \beta, \gamma) = P_j(\cos\beta), \tag{3.27}$$

where $P_j(x)$ is the Legendre polynomial of order j.

3.3 SPHERICAL-HARMONICS ADDITION THEOREM

Another useful relation that follows from the orthogonal property of the rotation matrix is the spherical-harmonics addition theorem. This theorem states that if (θ_1, ϕ_1) and (θ_2, ϕ_2) are the spherical coordinates that define points A_1 and A_2, respectively, on the unit sphere and if the angle between the factors $\mathbf{O}A_1$ and $\mathbf{O}A_2$ connecting the origin of the unit sphere to A_1 and A_2 is ω, then

$$P_l(\cos \omega) = \frac{4\pi}{2l+1} \sum_m Y^*_{lm}(\theta_1, \phi_1) Y_{lm}(\theta_2, \phi_2). \tag{3.28}$$

The theorem can be proved by considering the quantity given by

$$I = \sum_m Y^*_{lm}(\theta_1, \phi_1)\, Y_{lm}(\theta_2, \phi_2), \tag{3.29}$$

where (θ_1, ϕ_1) and (θ_2, ϕ_2) correspond to the spherical coordinates that define points A_1 and A_2.

We now perform a rigid rotation of vectors $\mathbf{OA}_1$ and $\mathbf{OA}_2$ about an arbitrary direction specified by $\hat{n}$ through angles γ to new positions specified by $\mathbf{OA}_1$ and $\mathbf{OA}_2$.

The functions $Y_{lm}(\theta_i'\phi_i')$, defined by points A_i', with $i = 1, 2$ are related to the original $Y_{lm}(\theta_i, \phi_i)$ according to Eq. (3.20) by

$$Y_{lm}(\theta_i', \phi_i') = D(\Omega)\, Y_{lm}(\theta_i, \phi_i) = \sum_{m'} D^{(j)}_{m'm}(\Omega)\, Y_{lm'}(\theta_i, \phi_i). \tag{3.30}$$

We now want to calculate the quantity given by

$$I' = \sum_m Y^*_{lm}(\theta_1', \phi_1')\, Y_{lm}(\theta_2', \phi_2'). \tag{3.31}$$

With the help of Eq. (3.30), we can write

$$\begin{aligned} I' &= \sum_m \left[\sum_{m'} D^{(j)}_{m'm}(\Omega)\, Y_{lm'}(\theta_1, \phi_1)\right]^* \left[\sum_{m''} D^{(j)}_{m''m}(\Omega)\, Y_{lm''}(\theta_2, \phi_2)\right] \\ &= \sum_m \sum_{m'} \sum_{m''} Y^*_{lm'}(\theta_1, \phi_1)\, Y_{lm''}(\theta_2, \phi_2)\, D^{*(j)}_{m'm}(\Omega)\, D^{(j)}_{m''m}(\Omega). \end{aligned} \tag{3.32}$$

Summation of the product of rotation matrices over m can now be carried out. The summation gives a Kronecker delta $\delta_{m'm''}$ because of Eq. (3.22). Further summation over m'' reduces Eq. (3.31) to

$$I' = \sum_{m'} Y^*_{lm'}(\theta_1, \phi_1)\, Y_{lm'}(\theta_2, \phi_2), \tag{3.33}$$

which is identical to I as given by Eq. (3.29). Therefore, we have shown that the quantity I is invariant under rotation, i.e.,

$$\sum_m Y^*_{lm}(\theta_1', \phi_1')\, Y_{lm}(\theta_2', \phi_2') = \sum_m Y^*_{lm}(\theta_1, \phi_1)\, Y_{lm}(\theta_2, \phi_2). \tag{3.34}$$

Equation (3.34) can now be used to show the spherical-harmonics addition theorem. For this purpose, we choose A' to be on the z axis (i.e., $\theta_1' = 0$ and

ϕ'_1 arbitrary) and A'_2 to be on the x–z plane with $\theta'_2 = \omega$ and $\phi'_2 = 0$. As a result, Eq. (3.34) becomes

$$\sum_m Y^*_{lm}(0, \phi'_1)\, Y_{lm}(\omega, 0) = \sum_m Y^*_{lm}(\theta_1, \phi_1)\, Y_{lm}(\theta_2, \phi_2). \tag{3.35}$$

But

$$Y_{lm}(0, \phi'_1) = \left(\frac{2l+1}{4\pi}\right)^{1/2} \delta_{m,0}, \tag{3.36}$$

and

$$Y_{lm}(\omega, 0) = \left(\frac{2l+1}{4\pi}\right)^{1/2} P_l(\cos\omega), \tag{3.37}$$

where $P_l(\cos\omega)$ is the Legendre polynomial of order l.

Substituting Eqs. (3.36) and (3.37) into (3.35) gives the spherical-harmonics addition theorem. For $l = 1$, we have

$$\cos\omega = \cos\theta_1 \cos\theta_2 + \sin\theta_1 \sin\theta_2 \cos(\phi_1 - \phi_2). \tag{3.38}$$

The spherical-harmonics addition theorem can be used to establish the identity given in Eq. (3.26a). For this purpose, we consider the rotation transformation of $Y_{lm}(\theta, \phi)$ to $Y_{lm}(\theta', \phi')$ according to Eq. (3.30)

$$Y_{lm}(\theta', \phi') = \sum_{m'} D^{(l)}_{m'm}(\alpha, \beta, \gamma)\, Y_{lm'}(\theta, \phi). \tag{3.39}$$

We carry out the rotation in such a way that the final point specified by (θ', ϕ') is on the xz plane. Then, $\phi' = 0$ and $\theta' = \omega$. Further, we set $m = 0$. As a result, Eq. (3.39) becomes

$$\begin{aligned} Y_{l0}(\omega) = \left(\frac{2l+1}{4\pi}\right)^{1/2} P_l(\cos\omega) &= \sum_{m'} D^{(l)}_{m'0}(\alpha, \beta, \gamma)\, Y_{lm'}(\theta, \phi) \\ &= \sum_{m'} D^{(l)}_{m'0}(\alpha, \beta, 0)\, Y_{lm'}(\theta, \phi). \end{aligned} \tag{3.40}$$

Comparison of Eq. (3.40) with the spherical-harmonics addition theorem gives

$$D^{(l)}_{m'0}(\alpha, \beta, 0) = \left(\frac{4\pi}{2l+1}\right)^{1/2} Y^*_{lm'}(\beta, \alpha), \tag{3.41}$$

which is identical to Eq. (3.26a).

From Eqs. (3.23) and (3.41) it follows that

$$Y^*_{lm}(\beta, \alpha) = (-1)^m\, Y_{l-m}(\beta, \alpha). \tag{3.42}$$

3.4 IRREDUCIBLE SPHERICAL TENSORS

The more familiar Cartesian tensors do not have simple transformation properties when subjected to rotation. Therefore, they are not convenient to use for treating physical problems involving rotation. On the other hand, the irreducible spherical-tensor operators, when subjected to rotation, yield simple transformation properties and are thus useful to describe rotational problems. In this section, we shall explore the various properties of irreducible spherical-tensor operators. An irreducible spherical tensor of rank j is defined as a set of $2j + 1$ operators given by T_{jm}, with $m = -j, -j + 1, \ldots, j - 1, j$, which is transformed according to

$$T'_{jm} = D(\alpha, \beta, \gamma)T_{jm} D^{-1}(\alpha, \beta, \gamma) = \sum_{m'} D^{(j)}_{m'm}(\alpha, \beta, \gamma)T_{jm'}. \tag{3.43}$$

This definition indicates that the T_{jm} transforms according to the spherical harmonics of order j and shares all the properties of spherical harmonics.

Like the Cartesian tensors, the sum of two spherical tensors of a given rank is another spherical tensor of the same rank. The multiplication and contraction of spherical tensors, on the other hand, are somewhat different from that of Cartesian tensors.

Irreducible spherical tensors $T_{j_1m_1}(A_1)$ and $T_{j_2m_2}(A_2)$ are multiplied in a manner analogous to the coupling of the eigenfunctions of angular momentum. We shall now describe how the eigenfunctions of angular momenta are coupled.

Let the eigenfunctions for angular momenta $\mathbf{J}_1$ and $\mathbf{J}_2$ be given by $\psi_{j_1m_1}$ and $\psi_{j_2m_2}$, respectively; i.e., $\psi_{j_1m_1}$ and $\psi_{j_2m_2}$ are the eigenfunctions of J_1^2, J_{1z}, and J_2^2, J_{2z}. The uncoupled representation corresponds to the direct product of $\psi_{j_1m_1}$ and $\psi_{j_2m_2}$. It is clear that J_1^2, J_{1z} and J_2^2, J_{2z} are diagonal in the uncoupled representation associated with eigenvalues given by $j_1(j_1 + 1)$, m_1 and $j_2(j_2 + 1)$, m_2, respectively. The total angular-momentum operator $\mathbf{J}$ is given by $\mathbf{J} = \mathbf{J}_1 + \mathbf{J}_2$. We seek a set of eigenfunctions $\{\psi_{jm}\}$ in which J^2 and J_z are diagonal with eigenvalues give by $j(j + 1)$ and m, respectively. The set of eigenfunctions $\{\psi_{jm}\}$ form the so-called coupled representation.

The coupled representation is related to the uncoupled representation by a unitary transformation:

$$\psi_{jm} = \sum_{m_1m_2} C(j_1j_2j|m_1m_2m)\psi_{j_1m_1}\psi_{j_2m_2}. \tag{3.44}$$

The coefficients for the transformation $C(j_1j_2j|m_1m_2m)$ are known as the Clebsch–Gordan (or Wigner) coefficients. For given j_1, j_2, m_1, m_2, and j values, the Clebsch–Gordan coefficient is a fixed number. Using the property of angular momentum operators, we can derive several symmetry relations as well as orthogonal relations for the Clebsch–Gordan coefficients. We shall omit the derivations because they can readily be found in standard books dealing with the theory of angular momentum.

As pointed out by Wigner, the maximum symmetry in the transformation coefficient is most conveniently expressed in terms of the so-called 3-j symbol. The 3-j symbol is related to the Clebsch–Gordan coefficient by

$$\begin{pmatrix} j_1 & j_2 & j_3 \\ m_1 & m_2 & m_3 \end{pmatrix} = (-1)^{j_1-j_2-m_3}(2j_3+1)^{-1/2}C(j_1j_2j_3|m_1m_2-m_3). \tag{3.45}$$

Several useful properties associated with the 3-j symbol are as follows:

(1) The symbol vanishes unless $m_1 + m_2 + m_3 = 0$.

(2) The value of the 3-j symbol is zero unless j_1, j_2, and j_3 satisfy the triangle condition.

(3) An even permutation of the columns leaves the numerical value of the 3-j symbol unchanged, whereas an odd permutation is equivalent to multiplying its value by $(-1)^{j_1+j_2+j_3}$, i.e.,

$$\begin{pmatrix} j_1 & j_2 & j_3 \\ m_1 & m_2 & m_3 \end{pmatrix} = (-1)^{j_1+j_2+j_3}\begin{pmatrix} j_2 & j_1 & j_3 \\ m_2 & m_1 & m_3 \end{pmatrix} = \begin{pmatrix} j_2 & j_3 & j_1 \\ m_2 & m_3 & m_1 \end{pmatrix}. \tag{3.46}$$

(4) Changing the sign of m_1, m_2, and m_3 is equivalent to multiplying its value by $(-1)^{j_1+j_2+j_3}$, i.e.,

$$\begin{pmatrix} j_1 & j_2 & j_3 \\ -m_1 & -m_2 & -m_3 \end{pmatrix} = (-1)^{j_1+j_2+j_3}\begin{pmatrix} j_1 & j_2 & j_3 \\ m_1 & m_2 & m_3 \end{pmatrix}. \tag{3.47}$$

(5) The 3-j symbols satisfy the following orthogonality properties:

$$\sum_{j_3m_3}(2j_3+1)\begin{pmatrix} j_1 & j_2 & j_3 \\ m_1 & m_2 & m_3 \end{pmatrix}\begin{pmatrix} j_1 & j_2 & j_3 \\ m_1' & m_2' & m_3 \end{pmatrix} = \delta_{m_1m_1'}\delta_{m_2m_2'} \tag{3.48}$$

and

$$\sum_{m_1m_2}(2j_3+1)\begin{pmatrix} j_1 & j_2 & j_3 \\ m_1 & m_2 & m_3 \end{pmatrix}\begin{pmatrix} j_1 & j_2 & j_3' \\ m_1 & m_2 & m_3' \end{pmatrix} = \delta_{j_3j_3'}\delta_{m_3m_3'}. \tag{3.49}$$

Wigner's 3-j symbols have been computed and tabulated by many authors. The most extensive tabulation is found in a book by Rotenberg *et al.* (1959), wherein the 3-j symbols with $j \leq 3$ are given in Table I.

We now return to the problem of the coupling of spherical-tensor operators. Analogous to the coupling of wave functions involving angular momenta, the irreducible spherical-tensor operators $T_{j_1m_1}(A_1)$ and $T_{j_2m_2}(A_2)$ are coupled according to

$$T_{jm}(A_1, A_2) = \sum_{m_1m_2} C(j_1j_2j|m_1m_2m)\, T_{j_1m_1}(A_1)T_{j_2m_2}(A_2)$$

$$= \sum_{m_1m_2} (-1)^{-j_1+j_2-m}(2j+1)^{1/2} \begin{pmatrix} j_1 & j_2 & j \\ m_1 & m_2 & -m \end{pmatrix}$$

$$\times\, T_{j_1m_1}(A_1)T_{j_2m_2}(A_2), \tag{3.50}$$

where A_1 and A_2 are the arguments of tensors $T_{j_1m_1}$ and $T_{j_2m_2}$, respectively.

The arguments A_1 and A_2 need not be the same. With the help of the orthogonality relationship of 3-j symbols [Eq. (3.49)], the inverse of Eq. (3.50) is immediately found to be

$$T_{j_1m_1}(A_1)T_{j_2m_2}(A_2) = \sum_{jm} (<1)^{j_1-j_2-m}\,(2j+1)^{1/2} \begin{pmatrix} j_1 & j_2 & j \\ m_1 & m_2 & -m \end{pmatrix} T_{jm}(A_1, A_2). \tag{3.51}$$

It is left to the reader to verify that $T_{jm}(A_1, A_2)$ also transform as irreducible spherical tensors.

In Eq. (3.42), the irreducible tensor operators are defined according to transformation under rotation. An alternative definition of irreducible tensor operators was provided by Racah using the commutation rules involving angular-momentum operators (Rose, 1957), i.e., if T_{jm} is an irreducible spherical tensor of rank j, then it satisfies the commutation relation

$$[J_i, T_{jm}] = \sum_{m'} \langle jm'|J_i|jm\rangle\, T_{jm'}, \tag{3.52}$$

where J_i, with $i = x, y, z$, is the ith component of the angular-momentum operator. From Eq. (3.52) it follows that

$$[J_z, T_{jm}] = mT_{jm}, \tag{3.53}$$

and

$$[J_\pm, T_{jm}] = [(j \mp m)(j \pm m + 1)]^{1/2}\, T_{jm\pm1}, \tag{3.54}$$

where $J\pm = (J_x \pm iJ_y)$. The Racah definition follows directly from Eq. (3.42) by considering infinitesimal rotation.

To give examples of irreducible tensors, we consider a vector $\mathbf{r}$ in terms of a spherical basis:

$$\mathbf{r} = \sum_{q=-1,0,1} (-1)^q r_q \hat{e}_{-q}, \tag{3.55}$$

where the spherical-basis vectors are

$$\hat{e}_0 = \hat{z}, \qquad \hat{e}_{\pm 1} = (\mp 1/\sqrt{2})(\hat{x} \pm i\hat{y}), \tag{3.56}$$

and the spherical components of **r** are

$$r_0 = z, \qquad r_{\pm 1} = (\mp 1/\sqrt{2})(x \pm iy). \tag{3.57}$$

From the Racah definition, it is easy to verify that r_+, r_{-1}, and r_0 are three components of an irreducible spherical-tensor operator of rank 1. Using this result, we can construct irreducible spherical tensors of higher rank. For example, in Eq. (3.50), if $T_{j_1 m_1}$ is equated to r_{m_1}, and $T_{j_2 m_2}$ to r_{m_2}, we can construct a tensor of rank 2 according to

$$T_{2m}(\mathbf{r}) = \sum_{m_1 m_2} (-1)^m \sqrt{5} \begin{pmatrix} 1 & 1 & 2 \\ m_1 & m_2 & -m \end{pmatrix} r_{m_1} r_{m_2}. \tag{3.58}$$

Evaluating the various 3-j symbols, we have found that the five tensor components are given by

$$\begin{aligned} T_{20}(\mathbf{r}) &= (1/\sqrt{6})(3z^2 - r^2) = \sqrt{(8\pi/15)}\, y_{20}(\mathbf{r}), \\ T_{2\pm 1}(\mathbf{r}) &= \pm z(x \pm iy) = \sqrt{(8\pi/15)}\, y_{2\pm 1}(\mathbf{r}), \\ T_{2\pm 2}(\mathbf{r}) &= \tfrac{1}{2}(x \pm iy)^2 = \sqrt{(8\pi/15)}\, y_{2\pm 2}(\mathbf{r}) \end{aligned} \tag{3.59}$$

where $y_{2m}(\mathbf{r})$ are solid harmonics of order 2. Solid harmonics are related to spherical harmonics by $y_{lm}(\mathbf{r}) = r\, Y_{lm}(\theta, \phi)$. It is obvious that starting with $T_{2m}(\mathbf{r})$ and $T_{1m}(\mathbf{r})$, we can construct spherical tensors of the third rank, the fourth rank, and so on.

Another example of the decomposition of a Cartesian tensor into components of irreducible spherical tensors is the resolution of the polarizability tensor of molecule m, given by $\alpha_{ij}^{(m)}$, using the coupling formula of spherical tensors as given by Eq. (3.51). By definition, $\alpha_{ij}^{(m)}$ is given as

$$\alpha_{ij}^{(m)} = \hat{n}_i \cdot \boldsymbol{\alpha}^{(m)} \cdot \hat{n}_j \tag{3.60}$$

where $\hat{n}_i$ and $\hat{n}_j$ represent unit vectors for the polarization of incident and scattered light respectively.

Clearly, $\alpha_{ij}^{(m)}$ is expressed in terms of the space-fixed (or laboratory) coordinate system. However, to utilize the symmetry of the molecule, it is more convenient to express $\alpha_{ij}^{(m)}$ in terms of the coordinate system fixed in the molecule (i.e., the body-fixed coordinate system).

To express $\boldsymbol{\alpha}$ (we drop the superscript m for simplicity of notation) in terms of the body-fixed system, we introduce the spherical-basis vectors

$\hat{e}_q$, which define the body-fixed coordinate system. The polarization vectors $\hat{n}_i$ and $\hat{n}_j$ expressed in terms of $\hat{e}_q$ are given by

$$\hat{n}_l = \sum_q (-1)^q n_l^{(q)} \hat{e}_{-q}, \qquad l = i, j. \tag{3.61}$$

Using this expression, we can rewrite α_{ij} as

$$\alpha_{ij} = \sum_{q,q'} (-1)^{q-q'} n_i^{(q)} n_j^{(-q')} \alpha_{-qq'}, \tag{3.62}$$

where

$$\alpha_{-qq'} = \hat{e}_{-q} \cdot \boldsymbol{\alpha} \cdot \hat{e}_{q'}, \tag{3.63}$$

and $n_i^{(q)}$ and ${}_j^{(q')}$ are spherical tensors of rank 1 and can be combined to yield spherical tensors of rank 0, 1, and 2, according to the expression [compare with Eq. (3.50)]

$$T_{km}(ij) = \sum_{q,q'} (-1)^m (2k+1)^{1/2} \begin{pmatrix} 1 & 1 & k \\ q & -q' & -m \end{pmatrix} n_i^{(q)} n_j^{(-q')}, \tag{3.64}$$

where the T_{km} vanish unless $k = 0$, 1, and 2. The inverse of Eq. (3.64) is given by

$$n_i^{(q)} n_j^{(-q')} = \sum_{k,m} (-1)^m (2k+1)^{1/2} \begin{pmatrix} 1 & 1 & k \\ q & -q & -m \end{pmatrix} T_{km}(ij). \tag{3.65}$$

Substituting Eq. (3.65) into (3.62) gives

$$\alpha_{ij} = \sum_{km} (-1)^m \alpha(k, -m) T_{km}(ij), \tag{3.66}$$

where

$$\alpha(k, -m) = \sum_{q,q'} (-1)^{q-q'} (2k+1)^{1/2} \begin{pmatrix} 1 & 1 & k \\ q & -q' & -m \end{pmatrix} \alpha_{-qq'}. \tag{3.67}$$

By expressing α_{ij} in terms of Eq. (3.66), we have succeeded in decomposing a Cartesian tensor into the sum of irreducible spherical tensors of rank, 0, 1, and 2.

Evaluation of $\alpha(k, m)$ gives [using Eqs. (3.63) and (3.67)]

$$\begin{aligned}
\alpha(0,0) &= (-1/\sqrt{3}) \operatorname{Tr} \boldsymbol{\alpha}, \\
\alpha(1,0) &= (-i/\sqrt{2})(\alpha_{xy} - \alpha_{yx}), \\
\alpha(1, \pm 1) &= \tfrac{1}{2}[(\alpha_{zx} - \alpha_{xz}) \pm i(\alpha_{zy} - \alpha_{yz})], \\
\alpha(2,0) &= (-1/\sqrt{6})(\operatorname{Tr} \boldsymbol{\alpha} - 3\alpha_{zz}), \\
\alpha(2, \pm 1) &= \mp \tfrac{1}{2}[(\alpha_{xz} + \alpha_{zx}) \pm i(\alpha_{yz} + \alpha_{zy})], \\
\alpha(2, \pm 2) &= \tfrac{1}{2}[\alpha_{xx} - \alpha_{yy}) \pm i(\alpha_{xy} + \alpha_{yx})].
\end{aligned} \tag{3.68}$$

When the molecular polarizability tensor is symmetric, all components of $\alpha(1, m)$ vanish. The expressions for $\alpha(k, m)$ are further simplified if the molecule possesses symmetry. For example, for a symmetric-top molecule, there are only two nonvanishing elements for $\alpha(k, m)$:

$$\alpha(0, 0) = (-1/\sqrt{3})(2\alpha_\perp + \alpha_\parallel) = -\sqrt{3}\,\bar{\alpha},$$
$$\alpha(2, 0) = (2/\sqrt{6})(\alpha_\parallel - \alpha_\perp) = (2/\sqrt{6})\,\beta, \tag{3.69}$$

where $\alpha_\parallel$ and $\alpha_\perp$ are the principal polarizability values parallel and perpendicular to the symmetry axis of the molecule, respectively; and $\bar{\alpha}$ and β correspond to the average value and the anisotropy of the polarizability tensor, respectively. Scattering caused by the average and the anisotropy of the polarizability tensor in a molecular liquid was discussed in Chapter 2.

In Eq. (3.66), both $\alpha(k, -m)$ and $T_{km}(ij)$ are given in terms of the molecule-fixed coordinate frame. The quantities $T_{jm}(i, j)$, which are defined by the polarization vectors of the incident ($\hat{n}_i$) and scattered ($\hat{n}_j$) light, are related to the laboratory-fixed coordinate frame by

$$T_{km}(ij) = \sum_q F_{kq}(ij) D_{mq}^{(k)*}(\Omega), \tag{3.70}$$

where the quantities $F_{kq}(ij)$ are defined with respect to the laboratory-fixed coordinate frame and can be evaluated according to Eq. (3.64).

Substituting Eq. (3.70) into (3.66), we can write the polarizability tensor α_{ij}, defined according to the laboratory-fixed coordinate system, in terms of $\alpha(k, m)$, defined with respect to the molecule-fixed coordinate system by

$$\alpha_{ij} = \sum_{k,m,q} (-1)^m F_{kq}(ij) D_{mq}^{(k)*}(\Omega) \alpha(k, -m). \tag{3.71}$$

This equation provides a useful connection between light-scattering spectroscopy and the orientational behavior of the molecules in the scattering medium. The effect of molecular reorientation is now compactly included in Wigner rotation matrices $D_{qm}^{(k)}(\Omega)$. The quantities $\alpha(k, -m)$ are defined with reference to the molecule-fixed coordinate system and are affected only by molecular vibration, but they are independent of molecular orientation coordinates. Molecular reorientation will modulate the rotation matrices and make these quantities functions of time. We shall describe in a later section how the time dependence in the rotation matrices is reflected in the light-scattering spectrum.

For the polarized-scattering configuration we have $\hat{n}_i = \hat{n}_j = \hat{z}$, and, according to Eq. (3.64), the only nonvanishing elements for $F_{kq}(ij)$ are

$$F_{00}(z, z) = -\sqrt{\frac{1}{3}}, \qquad F_{20}(z, z) = \sqrt{\frac{2}{3}}, \tag{3.72}$$

and Eq. (3.71) becomes

$$\alpha_{zz} = -\sqrt{\frac{1}{3}}\,\alpha(0, 0) + \sqrt{\frac{2}{3}}\sum_{m}(-1)^m D_{m0}^{(2)*}(\Omega)\,\alpha(2, -m). \tag{3.73}$$

On the other hand, for the depolarized-scattering configuration with $\hat{n}_i = \hat{z}$ and $\hat{n}_i = \hat{y}$, the nonvanishing $F_{kq}(y, z)$ are

$$\begin{aligned} F_{1,1}(y, z) &= F_{1,-1}(y, z) = -i/2, \\ F_{2,-1}(y, z) &= F_{2,1}(y, z) = -i/2, \end{aligned} \tag{3.74}$$

and Eq. (3.71) is reduced to

$$\begin{aligned} \alpha_{zy} = \frac{i}{2}\sum(-1)^m\{&\alpha(1, m)[(D_{m1}^{(1)*}(\Omega) + D_{m-1}^{(1)*}(\Omega))] \\ &+ \alpha(2, -m)[-D_{m-1}^{(2)*}(\Omega) - D_{m1}^{(2)*}(\Omega)]\}. \end{aligned} \tag{3.75}$$

In the event that the molecule has a center of inversion or a plane of symmetry (i.e., an optically inactive molecule), the Cartesian polarizability tensor is symmetric. In this case, $\alpha(1, m) = 0$ for all m, and Eq. (3.75) is reduced to

$$\alpha_{zy} = \frac{i}{2}\sum_{m}(-1)^m\alpha(2, -m)[D_{m-1}^{(2)*}(\Omega) + D_{m1}^{(2)*}(\Omega)]. \tag{3.76}$$

We now apply Eqs. (3.73) and (3.76) to the problem relating the integrated Raman-scattering light intensity to the orientation parameters of a partially oriented system, such as a liquid crystal in the nematic or smectic phase or a partially oriented polymer chain.

3.5 RAMAN-SCATTERING INTENSITY AND THE ORIENTATION PARAMETERS

Since the Raman polarizability tensor is orientation dependent, measurements of the Raman-scattering intensities of the various polarization components associated with a vibrational band can provide useful information about the orientation-distribution function in a partially oriented system.

To illustrate this, we consider for simplicity that the Raman tensor of a vibrational mode is given by

$$\boldsymbol{\alpha} = \begin{bmatrix} \alpha_\perp & 0 & 0 \\ 0 & \alpha_\perp & 0 \\ 0 & 0 & \alpha_\parallel \end{bmatrix}. \tag{3.77}$$

The intensity detected using the zz scattering geometry is proportional to [Eq. (3.73)]

$$\langle|\alpha_{zz}|^2\rangle = \langle|(-1/\sqrt{3})\alpha\,(0,0) + (2/\sqrt{6})\,D^{(2)}_{00}(\Omega)\,\alpha(2,0)|^2\rangle \tag{3.78}$$

where $\alpha(0, 0) = (2\alpha_\perp + \alpha_\parallel)/\sqrt{3} = \sqrt{3}\alpha$, and $\alpha(2, 0) = (2/\sqrt{6})(\alpha_\parallel - \alpha_\perp) = (2/\sqrt{6})\,\beta$. The quantities α and β depend only on the vibrational coordinates. The angular brackets denote an average over the various orientations of the partially oriented scattering element. That is, $\langle A\rangle$ is given by

$$\langle A\rangle = \int d\Omega\; Af(\Omega), \tag{3.79}$$

where $f(\Omega)$ is the distribution function describing the orientational ordering of the molecules.

Expanding Eq. (3.78) gives

$$\langle|\alpha_{zz}|^2\rangle = \alpha^2 + \tfrac{4}{3}\alpha\beta\langle D^{(2)}_{00}(\Omega)\rangle + \tfrac{4}{9}\beta^2\langle|D^{(2)}_{00}(\Omega)|^2\rangle. \tag{3.80}$$

The third term on the right-hand side of this equation can be written in terms of the Clebsch–Gordan series, which reduces the product of the Wigner matrices to a single Wigner matrix:

$$D^{(j)}_{m_1n_1}(\Omega)\,D^{(j)}_{m_2n_2}(\Omega) = \sum_{jmn}(2j+1)\begin{pmatrix} j_1 & j_2 & j \\ m_1 & m_2 & m \end{pmatrix} D^{(j)*}_{mn}(\Omega)\begin{pmatrix} j_1 & j_2 & j \\ n_1 & n_2 & n \end{pmatrix}. \tag{3.81}$$

Using this formula and evaluating the 3-j symbols, we find that

$$\langle D^{(2)}_{00}(\Omega)\,D^{(2)}_{00}(\Omega)\rangle = \tfrac{1}{5} + \tfrac{2}{7}\,\langle D^{(2)}_{00}(\Omega)\rangle + \tfrac{18}{35}\,\langle D^{(4)}_{00}(\Omega)\rangle. \tag{3.82}$$

Substituting Eq. (3.82) into Eq. (3.80) gives

$$\langle|\alpha_{zz}|^2\rangle = \alpha^2 + \tfrac{4}{45}\beta^2 + (\tfrac{4}{3}\alpha\beta + \tfrac{8}{63}\beta^2)\langle P_2\rangle + \tfrac{8}{35}\beta^2\langle P_4\rangle, \tag{3.83}$$

where the orientation parameters $\langle P_l\rangle$ are given by

$$\langle P_l\rangle = \langle P_l(\cos\theta)\rangle = \int d\Omega\; D^{(l)}_{00}(\Omega)\,f(\Omega). \tag{3.84}$$

For a partially oriented system, the quantities $\langle P_l\rangle$ do not vanish. Equation (3.83) indicates that the I_{zz} intensity is affected by $\langle P_2\rangle$ and $\langle P_4\rangle$.

The I_{zy}, I_{xy}, and I_{yy} scattering intensities can also be related to the orientation parameters. According to Eqs. (3.76) and (3.77), we have

$$\begin{aligned}\langle|\alpha_{yz}|^2\rangle &= \langle|(i/\sqrt{6})\beta[D^{(2)}_{10}(\Omega) + D^{(2)}_{-10}(\Omega)]|^2\rangle \\ &= (\beta^2/6)\langle D^{(2)}_{10}(\Omega)\, D^{(2)*}_{10}(\Omega) + D^{(2)}_{-10}(\Omega)\, D^{(2)*}_{-10}(\Omega) \\ &\qquad + 2D^{(2)}_{10}(\Omega)D^{(2)*}_{-10}(\Omega)\rangle. \end{aligned} \tag{3.85}$$

Again using Eq. (3.81), evaluating the various 3-j symbols, and assuming that the distribution has a cylindrical symmetry about the z-axis, we obtain

$$\langle|\alpha_{yz}|^2\rangle = \beta^2(\tfrac{1}{15} + \tfrac{1}{21}\langle P_2\rangle - \tfrac{4}{35}\langle P_4\rangle). \tag{3.86}$$

In a similar manner, the intensities I_{xy} and I_{yy} can also be calculated. The results are

$$\langle|\alpha_{xy}|^2\rangle = \beta^2(\tfrac{1}{15} - \tfrac{1}{21}\langle P_2\rangle + \tfrac{1}{35}\langle P_4\rangle), \tag{3.87}$$

$$\langle|\alpha_{yy}|^2\rangle = (\alpha^2 + \tfrac{4}{45}\beta^2) - (\tfrac{2}{3}\alpha\beta + \tfrac{4}{63}\beta^2)\langle P_2\rangle + \tfrac{3}{35}\beta^2\langle P_4\rangle. \tag{3.88}$$

It is clear from these equations that, given the parameter α/β, the orientation parameters $\langle P_2\rangle$ and $\langle P_4\rangle$ can be determined by measuring the intensity ratios of scattered light (such as the ratios, I_{zy}/I_{zz}, I_{zy}/I_{yy}, and I_{xy}/I_{zy}). There is no need to make absolute intensity measurements.

It should be pointed out that in a rotationally isotropic system, such as a viscous liquid or an amorphous solid, the orientation paramenters $\langle P_2\rangle$ and $\langle P_4\rangle$ vanish identically, and we obtain from Eqs. (3.83) and (3.88)

$$\langle|\alpha_{zz}|^2\rangle = \langle|\alpha_{yy}|^2\rangle = \alpha^2 + \tfrac{4}{45}\beta^2 \tag{3.89}$$

and from Eqs. (3.82) and (3.83)

$$\langle|\alpha_{yz}|^2\rangle = \langle|\alpha_{xy}|^2\rangle = \tfrac{1}{15}\beta^2, \tag{3.90}$$

which gives the well-known result for the depolarization ratio for an isotropic liquid

$$\mathrm{R}_{\mathrm{iso}} = 3\beta^2/(45\alpha^2 + 4\beta^2). \tag{3.91}$$

Equations (3.86)–(3.88) have been used to extract the orientation parameters of the polypropylene chain backbone (Satija and Wang, 1973). This method deals with measurements of the Raman intensity ratios I_{xy}/I_{zz} and I_{xy}/I_{yy} for the CH_2 twisting mode at 1256 cm^{-1} of polypropylene as a function of the extrusion ratio. If it is assumed that the α/β ratio is insensitive to temperature variation and remains unchanged in the polymer melt, then, knowing the α/β value and the intensity ratios, we can determine

$\langle P_2 \rangle$ and $\langle P_4 \rangle$ with the help of Eqs. (3.86)–(3.88). These $\langle P_2 \rangle$ and $\langle P_4 \rangle$ data are very useful because together they provide a more stringent test of the model that is used to calculate the orientation-distribution function than a set of separate $\langle P_2 \rangle$ data alone. Studies of this type were applied to other polymer systems and to liquid crystals.

3.6 STUDIES OF REORIENTATIONAL MOTION BY DEPOLARIZED RAYLEIGH AND RAMAN SCATTERING

We shall next consider the time-correlation function associated with molecular reorientation for isotropic liquids or amorphous solids. Only two experimental geometries are needed to obtain all the useful information needed concerning the molecular reorientation. We denote those geometries by VV and VH respectively. The polarizability associated with the VV geometry is α_{zz}, which is given by Eq. (3.73). The polarizability for VH scattering is α_{zy}, given in Eq. (3.76) (assuming that the Cartesian polarizability tensor is symmetric). The time-correlation function for VV scattering is thus

$$G_{VV}(t) = \langle \bar{\alpha}(t)\, \bar{\alpha}(0) \rangle + \tfrac{2}{3} \sum_{m} \sum_{m'} \langle \alpha(2, m; t) \alpha^*(2, m')\, D_{m0}^{(2)}[\Omega(t)]\, D_{m'0}^{(2)*}(\Omega) \rangle, \quad (3.92)$$

where $\bar{\alpha} = Tr\, \boldsymbol{\alpha}$. The cross terms linear in D vanish because there is no macroscopic order in the isotropic system.

The time-correlation function for VH scattering is

$$G_{\mathrm{VH}}(t) = \tfrac{1}{4} \sum_{m} \sum_{m'} \langle \alpha(2, m; t) \alpha^*(2, m') \{ D_{m1}^{(2)}[\Omega(t)] + D_{m(-1)}^{(2)}[\Omega(t)] \} \times \{ D_{m'1}^{(2)*}(\Omega) + D_{m'(-1)}^{(2)*}(\Omega) \} \rangle. \quad (3.93)$$

These time-correlation functions that contain rotation matrices can be simplified. Since the rotation matrices given by $D_{mm'}^{(2)}[\Omega/t)]$ result from reorientation of $\delta\Omega = \delta\alpha$, $\delta\beta$, $\delta\alpha$ in time t from its initial orientation Ω at $t = 0$, they can be written in terms of a rotational transformation as [compare with Eq. (3.20)] (Steele, 1969)

$$D_{nn'}^{(2)}[\Omega(t)] = \sum_{r} D_{nr}^{(2)}(\Omega)\, D_{rn'}^{(2)}(\delta\Omega). \quad (3.94)$$

Substituting Eq. (3.94) into Eqs. (3.92) and (3.93) and averaging over the initial orientation Ω [neglecting the angular dependence that may be present in the $\alpha(2, m)$ terms owing to the rotation–vibration coupling] we obtain, with the help of Eq. (3.24),

$$G_{VV}(t) = \langle \bar{\alpha}(t)\bar{\alpha}(0)\rangle + \tfrac{2}{15}\sum_{m}\sum_{m'}\langle \alpha(2, m; t)\alpha^*(2, m')\rangle\langle D^{(2)}_{mm'}[\delta\Omega(t)]\rangle, \quad (3.95)$$

$$G_{VH}(t) = \tfrac{1}{10}\sum_{m}\sum_{m'}\langle \alpha(2, m; t)\alpha^*(2, m')\rangle\langle D^{(2)}_{mm'}[\delta\Omega(t)]\rangle. \quad (3.96)$$

Note that $G_{VV}(t)$ splits into an isotropic and an anisotropic part; if we denote the VH time-correlation function owing to anisotropic scattering as

$$G_{\text{ani}}(t) = \tfrac{1}{10}\sum_{m}\sum_{m'}\langle \alpha(2, m; t)\alpha^*(2, m')\rangle\, C^{(2)}_{mm'}(t), \quad (3.97)$$

then we can write

$$G_{VV}(t) = G_{\text{iso}}(t) + \tfrac{4}{3}G_{\text{ani}}(t), \quad (3.98)$$

$$G_{VH}(t) = G_{\text{ani}}(t), \quad (3.99)$$

where

$$G_{\text{iso}}(t) = \langle \bar{\alpha}(t)\bar{\alpha}\rangle. \quad (3.100)$$

In Eq. (3.97), $C^{(2)}_{mm'}(t)$ is the orientational time-correlation function associated with the second-rank rotation matrix and is given by

$$C^{(2)}_{mm'}(t) = \langle D^{(2)}_{mm'}[\delta\Omega(t)]\rangle. \quad (3.101)$$

Equations (3.98) and (3.99) were utilized to study the dynamics of molecular reorientation in dense gases and liquids composed of linear or symmetric molecules from the line shape of the polarized (VV) and depolarized (VH) Raman spectra. We shall now discuss the methodology in more detail. As pointed out previously, if the collision-induced effect is ignored, then the quantity $\alpha(2, m)$ depends on the vibrational motion of the molecule. If we consider only first-order Raman scattering, then $\alpha(2, m)$ can be written as

$$\alpha(2, m) = \alpha^{(0)}(2, m) + \sum_{\mu}\alpha^{(\mu)}(2, m)\, Q_{\mu}, \quad (3.102)$$

where $\alpha^{(0)}(2, m)$, the polarizability of the molecule in its equilibrium nuclear configuration, is a constant in the molecule-fixed coordinate frame; Q_μ the coordinate of the μth normal mode; and $\alpha^{(\mu)}(2, m) = \partial\alpha(2, m)/\partial Q_\mu$ evaluated at the equilibrium configuration. The values of $\alpha^{(\mu)}(2, m)$ are determined by the symmetry of the normal modes in a molecule. In most cases, no attempt is made to evaluate the $\alpha^{(\mu)}(2, m)$ values explicitly but rather group theory is used to determine which elements are nonvanishing. For example, the CH_3I molecule belongs to the C_{3v} point group and has nine normal modes.

The Raman polarizability tensor for the A_1 mode has axial symmetry and is given by

$$\boldsymbol{\alpha}^{(A_1)} = \begin{bmatrix} a & 0 & 0 \\ 0 & a & 0 \\ 0 & 0 & b \end{bmatrix}. \tag{3.103}$$

Thus, according to Eq. (3.68), the only nonvanishing element among $\alpha(2, m)$ is $\alpha(2, 0)$, which is easily proven to be given by $(2/\sqrt{6})\,(b - a) = (2/\sqrt{6})\beta$. The average Raman polarizability $\bar{\alpha}$ is given by $\frac{1}{3}(2a + b)$. Thus, we have

$$G_{\mathrm{VV}}(t) = \langle \bar{\alpha}(t)\alpha \rangle + \tfrac{4}{45}\langle \beta(t)\beta \rangle C_{00}^{(2)}(t), \tag{3.104}$$

$$G_{\mathrm{VH}}(t) = \tfrac{1}{15}\langle \beta(t)\beta \rangle C_{00}^{(2)}(t). \tag{3.105}$$

Since

$$\bar{\alpha} = (\partial\bar{\alpha}/\partial Q_{A_1})_0\, Q_{A_1} = l_{A_1} Q_{A_1}, \tag{3.106}$$

$$\beta = (\partial\beta/\partial Q_{A_1})_0\, Q_{A_1} = t_{A_1} Q_{A_1}, \tag{3.107}$$

it follows that

$$G_{\mathrm{iso}}(t) = l_{A_1}^2 \,\langle Q_{A_1}(t) Q_{A_1} \rangle, \tag{3.108}$$

$$G_{\mathrm{ani}}(t) = \tfrac{1}{15}\, t_{A_1}^2 \,\langle Q_{A_1}(t)\, Q_{A_1} \rangle C_{00}^{(2)}(t). \tag{3.109}$$

The orientational time-correlation function can thus be obtained from Eqs. (3.108) and (3.109) according to the expression

$$\frac{G_{\mathrm{ani}}(t)/G_{\mathrm{ani}}(0)}{G_{\mathrm{iso}}(t)/G_{\mathrm{iso}}(0)} = C_{00}^{(2)}(t). \tag{3.110}$$

Therefore, to obtain $C_{00}^{(2)}(t)$ from the Raman-scattering spectra, we first calculate $I_{\mathrm{iso}}(\omega)$, using the relation

$$I_{\mathrm{iso}}(\omega) = I_{\mathrm{VV}}(\omega) - \tfrac{4}{3} I_{\mathrm{VH}}(\omega), \tag{3.111}$$

where $I_{\mathrm{VV}}(\omega)$ and $I_{\mathrm{VH}}(\omega)$ are the observed polarized and depolarized components that have been normalized to the same gain setting. The $I_{\mathrm{VH}}(\omega)$ and $I_{\mathrm{iso}}(\omega)$ spectra are then Fourier transformed to obtain the time-correlation functions $G_{\mathrm{ani}}(t)$ and $G_{\mathrm{iso}}(t)$, respectively. Finally, Eq. (3.110) is used to obtain $C_{00}^{(2)}(t)$.

The validity of this procedure lies in the assumption that the nonreorientational-relaxation mechanism is statistically independent of the reorientational relaxation. Under this assumption, Eqs. (3.95) and (3.96) become valid. However, a variety of intramolecular vibration–rotation coupling mechanisms are present in real molecules. Furthermore, note that the spectral line shapes in dense fluids are determined not only by intramolecular processes but are strongly affected by the intermolecular interactions that may also couple vibrational and rotational motion. Thus, although Raman scattering seems to be the only way to obtain $C_{00}^{(2)}(t)$, it is necessary to use this procedure with care to extract the orientational time-correlation function.

In Fig. 3.2, $C_{00}^{(2)}(t)$ as well as $G_{\text{iso}}(t)$ and $G_{\text{ani}}(t)$ obtained from the A_1 band in liquid CH_3I are shown at two temperatures (Wright *et al.*, 1973).

Note, in Fig. 3.2, that the time-correlation functions $G_{\text{iso}}(t)$ and $C_{00}^{(2)}(t)$ differ in temperature dependence. Studies of the temperature dependences of $C_{00}^{(2)}(t)$ and $G_{\text{iso}}(t)$ should provide useful information about the reorientational and nonreorientational mechanisms in a molecular liquid. Note also that $C_{00}^{(2)}(t)$ does not decay exponentially over a short time (<0.5 psec) but rather approaches an exponential decay pattern over a longer time. Comparison of the theoretically calculated time-correlation function, either by a physical theory or by computer-dynamics simulation with the experimentally determined $C_{00}^{(2)}(t)$, is essential to improve our knowledge about the molecular-reorientational process in the dense fluid.

We have used the A_1 mode in CH_3I as an illustration to demonstrate the utility of Raman scattering as a method for determining the second-rank orientational time-correlation function $C_{00}^{(2)}(t)$. A different time-correlation function is obtained if a different Raman band is used. For example, the Raman polarizability tensors for the E band in CH_3I have the form (Wilson *et al.*, 1955)

$$\boldsymbol{\alpha}^{(E_y)} = \begin{bmatrix} c & 0 & 0 \\ 0 & -c & d \\ 0 & d & 0 \end{bmatrix}, \qquad \boldsymbol{\alpha}^{(E_x)} = \begin{bmatrix} 0 & c & d \\ c & 0 & 0 \\ d & 0 & 0 \end{bmatrix}. \tag{3.112}$$

Thus for the E modes $\bar{\alpha} = 0$, and the nonvanishing elements for the E_y modes are $\alpha(2, \pm 1) = \pm id$ and $\alpha(2, \pm 2) = c$; for the E_x mode, $\alpha(2, \pm 1) = \pm d$ and $\alpha(2, \pm 2) = \pm ic$. Clearly, the E_y and E_x modes are degenerate.

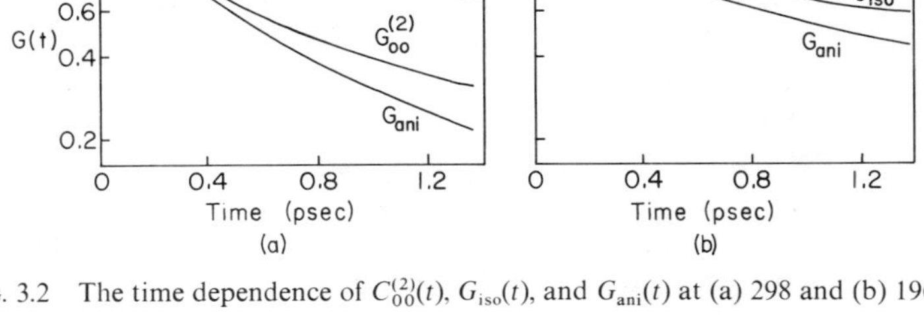

FIG. 3.2 The time dependence of $C_{00}^{(2)}(t)$, $G_{\text{iso}}(t)$, and $G_{\text{ani}}(t)$ at (a) 298 and (b) 196 K.

From Eqs. (3.97) and (3.100), we find that the time-correlation functions are given by

$$G_{\text{iso}}(t) = 0, \tag{3.113}$$

$$G_{\text{ani}}(t) = \tfrac{1}{5}[\langle d(t)\, d\rangle\, C_{11}^{(2)}(t) + \langle c(t)c\rangle\, C_{22}^{(2)}(t)]. \tag{3.114}$$

Other terms present in $G_{\text{ani}}(t)$ vanish because of symmetry. In the E band, unless the nonreorientational relaxation process is negligible, the extraction of the orientational time-correlation functions from the VV and VH scattering spectral components is impossible because in this case we have $G_{\text{VV}}(t) = \frac{4}{3}G_{\text{VH}}(t)$. However, if we combine additional information from the overtone bands in the Raman and infrared hyer-Raman or nonlinear spectroscopic techniques, then it may be possible to obtain the different components of the orientational time-correlation functions. Nevertheless, a variety of experimental factors (i.e., instrumental broadening, isotropic splitting can affect the line shape of the vibration–rotation band. Thus, for obtaining the orientational time-correlation function, it is not likely that analysis of the line shape of the E band can yield any additional useful information than that obtainable from the A band.

We discussed in detail the technique utilized to obtain the second-rank orientational time-correlation function using Raman-scattering spectra. The first-rank correlation function can be obtained from the line-shape analysis of the infrared-absorption vibration–rotation spectrum.

In Chapter 1 it was shown that if the vibration–rotation coupling is neglected, then the time-correlation function associated with the absorption of the vibration–rotation band is given by [see Eq. (1.173)]

$$C_{\text{IR}}(t) = \tfrac{1}{3} \sum_{i,j} \langle \mathbf{m}_i(t)\mathbf{m}_j\rangle\langle Q_i(t)Q_j\rangle, \tag{3.115}$$

where $\mathbf{m}_i$ is the derivative of the electric dipole moment of the ith molecule at the equilibrium position.

Since the orientational motion of molecules of physical–chemical interest can be described almost invariably in terms of classical statistical mechanics, it is not important to differentiate between the left-sided and the right-sided time-correlation function. Furthermore, if the intermolecular vibrational coupling is weak, we can neglect the cross terms with $i \neq j$ in Eq. (3.115) and approximate $C(t)$ by the single-particle time-correlation term. Thus, we have

$$C_{\text{IR}}(t) = \tfrac{1}{3}N\, \langle \mathbf{m}(t)\cdot\mathbf{m}\rangle\langle Q(t)Q\rangle, \tag{3.116}$$

where N is the number of absorbing molecules and $\mathbf{m}$ the transition dipole moment of a representative molecule for the particular vibration–rotation band under consideration. We now express the Cartesian vector to the Euler-angle representation. According to Eq. (3.55), the vector **m** can be written as

$$\mathbf{m} = \sum_{q=-1} (-1)^q m_q \hat{e}_{-q}, \tag{3.117}$$

where m_q is the body-fixed spherical component and $\hat{e}_{-q}$ is given by Eq. (3.56).

The body-fixed vector component is related to the space-fixed component (m'_q) by

$$m_q = \sum_{q'} m_{q'} D^{(1)*}_{qq'}(\Omega), \tag{3.118}$$

and the D function for the molecular orientation $\Omega(t)$ is related to the reorientation angle $\delta\Omega(t)$ in time t by (Steele, 1969)

$$D^{(1)}_{mm'}(\Omega(t)) = \sum_k D^{(1)}_{mk}(\Omega)\, D^{(1)}_{km'}[\delta\Omega(t)]. \tag{3.119}$$

Using Eqs. (3.117)–(3.119) and carrying out the average over Ω, we can rewrite $C_{\mathrm{IR}}(t)$ as

$$C_{\mathrm{IR}}(t) = \tfrac{1}{9} N m_n m_{n'} \langle Q(t)Q\rangle\, C^{(1)}_{nn'}(t),$$

where $C^{(1)}_{nn'}(t)$ is the first-rank orientational time-correlation function given by

$$C^{(1)}_{nn'}(t) = \langle D^{(1)}_{nn'}[\delta\Omega(t)]\rangle. \tag{3.120}$$

It is not evident that the absorption line shape of the vibration–rotation spectrum contains information about the first-rank orientational time-correlation function. For example, for the molecule CH_3I, the vibrational dipole moment that is parallel to the molecular symmetry axis generates the A_1 band. In this case, only m_0 is nonvanishing and

$$C_{\mathrm{IR}}(t) = \tfrac{1}{9} N m_0^2 \langle Q(t)Q\rangle\, C^{(1)}_{00}(t). \tag{3.121}$$

However, the difficulty in obtaining the orientational time-correlation function from the infrared-absorption spectral line shape is similar to that associated with the E band in the Raman spectrum. Experimentally, only a single spectrum (e.g., the parallel band) is obtained, but unless the vibrational-relaxation process is known there will be no way to extract $C^{(1)}_{00}(A)$ from Eq. (3.121).

3.7 MODELS FOR CALCULATING THE REORIENTATIONAL TIME-CORRELATION FUNCTION

Although the orientational time-correlation function can be determined experimentally (with the approximations used), theoretical calculation of the orientational-correlation function (or any other time-correlation function) in a dense system is a difficult, many-body problem. Two general types of approaches are used to calculate the orientational time-correlation function $C_r(t)$. One method deals with molecular dynamics; it uses a high-speed computer to solve the equations of motion for a system consisting of a small number of reorienting particles confined in a box. The particles are assumed to be described by classical mechanics. The size of the box is scaled so that the density of the particles in the box corresponds to the density of the fluid. The equation of motion of each molecule in the system is solved numerically by the computer. After the orientational dynamics of each particle is obtained, the computer is then programmed to perform the ensemble (thermal) average to obtain the time-correlation function $G_r(t)$. The degree of accuracy of this method increases with the number of particles in the box. However, the computation time increases disproportionately with the increase of particles. Although the molecular-dynamics calculation is a brute-force method, it is very useful to test the validity of the assumptions used in deriving a theory.

The second method is based on the stochastic approach. The orientational-correlation function is formulated in terms of a conditional probability for reorientation. The conditional-probability function is then treated in terms of a stochastic equation in the form of a diffusion equation, a Langevin equation, or a Fokker–Planck equation. The stochastic approach is valid, in general, only for large values of time, i.e., for time intervals long compared with the duration of the intermolecular collisions. For small values of time, $C_r(t)$ can be treated by the moment method.

We shall now find the second and fourth moments of $C_{00}^{(2)}(t)$. The moment expansion of the time-correlation function was described in Chapter 1. According to Eq. (1.184), we can write

$$C_r^{(t)} = C_{00}^{(2)}(t) = 1 - (t^2/2!)\, M_2^{(2)} + (t^4/4!)\, M_4^{(2)} + \cdots \qquad (3.122)$$

The odd moments vanish identically. The second is given by

$$\begin{aligned} M_2^{(2)} &= -(\partial^2/\partial t^2)\, C_r(t)|_{t=0} = -\tfrac{3}{2}\langle(\partial^2/\partial t^2)\cos^2\theta(t)\rangle \\ &= \langle 3\omega^2(t)\cos 2\theta(t) + \tfrac{3}{2}\dot{\omega}(t)\sin 2\theta\,(t)\rangle|_{t=0} = 3\langle\omega^2\rangle, \qquad (3.123) \end{aligned}$$

where $\theta = \delta\Omega$ is the reorientation angle of the symmetry axis in time t, $D_{00}^{(2)}(\theta) = \frac{1}{2}(3\cos^2\theta - 1)$ and $\omega = \partial\theta(t)/\partial t|_{t=0}$ is the initial angular velocity. The last step in Eq. (3.123) follows because at $t = 0$, $\theta = 0$. From the principle of equipartition of energy, we have

$$\langle\omega^2\rangle = \langle\omega_x^2 + \omega_y^y\rangle = 2kT/I_\perp, \tag{3.124}$$

where $I_\perp$ is the principal moment of inertia perpendicular to the symmetry axis. Thus, we obtain

$$M_2^{(2)} = 3\,\langle\omega^2\rangle = 6kT/I_\perp. \tag{3.125}$$

Clearly, the second moment of $C_r(t)$ depends only on the moment of inertia and the temperature of the system (Gordon, 1968). This is expected, as was shown in Chapter 1. The second moment is caused entirely by the kinetic effect, which in the classical system is determined entirely by the temperature of the thermodynamic system. Thus, the intermolecular-interaction potential plays no role in causing the initial decay of $C_r(t)$.

The fourth moment will be affected by the intermolecular-interaction potential (Gordon, 1968). From the definition of the fourth moment, we have

$$M_4^{(2)} = \partial^4 C_r(t)/\partial t^4|_{t=0} = \tfrac{3}{2}\,\partial^4/\partial t^4\langle\cos^2\theta(t)\rangle|_{t=0}. \tag{3.126}$$

Now, taking the fourth time derivative and evaluating the result at $t = 0$ (when $\theta = 0$), we obtain

$$M_4^{(2)} = 12\,\langle\omega^4\rangle - 9\,\langle\dot{\omega}^2\rangle - 12\,\langle\omega\ddot{\omega}\rangle = 12\,\langle\dot{\omega}^4\rangle + 3\,\langle\dot{\omega}^2\rangle, \tag{3.127}$$

where $\dot{\omega} = \partial^2\theta(t)/\partial t^2$, evaluated at $t = 0$. The second equality follows because of the identity

$$\langle\omega\ddot{\omega}\rangle = -\langle\dot{\omega}^2\rangle. \tag{3.128}$$

A straightforward thermodynamic calculation gives $\langle\omega^4\rangle = 8(kT/I_\perp)^2$. Thus,

$$M_4^{(2)} = 96\,(kT/I_\perp) + 3/I_\perp{}^2\,\langle\mathbf{L}^2\rangle \tag{3.129}$$

where $\mathbf{L}$ is the molecular torque and is equal to $I_\perp\dot{\omega}$.

In contrast to the second moment, we see that, in addition to the kinetic term given by $96(kT/I_\perp)^2$, the fourth moment also contains a term owing to the intermolecular interaction. This is consistent with the result we obtained in Chapter 1, where we found that the fourth moment is the first one to be affected by the dynamics of the intermolecular interaction.

The higher moments are increasingly sensitive to the intensity of the far wings of the band, which are further affected by overlap with other bands and by collision-induced effects. The data for the higher moments, being subject to significant experimental uncertainty, are not so meaningful as to warrant comparison with the theory. In conclusion, we have found that the rotational second moment is invariant with molecular motion. The second-moment calculation from experimental data thus gives a stringent test of the accuracy of the procedure used to extract orientational information from raw spectral data. The fourth moment depends on the mean-square torque exerted on a molecule and, in favorable cases, it can be used to provide useful estimates of this quality from spectra.

3.8 STOCHASTIC THEORY OF MOLECULAR REORIENTATION

In the previous section we discussed the behavior of the correlation function $C_r(t)$ over short time intervals. The results that followed Chapter 1 can be conveniently converted to an expression in terms of the cumulants by

$$C_r(t) = \exp[(K_2/2!)\, t^2 + (K_4/4!)\, t^4 + \cdots], \tag{3.130}$$

where K_2 and K_4 are the second and fourth cumulants given by

$$K_2 = M_2^{(2)} = 3\langle \omega^2 \rangle = 6kT/I_\perp = \Delta^2, \tag{3.131}$$

$$K_4 = M_4^{(2)} - 3(M_2^{(2)})^2 = 3\langle \dot{\omega}^2 \rangle. \tag{3.132}$$

The modulation coefficient ξ is given by

$$\xi = K_4/K_2^2 = 3\langle \dot{\omega}^2 \rangle/\Delta^4 = 1/\Delta^2 \tau_c'^2 \tag{3.133}$$

where the time constant τ_c' is given by

$$\tau_c' = (\Delta^2/3\langle \dot{\omega}^2 \rangle)^{1/2} \tag{3.134}$$

Using this equation, we can distinguish two cases:

(1) Slow modulation. The axis of molecular rotation is not interrupted by intermolecular torque; i.e., $\Delta\tau_c' \gg 1$, and thus ξ is small. The correlation function can be described approximately by a Gaussian function.

(2) Fast modulation. Molecular reorientation occurs rapidly, as the intermolecular torque constantly changes the axis of rotation. We then have $\Delta\tau_c' \ll 1$. Thus ξ is large, and the time-correlation function will have the shape of an exponential or a sum of exponentials.

We shall now consider the fast modulation case. Although the method to be developed here can be applied to the general type of orientation time-correlation functions, we shall first consider specifically the calculation of $C_r(t)$, to illustrate the various approximations used in developing the theory.

Using the spherical-harmonics theorem [Eq. 3.28], we can write $C_r(t)$ for a linear molecular system as

$$C_r(t) = \langle D^{(2)}_{00}[\delta\Omega(t)] \rangle = \langle P_2(\hat{U}(t) \cdot \hat{U}) \rangle = \frac{4\pi}{5} \left\langle \sum_{m=-2}^{2} Y_{2m}[\Omega(t)]\, Y^*_{2m}(\Omega) \right\rangle, \tag{3.135}$$

where $\Omega(t)$ and Ω are a set of polar and azimuthal angles that denote the orientation of the molecular axis at times t and 0, respectively. The calculation of $C_r(t)$ can be considered as a problem of calculating the conditional-probability function $P(\Omega', t \mid \Omega, 0)$. This is defined as the probability of finding a molecular axis with an orientation given by Ω' at time t, assuming that its orientation is at Ω at time 0. In terms of the conditional-probability function, $C_r(t)$ can be written as

$$C_r(t) = \frac{1}{4\pi} \sum_m \int_{\Omega'} \int_{\Omega} Y_{2m}(\Omega')\, P(\Omega', t \mid \Omega, 0)\, Y^*_{2m}\, d\Omega'\, d\Omega, \tag{3.136}$$

where the integrations are to be carried out over all angles specified by Ω' and Ω.

We want to find the equation that is obeyed by the probability function P. If this equation is found, then the calculation of $C_r(t)$ can be performed by first solving the equation, plugging the result into Eq. (3.315), and then integrating over the angles Ω' and Ω.

Consider first the probability-distribution function $F(\Omega', t)$. This is defined as the probability of finding the molecular axis $\hat{U}$ within the orientation specified by Ω' and $\Omega' + d\Omega'$, i.e.,

$$P(\Omega' < \Omega(t) < \Omega' + d\Omega') = F(\Omega', t). \tag{3.137}$$

Clearly, the time rate of change of $F[\Omega(t), t]$ can be written as

$$\partial F(\Omega', t)/\partial t' = \boldsymbol{\omega}(t) \cdot (\partial F(\Omega', t)/\partial \Omega') = \boldsymbol{\omega}(t) \cdot \nabla_{\Omega'} F(\Omega', t), \tag{3.138}$$

where $\boldsymbol{\omega}(t)$ is the angular velocity given by $\partial\Omega(t)/\partial t$, and $\nabla_{\Omega'}$ is a gradient operator over a unit sphere. The initial condition for $F(\Omega', t)$ is obvious, as shown here:

$$F(\Omega', 0) = \delta(\Omega' - \Omega). \tag{3.139}$$

When reorientation occurs in a very small step, approximation can be made by assuming that $\partial/\partial\Omega'$ (or $\nabla_{\Omega'}$) is independent of $\boldsymbol{\omega}(t)$ so that when Eq. (3.137) is integrated it gives

$$F(\Omega', t) = \exp\left[\int_0^t dt' \boldsymbol{\omega}(t') \cdot \nabla_{\Omega'} \right] \delta(\Omega' - \Omega). \tag{3.140}$$

This is the formal solution to Eq. (3.138) for a given $\boldsymbol{\omega}(t)$. The conditional-probability function is the expectation value of $F(\Omega', t)$ for a distribution of $\omega(t)$. Since $\omega(t)$ is assumed to be a stationary stochastic process, the probability of finding the molecule at Ω' at time t after it has been at Ω at time $t = 0$ is given by

$$P(\Omega', t\,|\,\Omega, 0) = \langle F(\Omega', t)\rangle = \left\langle \exp\left[\int_0^t dt'\, \boldsymbol{\omega}(t') \cdot \boldsymbol{\nabla}_{\Omega'}\right]\right\rangle \delta(\Omega' - \Omega). \quad (3.141)$$

We shall next assume that $\boldsymbol{\omega}(t)$ is a Gaussian process. The central-limit theorem then gives [compare with Eq. (1.200)]

$$\left\langle \exp\left[\int_0^t dt'\, \boldsymbol{\omega}(t') \cdot \boldsymbol{\nabla}_{\Omega'}\right]\right\rangle = \exp\left[\int_0^t dt'\, \langle \boldsymbol{\omega}(t')\rangle_c \cdot \boldsymbol{\nabla}_{\Omega'} + \frac{1}{2!}\int_0^t dt_1 \int_0^t dt_2\, \langle \boldsymbol{\omega}(t_1)\boldsymbol{\omega}(t_2)\rangle_c : \boldsymbol{\nabla}_{\Omega'}\boldsymbol{\nabla}_{\Omega'}\right], \quad (3.142)$$

where the dyad product is defined as

$$\mathbf{AB}:\mathbf{CD} = (\mathbf{A} \cdot \mathbf{C})(\mathbf{B} \cdot \mathbf{D}). \quad (3.143)$$

For simplicity, we shall assume that $\langle \boldsymbol{\omega}(t)\rangle = 0$. This is clearly valid in the condensed phase because free molecular rotation does not take place. Furthermore, since $\boldsymbol{\omega}(t)$ is a stationary stochastic process, we have [compare with Eq. (1.201)]

$$\int_0^t dt_1 \int_0^t dt_2 \langle \boldsymbol{\omega}(t_1)\boldsymbol{\omega}(t_2)\rangle_c = 2\int_0^t d\tau\,(t - \tau)\,\langle \boldsymbol{\omega}(\tau)\boldsymbol{\omega}\rangle. \quad (3.144)$$

The conditional-probability function then becomes

$$P(\Omega', t\,|\,\Omega, 0) = \exp[\mathbf{B}(t):\boldsymbol{\nabla}_{\Omega'}\boldsymbol{\nabla}_{\Omega'}]\,\delta(\Omega' - \Omega), \quad (3.145)$$

where

$$\mathbf{B}(t) = \int_0^t d\tau\,(t - \tau)\,\langle \boldsymbol{\omega}(\tau)\boldsymbol{\omega}\rangle. \quad (3.146)$$

3.9 ROTATIONAL-DIFFUSION MODEL

We shall now examine the rotational-diffusion model by assuming that the motion is caused by a series of random reorientational steps (which may be of a different magnitude around different axes). If the angular-velocity–

correlation function decays extremely rapidly, then the upper integration limit in Eq. (3.146) can be extended to ∞. In this eventuality, the tensor function $\mathbf{B}(t)$ approaches $\mathbf{R}t$, where $\mathbf{R}$ is a time-independent quantity given by

$$\mathbf{R} = \int_0^\infty d\tau \, \langle \boldsymbol{\omega}(\tau)\boldsymbol{\omega} \rangle, \tag{3.147}$$

which is known as the rotational-diffusion tensor. Using this result and then taking the time derivative of the conditional-probability function given in Eq. (3.145), we obtain the Debye rotational-diffusion equation as (Farro, 1960)

$$\partial P/\partial t = (\mathbf{R}:\nabla_{\Omega'}\nabla_{\Omega'})P. \tag{3.148}$$

The form of this equation depends on the coordinate system we choose to express $\mathbf{R}$. If the molecular-coordinate system is chosen so that the diffusion-constant tensor $\mathbf{R}$ is diagonal with three principal values given by $R_{ii}(i = 1, 2, 3)$, then Eq. (3.148) becomes

$$\partial P/\partial t = \sum_i R_{ii} \, (\partial^2 P/\partial \Omega_i^2). \tag{3.149}$$

Thus, the Debye equation has the same form as the Schrödinger equation for an asymmetric rotator and can also be written in terms of the dimensionless quantum-mechanical angular-momentum operator $\mathbf{J}$ by

$$\partial P/\partial t = -(R_{xx}\hat{\mathbf{J}}_x^2 + R_{yy}\hat{\mathbf{J}}_y^2 + R_{zz}\hat{\mathbf{J}}_z^2)P, \tag{3.150}$$

where the rotational-diffusion constants R_{ii} are related to the fluid's viscosity through the Stokes law.

Before proceeding to solve the rotational-diffusion equation, it should be emphasized that the equation holds only if the condition

$$\langle \omega_i^2 \rangle^{1/2} \tau_i (\partial P/\partial \Omega_{ii}) \ll 1, \qquad i = 1, 2, 3, \tag{3.151}$$

is valid. Here τ_i is the correlation time for the angular-velocity correlation function $\langle \omega_i(\tau)\omega_i \rangle$. Or, expressing it in another way, Eq. (3.151) means that

$$\varpi_i \tau_i < \delta\Omega_i, \tag{3.152}$$

where ϖ_i is the root-mean-square angular velocity, given by $\langle \omega_i^2 \rangle^{1/2}$, and $\delta\Omega_i$ is the total orientation change within the time interval that the diffusion process occurs. Therefore, if molecular reorientations occur continuously by infinitesimal rotations, as in the case of reorientations of sufficiently large molecules in solutions, then we expect that the rotational-diffusion equation given by Eq. (3.149) will be valid.

We shall now seek the solution of the Debye rotational-diffusion equation. The solution can be obtained by the expansion of P in terms of the eigenfunctions of the operator $\mathscr{H}$ defined by

$$\mathscr{H} = \mathbf{J} \cdot \mathbf{R} \cdot \mathbf{J} = R_{xx}\mathbf{J}_x^2 + R_{yy}\mathbf{J}_y^2 + R_{zz}\mathbf{J}_z^2. \tag{3.153}$$

Note that $\mathscr{H}$ becomes identical with the rotational kinetic-energy part of the Hamiltonian if R_{ii} is replaced by $-\hbar^2/2I_i$, I_i being the moment of inertia along the ith principal axis. Thus $\{\psi_{vm}^{(j)}(\Omega')\}$ is the set of orthonormal eigenfunctions of $\mathscr{H}$, with eigenvalues E_{jm} (the eigenvalues do not depend on v), i.e.,

$$\mathscr{H}\psi_{vm}^{(j)}(\Omega') = E_{jm}\psi_{vm}^{(j)}(\Omega'). \tag{3.154}$$

Thus we have

$$\sum_{jmv} \psi_{vm}^{(j)}(\Omega')\, \psi_{vm}^{(j)*}(\Omega) = \delta(\Omega' - \Omega), \tag{3.155}$$

where Eq. (3.155) is the closure relation for any complete set of orthonormal functions. The formal solution of Eq. (3.150) can be written as

$$P(\Omega', t\,|\,\Omega, 0) = e^{-t\mathscr{H}} P(\Omega', 0\,|\,\Omega, 0), \tag{3.156}$$

where the operator $\mathscr{H}$ acts only on Ω', and the initial condition is

$$P(\Omega', 0\,|\,\Omega, 0) = \delta(\Omega' - \Omega). \tag{3.157}$$

Substituting Eq. (3.155) into (3.156), we obtain

$$P(\Omega', t\,|\,\Omega, 0 = e^{-t\mathscr{H}} \sum_{jmv} \psi_{v,m}^{(j)}(\Omega')\, \psi_{v,m}^{(j)*}(\Omega) = \sum_{jmv} \psi_{v,m}^{(j)}(\Omega')\, \psi_{v,m}^{(j)*}(\Omega) \exp(-tE_{jm}). \tag{3.158}$$

The last step of Eq. (3.158) follows from Eq. (3.154); it can be seen that

$$\exp(-t\mathscr{H})\, \psi_{v,m}^{(j)}(\Omega') = \exp(-tE_{jm})\, \psi_{v,m}^{(j)}(\Omega').$$

In the special case when $R_{xx} = R_{yy} = R_{zz} = R$, the operator $\mathscr{H}$ becomes $R\hat{J}^2$ and the eigenfunctions $\psi_{v,m}^{(j)}(\Omega')$ reduce to the spherical harmonics $Y_{jm}(\Omega')$, with eigenvalues $E_{jm} = R_j(j+1)$. The conditional-probability function reduces to

$$P(\Omega', t\,|\,\Omega, 0) = \sum_{jm} Y_{jm}(\Omega')\, Y_{jm}^*(\hat{\Omega}) e^{-j(j+1)Rt}. \tag{3.159}$$

For a symmetric-top diffusor, $R_{xx} = R_{yy} = R_\perp$ and $R_{zz} = R_\parallel$, $(R_\parallel \neq R_\perp)$. The operator $\mathscr{H}$ can be written as $\mathscr{H} = R_\perp \hat{J}^2 + (R_\parallel - R_\perp)\hat{J}_z^2$, and the eigenfunctions of $\mathscr{H}$ are the rotational matrices $[(2j+1)/8\pi^2]^{1/2}\, D_{mv}^{(j)}(\Omega')$

associated with the eigenvalues given by $E_{jm} = R_\perp j(j+1) + (R_\parallel - R_\perp)m^2$.

The conditional-probability density is, therefore, equal to

$$P(\Omega', t\,|\,\Omega, 0) = \frac{1}{8\pi^2} \sum_{jmv} (2j+1) D_{mv}^{(j)}(\Omega') D_{mv}^{(j)*}(\Omega) \times \exp\{-t[R_\perp j(j+1) + (R_\parallel - R_\perp)m^2]\}, \tag{3.160}$$

where P is normalized to

$$\iint d\Omega\, d\Omega'\, P(\Omega', 0\,|\,\Omega, 0) = 8\pi^2. \tag{3.161}$$

The general asymmetric-top diffusor has $R_{xx} \neq R_{yy} \neq R_{zz}$. The eigenfunctions of $\mathscr{H}$ in this case cannot be written in terms of simple, known analytical functions for an arbitrary j. However, according to the standard scheme, the asymmetric-top eigenfunctions can be found by forming linear combinations of the symmetric-top eigenfunctions

$$\psi_{v,m}^{(j)}(\Omega') = \sum_k C_{km}^{(j)} \left(\frac{2j+1}{8\pi^2}\right)^{1/2} D_{kv}^{(j)}(\Omega'). \tag{3.162}$$

Except for small values of j, the coefficients $C_{km}^{(j)}$ and the associated eigenvalues of $\psi_{v,m}^{(j)}(\Omega')$ cannot be written in closed form. We now show how the C coefficients and the eigenvalues are obtained for small j values.

The operator $\mathscr{H}$ for the asymmetric-top diffusor can be written as (Wang, 1929)

$$\begin{aligned}\mathscr{H} = c[&-(a + b\cos 2\gamma)(\sin\beta)^{-1}(\partial/\partial\beta)(\sin\beta(\partial/\partial\beta)\\ &-(a - b\cos 2\gamma)(\sin^2\beta)^{-1}(\partial^2/\partial\alpha^2 - 2\cos\beta(\partial^2/\partial\gamma\,\partial\alpha + \partial^2/\partial\gamma^2)\\ &-(1 + b\cos 2\gamma)(\partial^2/\partial\gamma^2 - 2b\sin 2\gamma/\sin\beta)\\ &\times(\partial^2/\partial\beta\,\partial\gamma - \cos\beta\,\partial^2/\partial\beta\,\partial\gamma)\\ &+ 2b\cos 2\gamma\cot\beta(\partial/\partial\beta) - b\sin 2\gamma(1 + 2\cot^2\beta)(\partial/\partial\gamma)\\ &+ 2b\sin 2\gamma\cot\beta\csc\beta(\partial/\partial\alpha],\end{aligned} \tag{3.163}$$

where

$$\begin{aligned} a &= (R_{xx} + R_{yy})/2c\\ b &= (R_{xx} - R_{yy})/2c,\\ c &= R_{zz} - \tfrac{1}{2}(R_{xx} + R_{yy}).\end{aligned}$$

We want to find the eigenfunction of $\mathscr{H}$; this is accomplished by forming a linear combination of $D_{mk}^{(j)}(\Omega)$ according to

$$\psi(\Omega) = \sum_k C_k \left(\frac{2j+1}{8\pi^2}\right)^{1/2} D_{kv}^{(j)}(\Omega). \tag{3.164}$$

Now, taking the operator $\mathscr{H}$ to act on $\psi(\Omega)$ and using the recursion relation for the rotation matrices

$$(1/c)\mathscr{H}D_{kv}^{(j)} = [aj(j+1)+k^2]D_{kv}^{(j)} + bf(j,k+1)D_{k+2,v}^{(j)} + bf(j,k-1)D_{k-2,v}^{(j)} \tag{3.165}$$

with

$$f(j,k) = -\tfrac{1}{2}\{(j^2-k^2)[(j+1)^2-k^2]\}^{1/2}, \tag{3.166}$$

we obtain

$$\begin{aligned}\mathscr{H}\psi &= c\sum_k C_k\left(\frac{2j+1}{8\pi^2}\right)^{1/2}\{[aj(j+1)+k^2]D_{kv}^{(j)} \\ &\quad + bf(j,k+1)D_{k+2,v}^{(j)} + bf(j,k-1)D_{k-2,v}^{(j)}\} \\ &= E\sum_k C_k\left(\frac{2j+1}{8\pi^2}\right)^{1/2} D_{kv}^{(j)}.\end{aligned} \tag{3.167}$$

The equating for determining C_k can be obtained by eliminating the D functions. This is carried out by multiplying $(2j+1/8\pi^2)^{1/2}D_{kv}^{(j)*}(\Omega)$ and then integrating over $d\Omega$. With the help of the orthogonal property of the D functions given by Eq. (3.24), Eq. (3.167) is reduced to

$$[aj(j+1)+k^2-\varepsilon]C_k + bf(j,k-1)C_{k-2} + bf(j,k+1)C_{k+2} = 0, \tag{3.168}$$

where $\varepsilon = E/c$.

Equation (3.168) indicates that the coefficients C_k and the eigenvalues ε depend implicitly on the value of j. Let

$$ad(j+1) - \varepsilon = -w. \tag{3.169}$$

We rewrite (3.168) as

$$C_k(-w+k^2) + bf(j,k-1)C_{k-2} + bf(j,k+1)C_{k+2} = 0. \tag{3.170}$$

This is the desired equation for determining the eigenfunctions and the associated eigenvalues of $\mathscr{H}$. We shall now find the results for small values of j. For $j = 0$, the only k value is 0, and this gives $w = 0$. For $j = 1$, there are $+1, 0, -1$ for values k, and application of Eq. (3.170) gives a set of three equations:

$$\begin{aligned} C_1(-w+1) - bC_{-1} &= 0,\\ C_0(-w) &= 0,\\ C_{-1}(-w+1) + bC_1 &= 0.\end{aligned} \tag{3.171}$$

From these, we obtain $w = 0$, $w = 1 \pm b$. The corresponding eigenfunctions and eigenvalues of $\mathscr{H}$ are thus given by

$$\begin{aligned}
\psi_{v,1}^{(1)} &= (3/16\pi^2)^{1/2}(D_{1v}^{(1)} - D_{-1v}^{(1)}), \qquad & E_{1,1} &= 3\bar{R} - R_{xx};\\
\psi_{v,0}^{(1)} &= (3/8\pi^2)^{1/2}D_{0v}^{(1)}, & E_{1,0} &= R_{xx} + R_{yy}; \quad (3.172)\\
\psi_{v,-1}^{(1)} &= (3/16\pi^2)^{1/2}(D_{1v}^{(1)} + D_{-1v}^{(1)}), & E_{1,-1} &= 3\bar{R} - R_{yy};
\end{aligned}$$

where $\bar{R} = (R_{xx} + R_{yy} + R_{zz})/3$.

For $j = 2$, there are five allowed values of k: $k = 0, \pm 1, \pm 2$. From Eq. (3.169), we obtain two sets of linear functions:

$$\begin{bmatrix} -w+1 & -3b \\ -3b & -w+1 \end{bmatrix}\begin{bmatrix} C_1 \\ C_{-1} \end{bmatrix} = 0, \tag{3.173}$$

and

$$\begin{bmatrix} -w+4 & -\sqrt{6}b & 0 \\ -\sqrt{6}b & -w & -\sqrt{6}b \\ 0 & -\sqrt{6}b & -w+4 \end{bmatrix}\begin{bmatrix} C_2 \\ C_0 \\ C_{-2} \end{bmatrix} = 0. \tag{3.174}$$

Solving the secular equations obtained from these and then determining the C coefficients, we obtain the eigenfunctions and eigenvalues for the $j = 2$ case as follows:

$$\begin{aligned}
\psi_{v,2}^{(2)} &= (5/8\pi^2)^{1/2}[(1/\sqrt{2})\sin(\theta/2)(D_{2v}^{(2)} + D_{-2v}^{(2)})\\
&\quad + \cos(\theta/2)D_{0v}^{(2)}], \qquad E_{2,2} = 6\bar{R} + 2g;\\
\psi_{v,-2}^{(2)} &= (5/8\pi^2)^{1/2}[(1/\sqrt{2})\cos(\theta/2)(D_{2v}^{(2)} + D_{-2v}^{(2)})\\
&\quad - \sin(\theta/2)D_{0v}^{(2)}], \qquad E_{2,-2} = 6\bar{R} - 2g;\\
\psi_{v,0}^{(2)} &= (5/16\pi^2)^{1/2}(D_{2v}^{(2)} - D_{-2v}^{(2)}), \qquad E_{2,0} = 3(\bar{R} + R_{zz});\\
\psi_{v,1}^{(2)} &= (5/16\pi^2)^{1/2}(D_{1v}^{(2)} - D_{-1v}^{(2)}), \qquad E_{2,1} = 3(\bar{R} + R_{xx});\\
\psi_{v,-1}^{(2)} &= (5/16\pi^2)^{1/2}(D_{1v}^{(2)} + D_{-1v}^{(2)}), \qquad E_{2,-1} = 3(\bar{R} + R_{yy}); \quad (3.175)
\end{aligned}$$

where

$$\begin{aligned}
g &= (R_{xx}^2 + R_{yy}^2 + R_{zz}^2 - R_{xx}R_{yy} - R_{yy}R_{zz} - R_{zz}R_{xx})^{1/2}\\
&= \{\tfrac{1}{2}[(R_{xx} - R_{yy})^2 + (R_{yy} - R_{zz})^2 + (R_{zz} - R_{xx})^2]\}^{1/2},\\
\theta &= \tan^{-1}\sqrt{3}b.
\end{aligned} \tag{3.176}$$

3.10 ORIENTATIONAL TIME-CORRELATION FUNCTIONS OF THE ROTATIONAL-DIFFUSION MODEL

After obtaining the conditional-probability function using the rotational diffusion model, it is a simple matter to calculate the orientational time-correlation function. For example if coupling of the rotational motion to nonrotational degrees of function is neglected, then the expression of $G_{\mathrm{VH}}(t)$ given in Eq. (3.93) can be written as

$$G_{\mathrm{VH}}(t) = \tfrac{1}{4} \sum_{m} \sum_{m'} \langle \alpha(2, m; t)\alpha^*(2, m') \rangle C_{mm'}^{(\mathrm{VH})}(t), \tag{3.177}$$

where

$$C_{mm'}^{(\mathrm{VH})}(t) = \langle \{D_{m1}^{(2)}[\Omega(t)] + D_{m(-1)}^{(2)}[\Omega(t)]\}[D_{m'1}^{(2)*}(\Omega) + D_{m'(-1)}^{(2)*}(\Omega)] \rangle, \tag{3.178}$$

which, when expressed in terms of the conditional probability, becomes

$$C_{mm'}^{(\mathrm{VH})}(t) = \int d\Omega' \, d\Omega \, [D_{m1}^{(2)}(\Omega') + D_{m(-1)}^{(2)}(\Omega')] P(\Omega', t | \Omega, 0) \times [D_{m'1}^{(2)*}(\Omega) + D_{m'(-1)}^{(2)*}(\Omega)] P(\Omega), \tag{3.179}$$

where $P(\Omega)$ is the probability of finding a molecule at an initial orientation specified by Ω. Since the orientational distribution is uniform, we have

$$P(\Omega) = 1/8\pi^2. \tag{3.180}$$

Consider first the case of symmetric-top molecules. We substitute Eq. (3.160) into Eq. (3.179) and obtain

$$C_{mm'}^{(\mathrm{VH})}(t) = \left(\frac{1}{8\pi^2}\right)^2 \sum_{jnv} (2j + 1) \int d\Omega' \, d\Omega \, [D_{m,1}^{(2)}(\Omega') + D_{m',-1}^{(2)}(\Omega')] \times D_{n,v}^{(j)}(\Omega') \, D_{n,v}^{(j)}(\Omega')[D_{m',1}^{(2)*}(\Omega) + D_{m',-1}^{(2)*}(-\Omega)] \times \exp\{-t[R_\perp j(j + 1) + (R_\parallel - R_\perp)M^2]\}. \tag{3.181}$$

We next carry out the integrations over Ω and Ω', and with the help of Eqs. (3.23) and (3.24) we obtain

$$C_{mm'}^{(\mathrm{VH})}(t) = \tfrac{2}{5} \delta_{mm'} \exp\{-t[6R_\perp + (R_\parallel - R_\perp)m^2]\}. \tag{3.182}$$

Thus Eq. (3.177) [or Eq. (3.93)] becomes

$$G_{\mathrm{VH}}(t) = \tfrac{1}{10} \sum_{m} \langle \alpha(2, m; t)\alpha^*(2, m) \rangle \exp\{-t[6R_\perp + (R_\parallel - R_\perp)m^2]\}. \tag{3.183}$$

Carrying out a similar calculation for the reorientational-correlation function in $G_{VV}(t)$, we obtain

$$\langle D^{(2)}_{m0}[\Omega(t)]D^{(2)*}_{m'0}(\Omega)\rangle = \tfrac{1}{5}\,\delta_{mm'}\exp\{-t[6R_\perp + (R_\parallel - R_\perp)m^2]\}, \quad (3.184)$$

and

$$G_{VV}(t) \simeq \langle\bar{\alpha}(t)\bar{\alpha}(0)\rangle + \tfrac{2}{15}\sum_m \langle\alpha(2, m; t)\alpha^*(2, m)\rangle \times \exp\{-t[6R_\perp + (R_\parallel - R_\perp)m^2]\}. \quad (3.185)$$

From Eqs. (3.183) and (3.185), we note that

$$G_{\text{iso}}(t) = G_{VV}(t) - \tfrac{4}{3}G_{VH}(t) = \langle\bar{\alpha}(t)\bar{\alpha}(0)\rangle, \quad (3.186)$$

which was previously obtained using a method independent of the rotational-diffusion model.

Although it appears that $G_{VH}(t)$ consists of three exponentials corresponding to $m = 0, \pm 1, \pm 2$, according to Eq. (3.68) only the polarizability tensor element $\alpha(2, 0) = (2/\sqrt{6})(\alpha_\parallel - \alpha_\perp) = (2/\sqrt{6})\beta$ is nonvanishing. As a result, Eq. (3.183) is simplified to

$$G_{VH}(t) = \tfrac{1}{15}\langle\beta(t)\beta\rangle\exp(-6R_\perp t). \quad (3.187)$$

Measurements of $G_{VH}(t)$ will provide information about $R_\perp$, the rotational-diffusion constant for the perpendicular axis.

The case of asymmetric-top molecules is considerably more complicated. Consider again the calculation of $C^{(VH)}_{mm'}(t)$ for the VH scattering spectrum. Although the conditional-probability function includes the eigenfunctions $\psi^{(j)}_{v,m}$ for all j, v, and m [compare with Eq. (3.158)], to calculate $C^{(VH)}_{mm'}(t)$, we shall need only the set of functions $\{\psi^{(2)}_{v,m}\}$ given in Eq. (3.175), because those with $j \neq 2$ do not make a contribution to $G_{VH}(t)$. Thus, for practical purposes the conditional-probability function appropriate for the calculation of $C^{(VH)}_{mm'}(t)$ can be written as

$$P[\Omega; t(\Omega, 0)] = \sum_{v=2}^{2}\sum_{n=-2}^{2}\int \psi^{(2)}_{v,n}(\Omega')\psi^{(2)*}_{v,n}(\Omega)\exp(-E_{2,n}t). \quad (3.188)$$

The function $C^{(VH)}_{mm'}(t)$ for the asymmetric diffusion is then given by

$$C^{(VH)}_{mm'}(t) = \frac{1}{8\pi^2}\sum_{v=-2}^{2}\sum_{n=2}^{2}\int d\Omega'\,d\Omega\,[D^{(2)}_{m,1}(\Omega') + D^{(2)}_{m,-1}(\Omega')] \psi^{(2)}_{v,n}(\Omega')\psi^{(2)*}_{v,n}(\Omega)[D^{(2)*}_{m',1}(\Omega) + D^{(2)*}_{m',-1}(\Omega)]\exp(-E_{2,n}t). \quad (3.189)$$

Substituting the $\psi_{v,n}^{(2)}$ functions given in Eq. (3.175) into Eq. (3.189) and carrying out the integrations over Ω' and Ω, we obtain $C_{mm'}^{(\mathrm{VH})}(t)$; next, substituting $C_{mm'}^{(\mathrm{VH})}(t)$ into Eq. (3.177) and after some algebraic manipulation, we find

$$C_{\mathrm{VH}}(t) = \tfrac{1}{10} \sum_i A_i \exp(-R_i t), \tag{3.190}$$

where

$$\begin{aligned} A_1 &= \langle |\alpha(2,0)|^2 \rangle \cos^2(\theta/2) + (1/2)\langle [\alpha(2,2) + \alpha(2,-2)] \\ &\quad \times [\alpha^*(2,2) + \alpha^*(2,-2)] \rangle \sin^2(\theta/2) + (1/\sqrt{2})\langle \alpha(2,0) \\ &\quad \times [\alpha^*(2,2) + \alpha^*(2,-2)] \\ &\quad + [\alpha(2,2) + \alpha(2,-2)]\alpha^*(2,0)\rangle \sin(\theta/2)\cos(\theta/2) \\ &= \tfrac{1}{6}\langle |\mathrm{Tr}\,\alpha - 3\alpha_{zz}|^2 \rangle \cos^2(\theta/2) + \langle |\alpha_{xx} - \alpha_{yy}|^2 \rangle \sin^2(\theta/2) \\ &\quad - (1/2\sqrt{3})\langle (\mathrm{Tr}\,\boldsymbol{\alpha} - 3\alpha_{zz})(\alpha_{xx} - \alpha_{yy})\rangle \sin\theta, \end{aligned} \tag{3.190a}$$

$$R_1 = E_{2,2} = 6\bar{R} + 2g; \tag{3.190b}$$

$$\begin{aligned} A_2 &= \langle |\alpha(2,0)|^2 \rangle \cos^2(\theta/2) + \tfrac{1}{2}[\alpha(2,2) + \alpha(2,-2)] \\ &\quad \times [\alpha^*(2,2) + \alpha^*(2,-2)] \rangle \sin^2(\theta/2) - (1/\sqrt{2}) \\ &\quad \times \alpha(2,0)[\alpha^*(2,2) + \alpha^*(2,-2)] \\ &\quad + [\alpha(2,2) + \alpha(2,-2)]\alpha^*(2,0)\rangle \sin(\theta/2)\cos(\theta/2) \\ &= \tfrac{1}{6}\langle |\mathrm{Tr}\,\boldsymbol{\alpha} - 3\alpha_{zz}|^2 \rangle \cos^2(\theta/2) + \langle |\alpha_{xx} - \alpha_{yy}|^2 \rangle \sin^2(\theta/2) \\ &\quad + (1/2\sqrt{3})\langle (\mathrm{Tr}\,\boldsymbol{\alpha} - 3\alpha_{zz})(\alpha_{xx} - \alpha_{yy})\rangle \sin\theta, \end{aligned} \tag{3.191a}$$

$$R_2 = E_{2,2} = 6\bar{R} - 2g; \tag{3.191b}$$

$$A_3 = \tfrac{1}{2}\langle [\alpha(2,2) - \alpha(2,-2)][\alpha^*(2,2) - \alpha^*(2,-2)] \rangle = \tfrac{1}{2}\langle |\alpha_{xy}|^2 \rangle, \tag{3.192a}$$

$$R_3 = E_{2,0} = 3(\bar{R} + R_{zz}); \tag{3.192b}$$

$$A_4 = \tfrac{1}{2}\langle [\alpha(2,1) - \alpha(2,-1)][\alpha^*(2,1) - \alpha^*(2,-1)] \rangle = 2\langle |\alpha_{zy}|^2 \rangle, \tag{3.193a}$$

$$R_4 = E_{2,1} = 3(\bar{R} + R_{xx}); \tag{3.193b}$$

$$A_5 = \tfrac{1}{2}\langle [\alpha(2,1) + \alpha(2,-1)][\alpha^*(2,1) + \alpha^*(2,-1)] \rangle = 2\langle |\alpha_{yz}|^2 \rangle, \tag{3.194a}$$

$$R_5 = E_{2,-1} = 3(\bar{R} + R_{yy}). \tag{3.194b}$$

Thus, the correlation function $C_{\mathrm{VH}}(t)$ of an asymmetric-top molecule consists, in general, of five exponentials, and hence the VH spectrum consists of five Lorentzian functions. However, the result is considerably simplified if the principal axes of the diffusion tensor coincide with the principal axes of the polarizability tensor. In this case, $A_3 = A_4 = A_5 = 0$ because in the principal coordinate system the polarizability tensor has no off-diagonal elements and the VH spectrum will then consist of two lorentzians (Pecora, 1968). Asymmetric molecules with C_2 or C_{2v} symmetry will satisfy this condition.

We have considered the time-correlation function $C_{\mathrm{VH}}(t)$ in detail. The calculation serves as an example for the application of the anisotropic rotational-diffusion model. In other kinds of spectroscopy which are sensitive to the molecular-reorientational process, time-correlation functions of the type $\langle D_{mn}^{(j)}[\Omega(t)]\, D_{m'n'}^{(j)*}(\Omega)\rangle$ are encountered, and so long as the rotational-diffusion model is applicable, the technique developed in this section can be readily applied to carry out the calculations. The rotational-diffusion model has been extensively used to investigate the nuclear magnetic-relaxation spectroscopy of molecular-reorientational processes in the liquid state (Huntress, 1968).

3.11 OTHER MOLECULAR-REORIENTATION MODELS

Although the anisotropic rotational-diffusion model provides an explicit analytical expression of the time-correlation function for reorientation, this model is only applicable when molecules undergo many collisions before reorienting through an appreciable angle. Unfortunately, the rotational-diffusion model does not agree very well with experiments or with molecular-dynamics computer simulations. For example, for linear or spherical-top molecules, the rotational-diffusion model predicts that

$$\langle P_l[\hat{U}(t) \cdot \hat{U}]\rangle = \exp[-l(l+1)Rt] = \exp(-t/\tau_l), \tag{3.195}$$

where $\tau_l = 1/l(l+1)R$. So $\tau_2/\tau_1 = \frac{1}{3}$. In reality, the decay rates obtained for $\langle P_1[\hat{U}(t) \cdot \hat{U}]\rangle$ and for $\langle P_2[\hat{U}(t) \cdot \hat{U}]\rangle$ do not conform to this proportionality. Moreover, for short time intervals, the decay of the time-correlation function is not exponential (Steele, 1976, 1980).

Gordon (1966) proposed what is called "extended-diffusion" model for the reorientational processes of diatomic molecules in the liquid state. Gordon's model assumes that molecules in a liquid are in collision-interrupted free rotation. A collision is defined as an event that changes the angular momentum of a molecule. This model has several alternative forms. In M diffusion, a molecule terminates a free-rotational step by a collision that randomizes the components of the angular-momentum vector of the molecule. Both the magnitude of the angular-momentum vector and the spatial orientation of the molecule are assumed to be unchanged by collision. On the other hand, in J diffusion, a collision randomizes both the components and the amplitude of the angular-momentum vector. After collision, the distribution of the magnitudes of the angular-momentum vector is assumed to be given by a Boltzmann distribution function. In effect, in J diffusion, a collision destroys the memory of the orientational function. The extended-diffusion model for linear molecules was worked out by Gordon, and that for symmetric-top molecules was simultaneously worked out by McClung (1972) and by St. Pierre and Steele (1972).

We shall first consider the time-correlation function $\langle \hat{U}(t) \cdot \hat{U} \rangle$ of a system of linear molecules. Let $\hat{U}(0)$ denote the direction of the molecular axis at time $t = 0$, and $\hat{U}(t_1)$, that at time $t = t_1$. In the absence of collision, we have $\hat{U}(t_1) \cdot \hat{U} = \cos \omega_1 t_1$, where ω_1 is the angular velocity during the free-rotation interval $(0, t_1)$. If, however, the first free-rotational period is terminated by a collision that occurs at time t_1, then according to the spherical-harmonics addition theorem, the subsequent value of $\hat{U}(t) \cdot \hat{U}$ during the next free-rotation period is given by (see Fig. 3.3)

$$[\hat{U}(t) \cdot \hat{U}]_2 = \cos \omega_2(t - t_1) \cos \omega_1 t_1 + \cos \alpha \sin \omega_2(t - t_1) \sin \omega_1 t_1, \tag{3.196}$$

where ω_2 is the angular velocity during the second period of free rotation and α the angle between the angular-momentum vector of the first and second rotational steps. Since the direction of the angular-momentum vector is randomized by collisions at each collision step, we have $\langle \cos \alpha \rangle = 0$, and thus

$$\langle \hat{U}(t) \cdot \hat{U} \rangle_2 = \cos \omega_2(t - t_1) \cos \omega_1 t_1. \tag{3.197}$$

Since the collision that ended the first-rotation step could have occurred with equal probability at any time between 0 and t, instead of using Eq. (3.197), we should have

$$\langle \hat{U}(t) \cdot \hat{U} \rangle_2 = \frac{1}{t} \int_0^t dt_1 \cos \omega_2(t - t_1) \cos \omega_1 t_1. \tag{3.198}$$

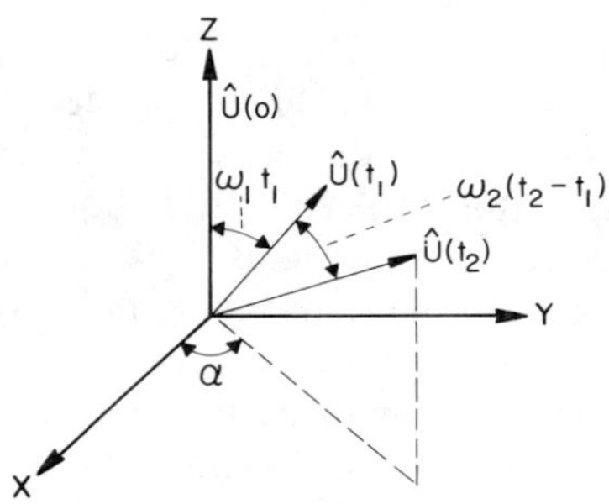

FIG. 3.3 Molecular reorientational geometry. $\hat{U}(0)$ represents the initial orientation of the molecule. The first collision occurs at t_1 and the second collision at t_2, when the orientation of the molecule is at $U(t_1)$ and at $U(t_2)$, respectively.

Following this reasoning, we can write down the general term for a molecule that has suffered n collisions as

$$\langle \hat{U}(t) \cdot \hat{U} \rangle_{n+1} = \frac{n!}{t^n} \int_0^t dt_n \cos \omega_{n+1}(t - t_n) \int_0^{t_n} dt_{n-1} \cos \omega_n(t_n - t_{n+1}) \cdots$$
$$\times \int_0^{t_3} dt_2 \cos \omega_3(t_3 - t_2) \int_0^{t_2} dt_1 \cos \omega_2(t_2 - t_1) \cos \omega_1 t_1. \tag{3.199}$$

Not all the collision steps considered within the time interval between 0 and t will occur with the same probability. If it is assumed that the probability of the occurrence of the $(n + 1)$ step is given by the Poisson distribution function $P(n, t)$,

$$P(n, t) = (1/n!)(t/\tau)^n \exp(-t/\tau), \tag{3.200}$$

where τ is the mean time between collisions, then the time-correlation function is the sum of all $(n + 1)$ steps weighted by the Poisson distribution function, i.e.,

$$C_1(t) \equiv \langle \hat{U}(t) \cdot \hat{U} \rangle = \sum_{n=0}^{\infty} P(n, t) \langle \hat{U}(t) \cdot \hat{U} \rangle_{n+1}. \tag{3.201}$$

For M diffusion, $\Omega_1 = \Omega_2 = \cdots = \Omega_n = \Omega$, and Ω is given by a Boltzmann distribution function. In this case,

$$C_1(t) = \int d\Omega\, f(\Omega) \exp(-t/\tau)$$
$$\times \sum_{n=0} \frac{1}{\tau^n} \int_0^t dt_n \cos \Omega(t - t_n) \int_0^t dt_{n-1} \cos \Omega(t_n - t_{n-1}) \cdots$$
$$\times \int_0^{t_3} dt_2 \cos \Omega(t_3 - t_2) \int_0^{t_2} dt_1 \cos \Omega(t_2 - t_1) \cos \Omega t_1, \tag{3.202}$$

where

$$f(\Omega) = (I/kT)\Omega \exp(-I\Omega^2/2kT), \tag{3.203}$$

I being the moment of inertia.

Consider next the J-diffusion model. In this case, both the amplitude and the direction of the angular velocity are randomized. Thus, instead of Eq. (3.198), we have

$$\langle \hat{U}(t) \cdot \hat{U} \rangle_2 = \frac{1}{t} \int_0^t dt_1 \overline{\cos \omega_2(t - t_1)}\; \overline{\cos \omega_1 t_1}$$
$$= \frac{1}{t} \int_0^t dt_1 F(t - t_1) F(t_1), \tag{3.204}$$

where

$$F(t) = \overline{\cos \omega t} = \int_0^\infty d\omega\, f(\omega) \cos \omega t. \tag{3.205}$$

Similarly, Eq. (3.202) is replaced by

$$C_1(t) = \sum_{n=0}^{\infty} \frac{1}{\tau^n} e^{-t/\tau} \left\{ \int_0^t dt_n\, F(t - t_n) \int_0^{t_n} dt_{n-1} F(t_n - t_{n-1}) \right.$$
$$\left. \times \int_0^{t_2} dt_1\, F(t_2 - t_1) F(t_1) \right\}. \tag{3.206}$$

The time-correlation function given here is expressed in terms of a multiple-time convolution integral. According to the convolution theorem, the Laplace transform of Eq. (3.206) is

$$\tilde{C}_1(s) = \sum_{n=0}^{\infty} \frac{1}{\tau^n} [F(s + \tau^{-1})]^{n+1} = \frac{\tilde{F}(s + \tau^{-1})}{1 - \tau^{-1} \tilde{F}(s + \tau^{-1})}, \tag{3.207}$$

where, to obtain the second step, we have summed the geometric series. The functions $\tilde{C}_1(s)$ and $\tilde{F}(s + \tau^{-1})$ are given by

$$\tilde{C}_1(s) = \int_0^\infty dt\, e^{-st} C_1(t) \tag{3.208a}$$

and

$$\tilde{F}(s + \tau^{-1}) = \int_0^\infty dt \exp[-(s + \tau^{-1})t] F(t). \tag{3.208b}$$

The correlation time τ_1 is the area under the correlation function $C_1(t)$, and is given by

$$\tau_1 = \int_0^\infty dt\, C_1(t) = \lim_{s \to 0} \tilde{C}_1(s) = \frac{\tilde{F}(\tau^{-1})}{[1 - \tau^{-1} \tilde{F}(\tau^{-1})]}. \tag{3.209}$$

Thus, by finding $\tilde{F}(\tau^{-1})$ using Eqs. (3.208b) and (3.205) we can evaluate τ_1.

The correlation function $C_l(t)$ given by

$$C_l(t) = \langle P_l[\hat{U}(t) \cdot \hat{U}] \rangle \tag{3.210}$$

can also be evaluated using the M or J-diffusion model. We shall now consider the J diffusion model. If we let $C_l^{(0)}(t)$ denote the free-rotation correlation function and $C_l^{(1)}(t)$ the correlation function for molecules that have suffered one collision, then

$$C_l^{(1)}(t) = e^{-t/\tau} C_l^{(0)}(t). \tag{3.211}$$

We next note that $C_l^{(n)}(t)$ can be written as a convolution integral.

$$C_l^{(n)}(t) = \tau^{-1} \int_0^t dt' \, C_l^{(1)}(t - t')C_l^{(n-1)}(t'), \tag{3.212}$$

where $C_l^{(n)}(t)$ represents the correlation function for those molecules that have suffered n collisions from time 0 to t, and dt'/τ is the fraction of the molecules that undergo one collision during the time between t' and $t' + dt'$. Equation (3.212) represents a generalization of Eq. (3.204) and is correct because, in the extended-diffusion model, successive collisions are assumed to be uncorrelated.

The Laplace transform of Eq. (3.212) gives

$$\begin{aligned} \tilde{C}_l^{(n)}(s) &= (1/\tau)\tilde{C}_l^{(1)}(s)C_l^{(n-1)}(s) \\ &= (1/\tau)\tilde{C}_l^{(n-1)}(s)\tilde{C}_l^{(0)}(s + \tau^{-1}), \end{aligned} \tag{3.213}$$

where the second step follows from Eq. (3.211).

Interaction with Eq. (3.213) gives

$$\begin{aligned} \tilde{C}_l^{(1)}(s) &= (1/\tau)[\tilde{C}_l^{(0)}(s + \tau^{-1})]^2, \\ \tilde{C}_l^{(2)}(s) &= (1/\tau^2)[\tilde{C}_l^{(0)}(s + \tau^{-1})]^3, \\ \tilde{C}_l^{(n)}(s) &= (1/\tau^n)[\tilde{C}_l^{(0)}(s + \tau^{-1})]^{n+1}. \end{aligned} \tag{3.214}$$

The value of $\tilde{C}_l(s)$ can be obtained by summing all of $\tilde{C}_l^{(n)}(s)$, $n = 1, 2, \ldots, \infty$, i.e.,

$$\tilde{C}_l(s) = \sum_{n=1}^{\infty} \tilde{C}_l^{(n)}(s) = \frac{\tilde{C}_l^{(0)}(s + \tau^{-1})}{1 - \tau^{-1}\tilde{C}_l^{(0)}(s + \tau^{-1})}, \tag{3.215}$$

which is similar to Eq. (3.207) for $\tilde{C}_1(s)$. The correlation time τ_l is thus given by

$$\tau_l = \lim_{s \to 0} \tilde{C}_l(s) = \frac{\tilde{C}_l^{(0)}(\tau^{-1})}{1 - \tau^{-1}\tilde{C}_l^{(0)}(\tau^{-1})}. \tag{3.216}$$

Much work interpreting the infrared and Raman spectral line shapes of molecular liquids has utilized the extended-diffusion model (Steele, 1976, 1980). The procedure commonly used to test the model is as follows: Using the experimentally determined time-correlation function, one first calculates the correlation time τ_l for a particular l by integrating the area under the time-correlation function. Given τ_l, τ is evaluated using Eq. (3.216). From the τ value, $C_l(t)$ is computed and the result is compared with the experimental time-correlation function.

Extensive studies of the extended-diffusion model have indicated that this model is most realistic in dilute gases in which the molecules actually undergo collision-interrupted free rotation. The model does not appear to be adequate for describing the reorientational process of molecules in the liquid state when the hindrance to rotation is important.

It should be noted that $\tilde{C}_l(s)$, and hence the spectral power density, depend on the parameter τ as well as on the free-rotor time-correlation function $C_l^{(0)}(t)$. The free-rotor time-correlation function decays on a time scale determined by $(kT/I)^{1/2}$, which is the root-mean-square angular speed. It is useful to define a dimensionless parameter β given by

$$\beta = \tau_{\mathrm{f}}/\tau, \tag{3.217}$$

where $\tau_{\mathrm{f}} = (I/kT)^{1/2}$, is the free-rotor reorientation time.

When β is small ($\beta \ll 1$), the rotors undergo many free-rotation cycles before being interrupted by collisions. On the other hand, when β is large ($\beta \gg 1$), the rotors are constantly interrupted by collisions and undergo free rotation for only a small fraction of the free-rotation cycle. For this case, J diffusion is reduced to the rotational-diffusion limit but not necessarily to the Debye limit (Fixman and Rider, 1969).

It is evident that the extended-diffusion model makes a smooth transition to free-rotor motion as τ becomes very large or when $\beta \to 0$. From this model, we can also obtain the correct initial curvatures of $C_l(t)$ for finite β. Thus, in contrast to Debye's diffusion model, the inertial effect is properly included in the extended-diffusion model. However, the extended-diffusion model has not yet been applied to the problem of anisotropic reorientation, primarily because of the severe computational problems involved. A more severe difficulty in the extended-diffusion model is its inconsistency with the result of the rough-sphere molecular-dynamics calculation of O'Dell and Berne (1975). They have shown that the best-fit values of β to the time-correlation functions obtained by computer molecular-dynamics simulation bear no relationship to the mean free time τ in the rough-sphere fluid.

The weight of available evidence indicates that neither the extended-diffusion model nor Debye's rotation model is capable of quantitatively describing the reorienting motion of simple molecules in liquids. Consequently, many researchers have been exploring other approaches to this dynamic problem. One of the most promising lines of attack is devising approximations to the memory functions for the rotational time-correlation functions (Steele, 1976, 1980).

Following the memory-function approach, to be discussed in detail in Chapter 4, we can define a memory function $K_l(t)$ for each rotational time-correlation function $C_l(t)$ by

$$\frac{d}{dt} C_l(t) = -\int_0^t d\tau\, K_l(\tau) C_l(t - \tau), \tag{3.218}$$

where we are limited by the consideration of the time correlation for the reorientation of the molecular symmetry axis. The memory function $K_l(\tau)$ is given by $\langle \dot{D}^{(l)}_{0,0}[\Omega(t)]\dot{D}^{(l)}_{0,0}(\Omega)\rangle$, $\dot{D}^{(l)}_{0,0}(\Omega)$ being the time derivative of $D^{(l)}(\Omega)$. In Eq. (3.218), coupling between orientational and other variables is neglected.

Given the memory function $K_l(\tau)$, Eq. (3.218) can be solved numerically or by the Laplace transform technique. A quantitative calculation of $K_l(t)$ is extremely difficult. Therefore, to calculate $C_l(t)$, some kind of ansatz is often made for $K_l(t)$ so that the result obtained for $C_l(t)$ may be in agreement with the experimentally obtained data. Another approach is based on the definition of a hierarchy of memory functions for $K_l(t)$ by means of Mori's continued-fraction theory. The theory is given by the equation

$$\frac{d}{dt} K_l^{(n)}(t) = -\int_0^t d\tau\, K_l^{(n+1)}(t-\tau)K_l^{(n)}(\tau). \tag{3.219}$$

Thus, Eq. (3.218) can be viewed as a special case of Eq. (3.219), in which we have $K_l^{(0)}(t) = C_l(t)$. If we assume that there is no memory in one of the higher $K_l^{(n+1)}(t)$, i.e., that

$$K_l^{(n+1)}(t) = \alpha\, \delta(t) \tag{3.220}$$

for some $n > 1$, then the hierarchy of equations generates solutions for all the lower $K_l^{(n)}(t)$. In particular, substituting Eq. (3.220) into (3.219) gives

$$dK_l^{(n)}(t)/dt = aK_l^{(n)}(t), \tag{3.221a}$$

or

$$K_l^{(n)}(t) = K_l^{(n)}(0)e^{-at}, \tag{3.221b}$$

which gives

$$K_l^{(n-1)}(t) = K_l^{(n-1)}(0)e^{-at/2}[\cosh bt + (a/b)\sinh bt], \tag{3.222}$$

where $b = (a^2 - 4K_l^{(n)}(0))^{1/2}$. From $K_l^{(n-1}(t)$, we then calculate $K_l^{(n-2)}(t), \ldots,$ and $C_l(t)$. The Laplace transforms of these memory functions have particularly simple forms, and the Laplace transform of $C_l(t)$ can be represented by a continued fraction.

Although Eq. (3.220) is strictly incorrect, the continued-fraction theory can yield a useful approximation for $C_l(t)$, especially if the value of n is large enough. However, a number of adjustable parameters (the parameter a plus the initial value of each of the memory functions of order n to 1) is needed in this method. Thus, the fact that good fits between theory and experiment are obtained may well be a consequence of the increasing number of fitting parameters rather than a physical realization of Eq. (3.220). Nevertheless, this theory can be improved if the initial values of the memory functions $K_l^{(n)}(0)$, $n = 1, 2, \ldots$ generated by the theory can be rigorously evaluated.

For example, in the case of linear molecules, we find that

$$K_l^{(1)}(0) = l(l + 1)(kT/I), \tag{3.223}$$

$$K_l^{(2)}(0) = (kT/I)[2l(l + 1) - 2 + \langle N^2 \rangle / 2(kT)^2], \tag{3.224}$$

where $\langle N^2 \rangle$ is the mean square of the molecular torque that is exerted on the molecule. Thus, if the parameters are constrained to these equations, then fitting the time-correlation function in accordance with the continued-fraction approach can yield information about intermolecular forces that appear explicitly via the mean-square torque and higher derivatives of the molecular torque.

Another useful form of memory-function theory dealing with a perturbation approximation has been proposed by Steele (1981). This theory to describe $C_l(t)$ concerns modeling the memory function $K_l(t)$ by the product of the free-rotor memory function and a Gaussian function involving the mean-square torque on the reorienting axis in the molecule. This perturbation theory works quite well for small and moderate mean-square torques but breaks down when the torque is large enough to introduce some degree of torsional oscillation in the reorientation.

There are other theories for reorientation, but we shall not discuss them here except to mention that the memory effect and anisotropy in the reorientation dynamics play important roles in determining the shape of the orientational-correlation function. A detailed understanding of the vibration–rotation spectra bands of molecular liquids requires clarification of these effects, and much work, experimental and theoretical, remains to be done on this subject.

We shall now return to the discussion of an important theorem in the theory of angular momentum.

3.12 WIGNER–ECKART THEOREM

The Wigner–Eckart theorem is one of the most useful theorems in the theory of angular momentum. In atomic and molecular spectroscopy, there are important problems concerning the interaction of two separated parts of a physical system, and the interaction Hamiltonian can often be written as the scalar product of two spherical-tensor operators of the same rank,

$$\mathscr{H}' = \sum_{j,m} (-1)^m T_{jm}(A) U_{j-m}(B), \tag{3.225}$$

where the tensor operators $T_{jm}(A)$ and $U_{j-m}(B)$ are associated with separate parts of the system.

In solving problems, it is often necessary to calculate matrix elements of one of the tensor operators between eigenstates of angular momenta. The Wigner–Eckart theorem provides a simple method for the calculations and results in a great simplification of the problem. This theorem states that for an irreducible tensor operator of rank j, given by $T_{jm}(A)$, the matrix element between states of sharp angular momenta is given by

$$\begin{aligned}&\langle \alpha_1 j_1 m_1 | T_{jm}(A) | \alpha_2 j_2 m_2 \rangle \\ &\quad = (-1)^{j_1 - m_1} \begin{pmatrix} j_1 & j & j_2 \\ -m_1 & m & m_2 \end{pmatrix} \langle \alpha_1 j_1 \| T_j(A) \| \alpha_2 j_2 \rangle, \qquad (3.226)\end{aligned}$$

where α_1 and α_2 are quantum numbers that are not associated with angular momentum. The quantity $\langle \alpha_1 j_1 \| T_j(A) \| \alpha_2 j_2 \rangle$, known as the reduced matrix element, is independent of the projection quantum numbers m_1, m, and m_2 but depends only on the nongeometrical nature of the physical problem.

Thus, the Wigner–Eckart theorem reduces the matrix element of an irreducible spherical tensor to an explicit form with its dependence on the projection quantum numbers in a form of Wigner's 3-j symbol, with the rest of the nongeometrical aspect of the problem being included in the reduced matrix element. Therefore, the theorem provides a formal statement for the conservation law of angular momentum. Moreover, through the symmetry nature of the 3-j symbol, the Wigner–Eckart theorem also furnishes a natural expression for the derivation of selection rules involving transitions between states of well-defined angular momentum.

We shall not provide the proof of this theorem; its proof can be found in books dealing with the theory of angular momentum (Edmonds, 1960; Rose, 1957). However, we shall illustrate the application of the theorem by evaluating the matrix elements of the nuclear quadrupole operator $Q_{\alpha\beta}(r)$ between two nuclear states with angular momentum I.

The nuclear quadrupole operator $Q_{\alpha\beta}(r)$ is defined as

$$Q_{\alpha\beta}(r) = e \sum_k (3 r_{\alpha k} r_{\beta k} - \delta_{\alpha\beta} r_k^2), \qquad (3.227)$$

where e is the proton charge and the summation over k includes all protons.

The total angular momentum of the nucleus is given by

$$\mathbf{I} = \sum_k (\mathbf{l}_k + \mathbf{S}_k), \qquad (3.228)$$

where $\mathbf{l}_k$ and $\mathbf{S}_k$ are the orbital and spin momenta of a nucleon (either a proton or a neutron), respectively.

We shall now construct an operator function of total nuclear angular momentum, given by

$$A_{\alpha\beta}(\mathbf{I}) = \tfrac{3}{2}(I_\alpha I_\beta + I_\beta I_\alpha) - \delta_{\alpha\beta} I^2. \tag{3.229}$$

In this expression, we have used the symmetrized product $I_\alpha I_\beta + I_\beta I_\alpha$ because different components of the angular-momentum operator **I** do not commute.

Using Eqs. (3.53) and (3.54), it is easy to verify that both $Q_{zz}(\mathbf{r})$ and $A_{zz}(\mathbf{I})$ transform as the irreducible spherical tensor T_{20} does. Thus, according to the Wigner–Eckart theorem, the matrix elements of $Q_{zz}(\mathbf{r})$ and $A_{zz}(\mathbf{I})$ have the same geometric factor:

$$\langle \alpha I m | Q_{zz}(\mathbf{r}) | \alpha I m' \rangle = (-1)^{I-m} \begin{pmatrix} I & 2 & I \\ -m & 0 & m' \end{pmatrix} \langle \alpha I \| Q(\mathbf{r}) \| \alpha I \rangle, \tag{3.230}$$

$$\langle \alpha I m | A_{zz}(I) | \alpha I m' \rangle = (-1)^{I-m} \begin{pmatrix} I & 2 & I \\ -m & 0 & m' \end{pmatrix} \langle \alpha I \| A(I) \| \alpha I \rangle. \tag{3.231}$$

Thus, we can divide out the common Wigner 3-j symbol and obtain

$$\langle \alpha I m | Q_{zz}(\mathbf{r}) | \alpha I m' \rangle = C \langle \alpha I m | A_{zz}(I) | \alpha I m' \rangle, \tag{3.232}$$

where the constant C is given by

$$C = \frac{\langle \alpha I \| Q(r) \| \alpha I \rangle}{\langle \alpha I \| A(r) \| \alpha I \rangle}, \tag{3.233}$$

which can be evaluated by using Eq. (3.232) and setting $m = m' = I$:

$$\langle \alpha I I | Q_{zz}(\mathbf{r}) | \alpha I I \rangle = C \langle \alpha I I | 3I_z^2 - I^2 | \alpha I I \rangle = C I (2I - 1)$$

or

$$C = eQ/I(2I - 1) \tag{3.234}$$

where eQ, known as the nuclear quadrupole moment, is given by

$$eQ = \langle \alpha I I | Q_{zz}(\mathbf{r}) | \alpha I I \rangle. \tag{3.235}$$

According to Eq. (3.232), it is clear that within the $2I + 1$ manifold of the nuclear angular momentum **I**, the zz component of the nuclear quadrupole operator can be expressed in terms of the nuclear angular-momentum operator as

$$Q_{zz}(r) = [eQ/I(2I - 1)](3I_z^2 - I^2). \tag{3.236}$$

This result can be generalized to consider an operator function $F_j(A)$ of rank j; $F_j(A)$ can be expressed in terms of a linear combination of $T_{jm}(A)$:

$$F_j(A) = \sum_m A_m T_{jm}(A). \tag{3.237}$$

Replacing A by B, we obtain

$$F_j(B) = \sum_m A_m T_{jm}(B). \tag{3.238}$$

The Wigner–Eckart theorem allows us to relate the matrix element of $T_{jm}(A)$ to that of $T_{jm}(B)$ between the states $|\alpha_1 j_1 m_1\rangle$ and $|\alpha_2 j_2 m_2\rangle$ according to the expression given by

$$\langle\alpha_1 j_1 m_1 | T_{jm}(A) | \alpha_2 j_2 m_2\rangle = C\langle\alpha_1 j_1 m_1 | T_{jm}(B) | \alpha_2 j_2 m_2\rangle, \tag{3.239}$$

where

$$C = \frac{\langle\alpha_1 j_1 \| T_j(A) \| \alpha_2 j_2\rangle}{\langle\alpha_1 j_1 \| T_j(B) \| \alpha_2 j_2\rangle}. \tag{3.240}$$

Using the Wigner–Eckart theorem, we find that the matrix element of operator $F_j(A)$ can be written as

$$\begin{aligned}\langle\alpha_1 j_1 m_1 | F_j(A) | \alpha_2 j_2 m_2\rangle &= \sum_m A_m \langle\alpha_1 j_1 m_1 | T_{jm}(A) | \alpha_2 j_2 m_2\rangle \\ &= C \sum_m A_m \langle\alpha_1 j_1 m_1 | T_{jm}(B) | \alpha_2 j_2 m_2\rangle \\ &= \langle\alpha_1 j_1 m_1 | F_j(B) | \alpha_2 j_2 m_2\rangle \frac{\langle\alpha_1 j_1 \| T_j(A) \| \alpha_2 j_2\rangle}{\langle\alpha_1 j_1 \| T_j(B) \| \alpha_2 j_2\rangle}.\end{aligned} \tag{3.241}$$

3.13 ELECTRONIC QUADRUPOLE EFFECT

The example given in Eq. (3.236) in which operator $Q_{zz}(r)$ is replaced by operator $A_{zz}(I)$ provides fundamental background material for the discussion of the electronic quadrupole effect. Consider a molecule in a stable electronic and nuclear configuration. If parts of the constituting nuclei possess nuclear quadrupole moments, each of these nuclei can interact with the inhomogeneous electric field produced by the electron cloud surrounding the nuclei. This type of interaction results because of the electrostatic interaction of the charge distribution of the nucleus with the charge distribution of the electron cloud. We are interested in obtaining the Hamiltonian operator that is responsible for the electronic quadrupole effect. The availability of the Hamiltonian operator will allow us to determine its eigenstates and transitions between these states.

Consider first that the nucleus and the electron cloud in a molecule are described in terms of two classical charge distributions $\rho_n(r_n)$ and $\rho_e(r_e)$. Their mutual electrostatic-interaction energy is given by

$$E = \iint d^3r_e \, d^3r_n \frac{\rho_e(\mathbf{r}_e)\rho_n(\mathbf{r}_n)}{|\mathbf{r}_e - \mathbf{r}_n|}, \tag{3.242}$$

where $\mathbf{r}_e$ and $\mathbf{r}_n$ indicate, respectively, the electronic and nuclear positions with respect to some arbitrary reference.

If the small penetration of the electron inside the nucleus is neglected, we can assume that $r_e > r_n$ and write

$$\begin{aligned} 1/|\mathbf{r}_e - \mathbf{r}_n| &= (r_e^2 + r_n^2 - 2r_e r_n \cos\omega)^{-1/2} \\ &= r_e^{-1}[1 - 2(r_n/r_e)\cos\omega + (r_n/r_e)^2]^{-1/2} \\ &= r_e^{-1}\sum_{l=0}^{\infty} P_l(\cos\omega)(r_n/r_e)^l, \end{aligned} \tag{3.243}$$

where ω is the angle between vectors $\mathbf{r}_e$ and $\mathbf{r}_n$ and $P_l(\cos\omega)$ the Legendre polynomial of order l.

The spherical-harmonic addition theorem [Eq. (3.29)] allows us to express $P_l(\cos\omega)$ in terms of the spherical harmonics. Using this formula we can rewrite Eq. (3.243) as

$$\frac{1}{\mathbf{r}_e - \mathbf{r}_n} = \frac{4\pi}{\mathbf{r}_e}\sum_{l=0}^{\infty}\sum_{m=-l}^{\infty}\left(\frac{1}{2l+1}\right)^l\left(\frac{r_n}{r_e}\right)Y_{lm}^*(\theta_e, \phi_e)Y_{lm}(\theta_n, \phi_n). \tag{3.244}$$

Substituting Eq. (3.244) into (3.242), we obtain

$$E = \sum_{l,m}(-1)^m A_{lm} B_{l-m}, \tag{3.245}$$

where A_{lm} is a function associated only with the nucleus and B_{lm} only with the electron. These functions are given by

$$A_{lm} = \sqrt{\frac{4\pi}{2l+1}}\int d^3r_n r_n^l \rho(\mathbf{r}_n) Y_{lm}(\theta_n, \phi_n), \tag{3.246}$$

$$B_{lm} = \sqrt{\frac{4\pi}{2l+1}}\int d^3r_e r_e^{-(l+1)} \rho(\mathbf{r}_e) Y_{lm}(\theta_e, \phi_e). \tag{3.247}$$

Therefore, nuclear and electronic charge densities at point $\mathbf{r}_n$ and $\mathbf{r}_e$ can be written as

$$\rho(\mathbf{r}_n) = \sum_j \delta(\mathbf{r}_n - \mathbf{R}_j)\, q_j, \tag{3.248}$$

$$\rho(\mathbf{r}_e) = e\sum_j \delta(\mathbf{r}_e - \mathbf{r}_j), \tag{3.249}$$

where q_i is the nucleon charge, which is equal to q for a proton and zero for a neutron; e, the electronic charge equal to $-q$; and $\mathbf{R}_j$ and $\mathbf{r}_j$ denote the positions of nucleon j and electron j, respectively.

With the help of these equations, A_{lm} and B_{lm} can be written as

$$A_{lm} = \sqrt{\frac{4\pi}{2l+1}} \sum_j q_i R_j^l Y_{lm}(\theta_{nj}, \phi_{nj}), \tag{3.250}$$

$$B_{lm} = e\sqrt{\frac{4\pi}{2l+1}} \sum_j r_j^{-(l+1)} Y_{lm}(\theta_{ej}, \phi_{ej}). \tag{3.251}$$

The nuclear functions A_{lm} have $2l + 1$ components, which define the multiple moments of the nucleus. The zeroth moment, which corresponds to the $l = 0$, $m = 0$ term, is given by

$$A_{00} = \sqrt{4\pi} \sum_j q_j Y_{00}(\theta_{ni}, \phi_{ni}) = \sum_j q_j = Q_n, \tag{3.252}$$

which is simply the nuclear charge. The corresponding term for the electronic function,

$$B_{00} = e\sqrt{4\pi} \sum_j \frac{1}{r_j} Y_{00}(\theta_{ej}, \phi_{ej}) = \sum_j \frac{e}{r_j}, \tag{3.253}$$

is the electrostatic potential of all electrons at the position of the nucleus. Thus, the zeroth moment describes the Coulomb interaction between the point nuclear charge and a collection of electrons.

The $l = 1$ terms are associated with the electric dipole–dipole interaction. However, since a nucleus cannot have an electric dipole moment or any higher multipole moments with odd values of l, these odd multipole moments do not contribute to the electrostatic energy. This general result follows from the fundamental hypothesis of quantum mechanics.

According to the correspondence principle, the electrostatic interaction energy between the electrons and a nucleus corresponds to the expectation value of the Hamiltonian. Thus, the expression for E given in Eq. (3.245) can be considered as the expectation value of the Hamiltonian, given by

$$\mathscr{H} = \sum_{l,m} (-1)^m A_{lm} B_{l-m}, \tag{3.254}$$

where the functions A_{lm} and B_{l-m} are now regarded as operators. The operators A_{lm} have in principle both the diagonal and the off-diagonal matrix elements between nuclear angular-momentum states of different energies. However, in the field of molecular spectroscopy, the off-diagonal matrix elements are of no concern because of the wide energy separation of the nuclear states (in the range of γ ray), whereas for matrix elements between the nuclear states of the same nuclear angular-momentum manifold, the elements of A_{lm} with odd values of l vanish identically because of the parity restriction. Thus, beyond the $l - 0$ term, the next terms in the expansion of $\mathscr{H}$ start with those associated with $l = 2$.

According to Eq. (3.250), the five components of A_{zm} are given by

$$A_{2,\pm 2} = \sqrt{\frac{4\pi}{5}} \sum_j q_j R_j^2 Y_{2\pm 2}(\theta_{nj}, \phi_{nj}) = \frac{\sqrt{6}}{4} \sum_j q_j (X_j \pm iY_j)^2, \tag{3.255}$$

$$A_{2,\pm 1} = \sqrt{\frac{4\pi}{5}} \sum_j q_j R_j^2 Y_{2\pm 1}(\theta_{nj}, \phi_{nj}) = \mp \sqrt{\tfrac{3}{2}} \sum_j q_j Z_j (X_j \pm iY_j), \tag{3.256}$$

$$A_{2,0} = \sqrt{\frac{4\pi}{5}} \sum_j q_j R_j^2 Y_{20}(\theta_{nj}, \phi_{nj}) = \tfrac{1}{2} \sum_j q_j (3Z_j^2 - R_j^2). \tag{3.257}$$

Similarly, according to Eq. (3.251), the five components of B_{2m} are given by

$$B_{2\pm 2} = \frac{\sqrt{6}}{4} e \sum_j \frac{(x_j \pm iy_j)^2}{r_j^5} = \frac{1}{2\sqrt{6}} (V_{xx} - V_{yy} \pm 2iV_{xy}), \tag{3.258}$$

$$B_{2\pm 1} = \mp \frac{\sqrt{3}}{2} e \sum_j \frac{z_j(x_j \pm iy_j)}{r_j^5} = \mp \frac{1}{\sqrt{6}} (V_{xz} \pm iV_{yz}), \tag{3.259}$$

$$B_{20} = \tfrac{1}{2} e \sum_j \frac{(3z_j^2 - r_j^2)}{r_j^2} = \tfrac{1}{2} V_{zz}, \tag{3.260}$$

where $V_{\alpha\beta}$ is given by

$$V_{\alpha\beta} = e \sum_j \frac{\partial^2}{\partial r_{j\alpha}\, \partial r_{j\rho}} \left(\frac{1}{r_j}\right). \tag{3.261}$$

Here $\alpha, \beta = x, y, z$; $r_{jx} = x_j$; $r_{jy} = y_j$; and $r_{jz} = z_j$.

The A_{2m} operators transform as spherical tensors of rank 2, and according to the Wigner–Eckart theorem, within the $2I + 1$ manifold of the nuclear state with quantum number I they can be replaced by irreducible spherical tensors of the same rank, constructed by using nuclear angular-momentum operators. Thus, we can write

$$A_{2\pm 2} = [eQ/I(2I - 1)]\sqrt{6}/4 I_{\pm}^2, \tag{3.262}$$

$$A_{2\pm 1} = \mp [eQ/I(2I - 1)]\sqrt{6}/4 (I_2 I_{\pm} + I_{\pm} I_z), \tag{3.263}$$

$$A_{20} = [eQ/I(2I - 1)]\tfrac{1}{2}(3I_z^2 - I^2), \tag{3.264}$$

where eQ is the nuclear quadrupole moment, given by

$$eQ = I(2I - 1)\frac{\langle \alpha I \| \sum_j (3Z_j^2 - R_j^2)q_j \| \alpha I \rangle}{\langle \alpha I \| 3I_z^2 - I^2 \| \alpha I \rangle}. \tag{3.265}$$

Using the results obtained for the As and Bs, we can write the Hamiltonian associated with the electronic quadrupole effect as

$$\begin{aligned}\mathscr{H}_Q = {} & [eQ/4I(2I - 1)][V_0(3I_z^2 - I^2) + V_{+1}(I_- I_z + I_z I_-) \\ & + V_{-1}(I_+ I_z + I_z I_+) + (V_{+2}(I_-)^2 + V_{-2}(I_+)^2],\end{aligned} \tag{3.266}$$

where the V's are the spherical components of the electric-field gradient at the position of the nucleus, given by

$$\begin{aligned} V_0 &= V_{zz}, \\ V_{\pm 1} &= V_{zx} \pm iV_{zy} \\ V_{\pm 2} &= \tfrac{1}{2}(V_{xx} - V_{yy}) \pm iV_{xy}. \end{aligned} \tag{3.267}$$

Equation (3.266) is particularly useful for treating the problem associated with the relaxation effect resulting from the modulation of the quadrupole Hamiltonian owing to molecular reorientations. In the presence of molecular reorientation, the principal axes of the electric-field gradient are not fixed in space but rather are a function of time. Expressing $\mathscr{H}_Q$ in the form of Eq. (3.266) facilitates a systematic treatment of the nuclear quadrupole-relaxation problem.

It is an easy matter to show that Eq. (3.266) can be written in terms of the contraction of two types of Cartesian tensors $\mathbf{V}$ and $\mathbf{Q}$ in the form given by

$$\mathscr{H}_Q = \sum_{\alpha\beta} V_{\alpha\beta} Q_{\alpha\beta}, \tag{3.268}$$

where $V_{\alpha\beta}$ is the electric-field–gradient tensor given by Eq. (3.261) and $Q_{\alpha\beta}$ is given by

$$Q_{\alpha\beta} = eQ/6I(2I - 1)[\tfrac{3}{2}(I_\alpha I_\beta + I_\beta I_\alpha) - \delta_{\alpha\beta} I^2]. \tag{3.269}$$

Because there is no electronic charge at the nucleus, the electric-field–gradient tensor $V_{\alpha\beta}$ is traceless, in accordance with the Laplace equation

$$\nabla^2 \mathbf{V} = V_{xx} + V_{yy} + V_{zz} = 0. \tag{3.270}$$

Since $Q_{\alpha\beta}$ is also traceless in nuclear angular-momentum (or spin) space, $\mathscr{H}_Q$ vanishes when isotropically averaged in coordinate space or in spin space. It then follows that the first-order electronic–nuclear quadrupole-interaction effect is absent in gases, liquids, and rotationally isotropic systems such as amorphous solids. Furthermore, because of Eq. (3.270), the nuclear quadrupole effect also vanishes for the nucleus located at the center of the cubic system.

The electric-field–gradient tensor $V_{\alpha\beta}$ is symmetric and can thus be diagonalized to yield three real principal values. If we express $\mathscr{H}_Q$ in terms of the principal axes of $V_{\alpha\beta}$, then Eq. (3.268) becomes

$$\mathscr{H}_Q = [eQ/6I(2I - 1)][V_{xx}(3I_x^2 - I^2) + V_{yy}(3I_y^2 - I^2) + V_{zz}(3I_z^2 - I^2)]. \tag{3.271}$$

Because of the Laplace equation, not all principal values of $V_{\alpha\beta}$ are independent. If we define a quantity η, known as the symmetry parameter, as

$$\eta = (V_{xx} - V_{yy})/V_{zz}, \tag{3.272}$$

then, according to the Laplace equation, we express V_{xx} and V_{yy} as

$$\begin{aligned} V_{xx} &= \frac{-1}{2}(1 - \eta)V_{zz}, \\ V_{yy} &= \frac{-1}{2}(1 + \eta)V_{zz}, \end{aligned} \tag{3.273}$$

and Eq. (3.271) becomes

$$\mathscr{H}_Q = [eQ/4I(2I - 1)]V_{zz}[(3I_{zz}^2 - I^2) + \eta(I_x^2 - I_y^2)]. \tag{3.274}$$

This equation forms the basis for the discussion of nuclear quadrupole spectroscopy in a number of excellent books. [See, for example, Smith (1974).]

4
The Time-Correlation Function and the Memory-Function Formalism

4.1 INTRODUCTION

We have shown that spectral power densities are related to their appropriate time-correlation functions by Fourier transforms. To understand a spectral line shape we must study the time-correlation function of the dynamic variable associated with the particular type of spectroscopy. Calculations of the time-correlation functions from first principles in systems consisting of many interacting particles have, however, proved to be very difficult, if not impossible. Thus, the study of the time-correlation function has relied on approximating the physical system of interest in terms of simple models such as the hydrodynamic theory and the various rotational-diffusion equations introduced in previous chapters. However, exact formulations have been introduced in which the time-correlation functions are rigorously related to the memory functions. Although the complexity of molecular dynamics in the condensed phase still makes the calculation of the memory function infeasible, it is very useful to interpret experimentally observed spectra in terms of memory functions.

The memory-function description provides a clear display of the dynamic nature of the interacting system; it further demonstrates some of the powerful results in the structure of the time-correlation function. In addition, even though the calculation of the memory function requires a complete understanding of many-body dynamics, we can often obtain an accurate description of the time-correlation function using a fairly simple memory function. We shall demonstrate this in a later section when we consider specific examples.

In this chapter, we shall begin by describing a formal method for analyzing the structure of the time-correlation function. We shall derive an exact integrodifferential equation for the time-correlation function in which the memory function is a kernel, using the projection-operator technique originally introduced by Zwanzig (1961). We shall then describe the projection-operator technique that was subsequently extended by Mori (1965a,b) to derive the generalized Langevin equation. Applications of this formalism to spectroscopy will then follow.

4.2 THE MEMORY FUNCTION

We seek an equation that describes the time evolution of the (time) autocorrelation function $C_A(t)$ of the dynamic variable A. As defined before, the autocorrelation function of A is given by

$$C_A(t) = \langle A(t)A^* \rangle, \tag{4.1}$$

where for brevity we drop the argument 0 in A^*, understanding that $A^* = A^*(0)$; A^* is the Hermitian conjugate of A. We shall also assume $\langle A(t) \rangle = 0$ for arbitrary t. The angular bracket defines an average in the classical ensemble. Thus, $C_A(t) \to 0$ as t becomes large. If $A(t) \neq 0$, we can redefine another variable $a(t) = A(t) - \langle A(t) \rangle$ and then consider the correlation function of the dynamic variable $a(t)$.

It is understood that the dynamic variable $A = A(p, q)$ is a function of the phase-space coordinates (p, q), collectively represented by Γ_N. According to classical mechanics, $A(t)$ obeys the equation of motion

$$\frac{\partial A(t)}{\partial t} = -\{\mathscr{H}, A\}_{\mathbf{PB}} \equiv i\mathscr{L}A(t), \tag{4.2}$$

where $\mathscr{L}$ is the Liouville operator. The formal solution of Eq. (4.2) is $A(t) = e^{i\mathscr{L}t}A$. Inserting this result into Eq. (4.1), we have

$$C_A(t) = \langle A(t)A^* \rangle = \langle [e^{i\mathscr{L}t}A]A^* \rangle = \langle Ae^{-i\mathscr{L}t}A^* \rangle. \tag{4.3}$$

Here the last step follows from the time-reversal invariance property that always holds for a canonical ensemble.

If $C_A(t)$ is differentiated twice with respect to time t, we have

$$\frac{\partial^2 C_A(t)}{\partial t^2} = \langle i\mathscr{L}[e^{i\mathscr{L}t} i\mathscr{L} A]A^*\rangle = -\langle [e^{i\mathscr{L}t} i\mathscr{L} A][i\mathscr{L} A]^*\rangle. \tag{4.4}$$

The second step follows because the Liouville operator is Hermitian. This can be proved by partial integration, as was shown in Chapter 1.

Since the time evolution of $A(t)$ is given by Eq. (4.2), clearly the quantity $i\mathscr{L}A$ is equal to the time derivative of $A(t)$ evaluated at $t = 0$. Therefore,

$$\frac{\partial^2 C_A(t)}{\partial t^2} = -\langle (e^{i\mathscr{L}t}\dot{A})\dot{A}^*\rangle. \tag{4.5}$$

The operator $e^{i\mathscr{L}t}$ is the unitary time-evolution operator that, when applied to $\dot{A}$, displaces it in time to $\dot{A}(t)$. Thus, Eq. (4.5) can be written in the form

$$\frac{\partial^2 C_A(t)}{\partial t^2} = -\phi(t), \tag{4.6}$$

where

$$\phi(t) = \langle \dot{A}(t)\dot{A}^*\rangle = \langle (e^{i\mathscr{L}t}\dot{A})\dot{A}^*\rangle \tag{4.7}$$

Equation (4.6) is to be solved subject to the initial conditions

$$C_A(0) = \langle |A|^2\rangle \qquad \text{and} \qquad \dot{C}_A(0) = i\langle (\mathscr{L}A)A^*\rangle = 0, \tag{4.8}$$

where we assume that A and $\dot{A}$ have opposite parity. This will be the case if A has time-reversal symmetry.

Denoting the Laplace transform of $C_A(t)$ by $\tilde{C}_A(s)$, with s the Laplace variable

$$\tilde{C}_A(s) = \int_0^\infty dt\, e^{-st} C_A(t), \tag{4.9}$$

and denoting the Laplace transform of the autocorrelation function $\phi(t)$ by $\tilde{\phi}(s)$, we obtain the Laplace transform of Eq. (4.6) as

$$s^2\tilde{C}_A(s) - sC_A(0) = -\tilde{\phi}(s). \tag{4.10}$$

Rearranging the identity

$$s\tilde{C}_A(s)[s\tilde{C}_A(s) - C_A(0)] = [s^2\tilde{C}_A(s) - sC_A(0)]\tilde{C}_A(s) \tag{4.11}$$

by adding and subtracting $C_A(0)$ in the term on the extreme left, we obtain

$$[s\tilde{C}_A(s) + C_A(0) - C_A(0)][s\tilde{C}_A(s) - C_A(0)] = [s^2\tilde{C}_A(s) - sC_A(0)]\tilde{C}_A(s). \tag{4.12}$$

For $s \neq 0$, Eq. (4.12) can be rewritten as

$$\{[s\tilde{C}_A(s) - C_A(0)] + C_A(0)\}[s\tilde{C}_A(s) - C_A(0)] \\ = [s^2\tilde{C}_A(s) - sC_A(0)]\tilde{C}_A(s). \tag{4.13}$$

Equation (4.10) can now be substituted into Eq. (4.13) to yield

$$[C_A(0) - (1/s)\tilde{\phi}(s)][s\tilde{C}_A(s) - C_A(0)] = -\tilde{\phi}(s)\tilde{C}_A(s). \tag{4.14}$$

For the values of s such that

$$C_A(0) \neq (1/s)\tilde{\phi}(s), \tag{4.15}$$

eq. (4.14) can be written as

$$\tilde{C}_A(s) = C_A(0)/[s + K(s)], \tag{4.16}$$

where

$$K(s) = \tilde{\phi}(s)/[C_A(0) - (1/s)\phi(s)] = \tilde{\phi}(s)/[1 - (1/s)\tilde{\phi}(s)], \tag{4.17}$$

and

$$\tilde{\phi}(s) = \phi(s)/C_A(0). \tag{4.18}$$

When Eq. (4.16) is inverted, it gives

$$\frac{\partial C_A(t)}{\partial t} = -\int_0^t d\tau\, K(\tau)C_A(t - \tau), \tag{4.19}$$

where the kernel function $K(\tau)$ is known as the memory function, which describes the extent to which past values of $C(t)$ influence its time rate of change at present. The function $K(\tau)$ is related to $K(s)$ given in Eq. (4.17) by an inverse Fourier transform. Equation (4.19) is rigorous and is very useful for describing the time evolution of $C_A(t)$ for both conceptual and computational purposes. We shall show the power of the memory function when it is applied to spectral analysis. But before taking up the application, we shall show that $K(\tau)$ can be manipulated into an interesting form. To accomplish this, we need a projection operator P, first introduced into statistical mechanics by Zwanzig (1961). The projection operator is defined such that when P acts on any well-behaved phase-space function $G(p, q)$ it projects the phase-space function G onto the dynamic variable A. Thus, we have

$$PG(p, q) = \langle GA^* \rangle \langle AA^* \rangle^{-1} A(p, q), \tag{4.20}$$

where the angular bracket denotes the ensemble average, i.e.,

$$\langle Fg^* \rangle = \int dp\, dq\, F(p, q)g^*(p, q)\rho_e(p, q), \tag{4.21}$$

where $\rho_e(p, q)$ is the equilibrium distribution function in a canonical ensemble.

It is easy to show that $P^2G = PG$. This property is one of the requirements of a projection operator. Another property of P is that it is a Hermitian operator. We do not need this property at present.

Consider now the function $\tilde{\phi}(s)$. By Eq. (4.7) it can be written in terms of the resolvent operator $(s - i\mathscr{L})^{-1}$ as

$$\tilde{\phi}(s) = \int_0^\infty dt\, e^{-st}\langle(e^{i\mathscr{L}t}\dot{A})\dot{A}^*\rangle = \langle[(s - i\mathscr{L})^{-1}\dot{A}]\dot{A}^*\rangle. \tag{4.22}$$

Note that for the operators X and Y there exists the identity

$$X^{-1} = Y^{-1} + X^{-1}(Y - X)Y^{-1}, \tag{4.23}$$

with $X = s - i\mathscr{L}$ and $Y = s - i(1 - P)\mathscr{L}$; thus, we can rewrite $\tilde{\phi}(s)$ in Eq. (4.22) as

$$\begin{aligned}\tilde{\phi}(s) = &\langle\{[s - i(1 - P)\mathscr{L}]^{-1}\dot{A}\}\dot{A}^*\rangle \\ &+ i\langle\{(s - i\mathscr{L})^{-1}P\mathscr{L}[s - i(1 - P)\mathscr{L}]^{-1}\dot{A}\}\dot{A}^*\rangle.\end{aligned} \tag{4.24}$$

Now we use the definition of P to rewrite the second term on the right side of Eq. (4.24). We let the phase-space function G be $\mathscr{L}[s - i(1 - P)\mathscr{L}]^{-1}\dot{A}$. Application of Eq. (4.20) yields

$$\begin{aligned}P\mathscr{L}[s - i(1 - P)\mathscr{L}]^{-1}\dot{A} &= \langle\{\mathscr{L}[s - i(1 - P)\mathscr{L}^{-1}\dot{A}\}A^*\rangle\langle AA^*\rangle^{-1}A \\ &= i\langle\{[s - i(1 - P\mathscr{L}]^{-1}\dot{A}\}\dot{A}^*\rangle\langle AA^*\rangle^{-1}A,\end{aligned} \tag{4.25}$$

where to write the last step, we used the Hermitian property of $\mathscr{L}$; i.e., we used the fact that for any two phase-space functions f and g,

$$\langle(\mathscr{L}f)g^*\rangle = \langle f(\mathscr{L}g)^*\rangle = i\langle f\dot{g}^*\rangle. \tag{4.26}$$

Substituting Eq. (4.25) into (4.24), we obtain

$$\begin{aligned}\tilde{\phi}(s) = &\langle\{[s - i(1 - P)\mathscr{L}]^{-1}\dot{A}\}\dot{A}^*\rangle \\ &- \langle\{[s - i(1 - P)\mathscr{L}]^{-1}\dot{A}\}\dot{A}^*\rangle\langle AA^*\rangle^{-1}\langle[(s - i\mathscr{L})^{-1}A]\dot{A}^*\rangle.\end{aligned} \tag{4.27}$$

It is easily recognized that

$$\langle[(s - i\mathscr{L})^{-1}A]\dot{A}^*\rangle = (1/s)\langle[(s - i\mathscr{L})^{-1}\dot{A}]\dot{A}^*\rangle = (1/s)\tilde{\phi}(s). \tag{4.28}$$

When Eq. (4.28) is substituted into (4.27), and after rearrangement of terms, we find that

$$\langle\{[s - i(1 - P)\mathscr{L}]^{-1}\dot{A}\}\dot{A}^*\rangle\langle AA^*\rangle^{-1} = \tilde{\phi}(s)/[C_A(0) - (1/s)\tilde{\phi}(s)], \tag{4.29}$$

which, according to Eq. (4.17), is simply $\tilde{K}(s)$. The memory function $K(\tau)$ is obtained by the inverse Laplace transform of $\tilde{K}(s)$.

$$K(\tau) = \frac{\langle [e^{i(1-P)\mathscr{L}\tau}\dot{A}]\dot{A}\rangle}{C_A(0)} = \frac{1}{2\pi i}\oint ds \frac{e^{s\tau}\tilde{\phi}(s)}{[1-(1/s)\tilde{\phi}(s)]}. \tag{4.30}$$

Note that $K(\tau)$ itself has the structure of a time-correlation function of the velocity A (i.e., $\dot{A}$). However, there is one important difference from the ordinary time-correlation function: the dynamic operator $(1 - P)\mathscr{L}$, which determines the time evolution of $\dot{A}$, is not the full Liouville operator $\mathscr{L}$; it is a projection from $\mathscr{L}$ that determines the intrinsic fluctuations of the dynamic variable A. If we refer to the degree of freedom associated with the dynamic variable A as a system of interest and to all other degrees of freedom as the "bath," then the dynamic operator $(1 - P)\mathscr{L}$ describes the internal dynamics of the bath. The memory function thus describes the interaction of the bath with the system of interest. Clearly, the name "memory function" is appropriate because, owing to the dynamic coherence of the bath, an event that occurred in the past will feed back to impose its behavior on the system represented by the time-correlation function of the variable A.

The formal development presented in this section is for a classical system using the classical Liouville operator. However, the formalism can be applied to quantum systems as well. All we have to clarify is that the dynamic variable A^* is modified to be the Hermitian conjugate of A instead of its complex conjugate, that the ensemble average represented by the angular bracket means taking the trace with respect to the equilibrium density operator $\rho_e = e^{-\beta\mathscr{H}}/Z$, and that the Liouville operator $\mathscr{L}$ corresponds to the super operator $\mathscr{H}^\times/\hbar$, as defined in Eq. (1.78) of Chapter 1.

Equation (4.4) is also valid in the quantal case. The proof of the second step of this equation can be carried out by using the property of the trace,

$$\langle [\mathscr{H}, A]B\rangle = \mathrm{Tr}\{\rho_e[\mathscr{H}, A]B\} = -\langle A[\mathscr{H}, B]\rangle, \tag{4.31}$$

for any operators A and B.

4.3 THE GENERALIZED LANGEVIN EQUATION

We shall first describe the Langevin equation proposed for the theory of Brownian motion for the suspension of plant pollen in a fluid. In 1829, the British botanist Robert Brown observed that the tiny pollen particle executes an irregular fluctuating motion. This irregular motion is caused by constant collisions of the particle with molecules of the surrounding fluids. The theory of Brownian motion was first considered by Einstein and Smoluchowsky and later was further developed by Langevin. It continues to be a fascinating subject of statistical physics. It is asumed that the force on the Brownian particle can be divided into a fractional part and a random part. The frictional force results from frequent collisions with the fluid molecules, which appear as a continuous drag on a time scale long compared with the frequency and duration of individual collisions. If the drift velocity of the Brownian particle is $\mathbf{u}$, then the Brownian particle experiences a frictional force $\zeta\mathbf{u}$, where ζ is known as the friction coefficient and is given by $6\pi a\eta_s$ according to the Stokes–Einstein law. Here a is the radius of the Brownian particle and η_s the shear viscosity of the fluid. The other random force represents the instant molecular bombardment exerted by the fluid molecules. Since the time scale of the random force is very short compared with that of the frictional force, we can assume that the random force is independent of $\mathbf{u}$. Thus, the equation of motion for such a Brownian particle can be written as

$$m\, d\mathbf{u}/dt = -\xi\mathbf{u} + \mathbf{f}(t), \tag{4.32}$$

or, more conveniently,

$$d\mathbf{u}/dt = -\alpha\mathbf{u} + \mathbf{f}'(t), \tag{4.33}$$

where $\alpha = \xi/m$ and $\mathbf{f}'(t) = \mathbf{f}(t)/m$. The term $\mathbf{f}(t)$ is the random force. Equation (4.32), or (4.33), is known as the Langevin equation. This is a stochastic differential equation, which differs from the usual differential equation because it contains a randomly varying function. The solution of such a stochastic equation is given in terms of the time-correlation function or the conditional-probability density. In Chapter 1 we described the method for solving the Langevin equation given in Eq. (4.33). In a later section of this chapter, we shall provide the microscopic derivation of the Langevin equation. In such a derivation we shall again begin with the Liouville equation (either classical or quantal), which for an arbitrary dynamic variable $\mathbf{A}(t)$ is given by

$$\frac{\partial A(t)}{\partial t} = i\mathscr{L}A(t). \tag{4.34}$$

To be more general, we shall not assume the dynamic variable A to have a time-reversal symmetry. Thus, $\langle \dot{A}A^* \rangle$ will be finite. The generalization allows us to include the optical or acoustic phonons in the general dynamic variable A, because these phonon coordinates do not have simple time-reversal symmetry.

Since the formal solution of Eq. (4.34) is $A(t) = e^{i\mathscr{L}t}A$, the time derivative of $A(t)$ can be written as

$$\frac{\partial A(t)}{\partial t} = e^{i\mathscr{L}t}i\mathscr{L}A. \tag{4.35}$$

For the given dynamic operator A, we now introduce projection operator P, defined by

$$P \equiv A^+ \langle AA^+ \rangle^{-1} A. \tag{4.36}$$

We shall describe the property of the projection operator. The operator can project any function G onto the intial dynamic variable A according to

$$PG = \langle GA^+ \rangle \langle AA^+ \rangle^{-1} A. \tag{4.37}$$

Equation (4.37) is in agreement with Eq. (4.20), in which we consider only the classical case. The projection operator given by Eq. (4.36) is an appropriate generalization for both the classical and quantal cases.

It is a simple matter to verify that P has the following properties:

(1) $PA = A$,
(2) $P^2 G = PG$,
(3) P is a Hermitian operator, i.e., $\langle F(PG)^* \rangle = \langle (PF)G^* \rangle$.

Properties (2) and (3) are the necessary and sufficient conditions for P to be a projection operator.

It is convenient to define an operator complement to P as

$$Q = 1 - P. \tag{4.38}$$

Using Eq. (4.36), it is easy to verify that

(1) $QA = 0$,
(2) $Q^2 = Q$,
(3) Q is Hermitian.

Thus, Q is also a projection operator. Now we note that

$$\langle (QG)A^+ \rangle = \langle GA^+ \rangle - \langle (PG)A^+ \rangle = \langle GA^+ \rangle - \langle G(PA)^+ \rangle = 0. \tag{4.39}$$

The projection operator Q projects from the function G of a component orthogonal to A. We call the two functions f and g orthogonal if $\langle fg^+ \rangle = 0$. Clearly,

$$PQ = QP = 0. \tag{4.40}$$

Given A and the projection operator, we can transform the Louville equation given by Eq. (4.35) into the generalized Langevin equation following the method introduced by Mori (1965a,b).

Using the result that $P + Q = 1$, we write

$$\begin{aligned} i\mathscr{L}A &= i(P + Q)\mathscr{L}A = i\langle (\mathscr{L}A)A^+ \rangle \langle AA^+ \rangle^{-1} A + iQ\mathscr{L}A \\ &= i\Omega A + iQ\mathscr{L}A, \end{aligned} \tag{4.41}$$

where we have defined the frequency Ω as

$$\Omega = \langle(\mathscr{L}A)A^+\rangle\langle AA^+\rangle^{-1}. \tag{4.42}$$

The frequency Ω vanishes if A has time-reversal symmetry.

Substituting Eq. (4.4T) into (4.35) yields

$$\frac{\partial A(t)}{\partial t} = i\Omega A(t) + e^{i\mathscr{L}t}iQ\mathscr{L}A. \tag{4.43}$$

We next use the operator identity [this identity is the Laplace transform version of Eq. (4.23)]

$$\begin{aligned} e^{i\mathscr{L}t} &= e^{i(P+Q)\mathscr{L}t} \\ &= e^{iQ\mathscr{L}t} + \int_0^t d\tau\, e^{i\mathscr{L}(t-\tau)}iP\mathscr{L}e^{iQ\mathscr{L}\tau} \end{aligned} \tag{4.44}$$

and rewrite Eq. (4.43) as

$$\frac{\partial A(t)}{\partial t} = i\Omega A(t) + e^{iQ\mathscr{L}t}iQ\mathscr{L}A + \int_0^t d\tau\, e^{i\mathscr{L}(t-\tau)}iP\mathscr{L}e^{iQ\mathscr{L}\tau}iQ\mathscr{L}A. \tag{4.45}$$

For simplicity of notation, we define

$$F = iQ\mathscr{L}A \tag{4.46}$$

and

$$F(t) = e^{iQ\mathscr{L}t}F. \tag{4.47}$$

It should be noted that $F(t)$ does not develop in time according to the Liouville operator $\mathscr{L}$, but rather according to $Q\mathscr{L}$. That is, F is defined in a direction orthogonal to A, and $F(t)$ evolves in time in this orthogonal space. Naturally, we expect that

$$\langle F(t)A^+\rangle = 0. \tag{4.48}$$

This result is easily proved by recognizing that

$$F(t) = e^{iQ\mathscr{L}t}iQ\mathscr{L}A = Qe^{iQ\mathscr{L}t}iQ\mathscr{L}A = QF(t). \tag{4.49}$$

Thus, we have

$$\langle QF(t)A^+\rangle = \langle F(t)(QA)^+\rangle = 0, \tag{4.50}$$

because Q is Hermitian.

The quantity $F(t)$ given in Eq. (4.49) is known as the random force. Equation (4.48) states that the random force $F(t)$ does not correlate with A. This is an important result, as will be seen in a later discussion.

We shall now simplify the quantity associated with the integral in Eq. (4.45). Since

$$\begin{aligned} iP\mathscr{L}F(\tau) &= iP\mathscr{L}e^{iQ\mathscr{L}\tau}iQ\mathscr{L}A = iP\mathscr{L}F(\tau) = iP\mathscr{L}QF(\tau) \\ &= \langle i\mathscr{L}QF(\tau)A^+\rangle\langle AA^+\rangle^{-1}A \end{aligned} \tag{4.51}$$

and recognizing that both Q and $\mathscr{L}$ are Hermitian operators, we can rewrite Eq. (4.51) as

$$\begin{aligned} iP\mathscr{L}F(\tau) &= -\langle F(\tau)(Qi\mathscr{L}A)^+\rangle\langle AA^+\rangle^{-1}A \\ &= -\langle F(\tau)F^+\rangle\langle AA^+\rangle^{-1}A \equiv -K(\tau)A. \end{aligned} \tag{4.52}$$

Combining Eqs. (4.51) and (4.52) with (4.45), we obtain

$$\frac{dA(t)}{dt} = i\Omega A(t) - \int_0^t d\tau\, K(\tau)A(t-\tau) + F(t). \tag{4.53}$$

This equation is known as the generalized Langevin equation. The quantity $K(\tau)$, defined as

$$K(\tau) = \langle F(\tau)F^+\rangle\langle AA^+\rangle^{-1}, \tag{4.54}$$

is the memory function.

This memory function describes the extent to which the past value of $A(t)$ influences the value of $\partial A(t)/\partial t$ in the present. It can also be written as

$$K(\tau) = \langle (Q\dot{A})e^{-iQ\mathscr{L}\tau}(Q\dot{A})^+\rangle\langle AA^+\rangle^{-1}, \tag{4.55}$$

where $\dot{A} = i\mathscr{L}A$ is the initial velocity of A. It is clear that only the part of $\dot{A}$ that is orthogonal to A enters the expression of the memory function. Note that the memory function is not determined by the Liouville operator $\mathscr{L}$ but rather by the modified Liouville operator $Q\mathscr{L}$.

From Eq. (4.53) we can obtain an equation for the correlation function $C_A(t)$. This is obtained by multiplying A^+ and then taking an ensemble average. Thus,

$$\partial C_A(t)/\partial t = i\Omega C_A(t) - \int_0^t d\tau\, K(\tau)C_A(t-\tau), \tag{4.56}$$

where the random force $F(t)$ drops out because of Eq. (4.48). Equation (4.56) reduces to Eq. (4.19) if $\Omega = 0$.

4.4 TIME-CORRELATION FUNCTION MATRICES

The foregoing discussion was restricted to a consideration of a single dynamic variable **A**. However, when there are several coupled variables, we need to generalize this analysis to include the entire set of the coupled variables $\{A_1, A_2, \ldots, A_n\}$. In this case, the dynamic variable **A** is represented by the column matrix

$$\mathbf{A} = \begin{bmatrix} A_1 \\ A_2 \\ \vdots \\ A_n \end{bmatrix}. \tag{4.57}$$

The elements of the time-correlation–function matrix are defined as

$$C_{ij}(t) = \langle A_i(t)A_j^+ \rangle = \langle (e^{i\mathscr{L}t}A_i)A_j^+ \rangle, \tag{4.58}$$

where the diagonal elements correspond to the autocorrelation functions and the off-diagonal elements are the cross-correlation functions.

We shall show that the time dependence of the time-correlation function matrix also obeys Eq. (4.56). For convenience, we again assume that $\langle A_i \rangle = 0$ for $i = 1, \ldots, n$.

The projection operator P that projects any function onto the variable A_i and its complement Q is given by

$$P = 1 - Q = A_i^+\{\langle AA^+ \rangle^{-1}\}_{ij}A_j, \tag{4.59}$$

where the repeated index denotes summation over that index.

Following exactly the same arguments used to derive Eq. (4.53) for a single variable, we obtain a generalized Langevin equation for the dynamic variables A_j as

$$dA_j(t)/dt = i\Omega_{jk}A_k(t) - \int_0^t d\tau\, K_{jk}(\tau)A_k(t-\tau) + F_j(t). \tag{4.60}$$

Here Ω_{jk} is the jk component of the frequency matrix

$$\boldsymbol{\Omega} = \langle (\mathscr{L}\mathbf{A})\mathbf{A}^+ \rangle \langle \mathbf{A}\mathbf{A}^+ \rangle^{-1}, \tag{4.61}$$

where $\mathbf{A}^+$ is the conjugate matrix of **A** given in Eq. (4.57) and $K_{jk}(\tau)$ is the jk component of the memory-function matrix

$$\mathbf{K}(\tau) = \langle \mathbf{F}(\tau)\mathbf{F}^+ \rangle \langle \mathbf{A}\mathbf{A}^+ \rangle^{-1}. \tag{4.62}$$

The term $F_j(t)$ is the j component of the random force given by

$$F_j(t) = e^{iQ\mathscr{L}t}Qi\mathscr{L}A_j = e^{iQ\mathscr{L}t}Q\dot{A}_j. \tag{4.63}$$

We obtain from Eq. (4.60) an equation for the elements of the time-correlation function matrix by

$$dC_{jl}(t)/dt = i\Omega_{jk} C_{kl}(t) - \int_0^t d\tau \, K_{jk}(\tau) C_{kl}(t-\tau), \tag{4.64}$$

where we have again used the summation convention that a repeated index denotes summation over that index.

In matrix notation, Eq. (4.64) becomes

$$d\mathbf{C}(t)/dt = i\mathbf{\Omega} \cdot \mathbf{C}(t) - \int_0^t d\tau \, \mathbf{K}(\tau) \cdot \mathbf{C}(t-\tau), \tag{4.65}$$

where the bold dot denotes matrix multiplication. Equation (4.65) shows that if $\mathbf{\Omega}$ and $\mathbf{K}(\tau)$ are known, the time-correlation matrix can be obtained by manipulating the various matrices that enter into Eq. (4.65). Both Eqs. (4.53) and (4.65) are exact consequences of the Liouville equation and the projection operator. They are known as the Mori transforms of the Liouville equation.

4.5 MICROSCOPIC THEORY FOR BROWNIAN MOTION

We shall now use the memory-function formalism to set up a microscopic theory for Brownian motion. Consider a heavy Brownian particle moving with velocity $\mathbf{u}$ through a stationary fluid consisting of light particles; the position and momentum of the heavy particle are denoted by $\mathbf{r}$ and $\mathbf{p}$, respectively. The mass of the heavy particles is m and that of the fluid particle m_0. The positions and momenta of the fluid particles are given by $\mathbf{r}_i$ and $\mathbf{p}_i$ ($i = 1, 2, \ldots, N$). Using the Hamiltonian of the system (the Brownian particle plus the fluid) given by

$$\begin{aligned} \mathscr{H} &= \sum_i \frac{p_i^2}{2m_0} + \frac{p^2}{2m} + \sum_{i,j}{}' V(r_{ij}) + \sum_j U(|\mathbf{r}_j - \mathbf{r}|) \\ &= \mathscr{H}_0 + \frac{p^2}{2m} + \sum_j U(|\mathbf{r}_j - \mathbf{r}|), \end{aligned} \tag{4.66}$$

we shall examine the conditions under which the Langevin equation as given in Eq. (4.32) or (4.33) is valid. The analysis will be carried out using classical mechanics. In Eq. (4.66), $V(r_{ij})$ is the interaction potential owing to the fluid particles, and $U(|\mathbf{r}_j - \mathbf{r}|)$ is the interaction potential between the fluid particle j and the heavy Brownian particle. To apply the projection-operator formalism to study Brownian motion, we first choose the dynamic operator A to be the momentum of the Brownian particle $\mathbf{p}$:

$$\mathbf{A} = \begin{bmatrix} p_x \\ p_y \\ p_z \end{bmatrix} = \mathbf{p}. \tag{4.67}$$

The projection operator P is given by

$$P = \mathbf{p}^+\langle \mathbf{p}\mathbf{p}^+\rangle^{-1}\mathbf{p}, \tag{4.68}$$

which, when operating on a vector function $\mathbf{G}$, gives

$$P\mathbf{G} = \langle \mathbf{G}\mathbf{p}^+\rangle\langle \mathbf{p}\mathbf{p}^+\rangle^{-1}\mathbf{p} = (1/mkT)\langle \mathbf{G}\mathbf{p}^+\rangle\mathbf{p}. \tag{4.69}$$

To obtain the last step of this equation, we utilized the property that $\langle \mathbf{p}\mathbf{p}^+\rangle^{-1} = \mathbf{I}/mkT$, owing to the equipartition theorem.

Given the projection operator and the Hamiltonian, we can now calculate the various quantities that appear in the generalized Langevin equation [Eq. (4.60)]. We first recognize that the frequency matrix $\mathbf{\Omega}$ is

$$\mathbf{\Omega} = \langle(\mathscr{L}\mathbf{p})\mathbf{p}^+\rangle\langle \mathbf{p}\mathbf{p}^+\rangle^{-1} = \mathbf{0} \tag{4.70}$$

because $(\mathscr{L}\mathbf{p})$ and $\mathbf{p}$ have opposite time-reversal symmetry.

The random force is given by

$$\mathbf{F}(t) = e^{i(1-P)\mathscr{L}t}\mathbf{F} = e^{i(1-P)\mathscr{L}t}\dot{\mathbf{p}}, \tag{4.71}$$

because the random force at $t = 0$ is given by $\mathbf{F} = i(1 - P)\mathscr{L}\mathbf{p} = i\mathscr{L}\mathbf{p} = \dot{\mathbf{p}}$. To proceed with the evaluation of $\mathbf{F}(t)$, we need to have an explicit form of the Liouville operator $\mathscr{L}$. From the Hamiltonian given by Eq. (4.66), it is easy to find $\mathscr{L}$ in accordance with Eq. (4.50). We split $\mathscr{L}$ into two parts,

$$\mathscr{L} = \mathscr{L}_0 + \mathscr{L}_1, \tag{4.72}$$

where

$$i\mathscr{L}_0 = (1/m_0)\sum_i (\mathbf{p}_i \cdot (\partial/\partial \mathbf{r}_i) + \sum_i \mathbf{F}_i \cdot (\partial/\partial \mathbf{p}_i)) \tag{4.73}$$

and

$$i\mathscr{L}_1 = (1/m)\mathbf{p} \cdot \partial/\partial \mathbf{r} + \mathbf{F} \cdot \partial/\partial \mathbf{p}. \tag{4.74}$$

Here in Eqs. (4.73) and (4.74), $\partial/\partial \mathbf{r}_i$ denotes $(\hat{x}\, \partial/\partial x_i + \hat{y}\, \partial/\partial y_i + \hat{z}\, \partial/\partial z_i)$. Similar abbreviated notations are also used for $\partial/\partial \mathbf{p}_i$, $\partial/\partial \mathbf{r}$, and $\partial/\partial \mathbf{p}$. The term $\mathbf{F}_i$, the force acting on the ith fluid particle, is given by $-\partial \mathscr{H}/\partial \mathbf{r}_i$; and $\mathbf{F}$, the force acting on the Brownian particle, is given by $-\partial \mathscr{H}/\partial \mathbf{r}$.

In this division of the Liouville operator, $\mathscr{L}_0$ describes the fast dynamic process associated with the light fluid particles. On the other hand, the first term on the right-hand side of $\mathscr{L}_1$ [Eq. (4.74)] represents the free streaming of the Brownian particle, and the second term describes the interaction of the Brownian particle with all the fluid particles.

Since $\mathscr{L}_0$ acts only on the coordinates and momenta of the fluid particles, we have

$$\mathscr{L}_0 \mathbf{p} = 0 \tag{4.75}$$

The fluid particles will be considered to be the bath and the Brownian particle, the system of interest. The quantity denoted by

$$\mathbf{f}(t) \equiv e^{i\mathscr{L}_0 t}\dot{\mathbf{p}} \tag{4.76}$$

then represents the time evolution of the force $\dot{\mathbf{p}}$ acting on the Brownian particle by the random motion of the fluid particles. We shall now show that in the limit of $m_0/m \to 0$ the random force given by Eq. (4.71) is reduced to $\mathbf{f}(t)$.

The reason for splitting $\mathscr{L}$ into $\mathscr{L}_0$ and $\mathscr{L}_1$ will become clear from the following consideration. Note that the Brownian particle moves considerably slower than the particles constituting the fluid. This is obvious from the equipartition principle: $\langle|\mathbf{p}|^2\rangle = 3mkT$ and $\langle|\mathbf{p}_i|^2\rangle = 3m_0 kT$; thus

$$|\mathbf{p}_i|/|\mathbf{p}| \simeq (m_0/m)^{1/2}.$$

Using this we can scale the momenta $\mathbf{p}_i$ and $\mathbf{p}$ according to

$$\mathbf{p}_i = (m_0 kT)^{1/2}\mathbf{p}_i', \tag{4.77}$$

$$\mathbf{p} = (mkT)^{1/2}\mathbf{p}', \tag{4.78}$$

where $\mathbf{p}_i'$ and $\mathbf{p}'$ are dimensionless momenta.

In terms of $\mathbf{p}_i'$ and $\mathbf{p}'$, $\mathscr{L}_0$ and $\mathscr{L}_1$ can be written as

$$i\mathscr{L}_0 = (kT/m_0)^{1/2}\sum_i [\mathbf{p}_i'(\partial/\partial\mathbf{r}_i) + \mathbf{F}_i'(\partial/\partial\mathbf{p}_i') = i\mathscr{L}_0'/m_0^{1/2}, \tag{4.79}$$

$$i\mathscr{L}_1 = (kT/m_0)^{1/2}(\mathbf{p}'(\partial/\partial\mathbf{r}) + \mathbf{F}'(\partial/\partial\mathbf{p}')) = i\mathscr{L}'/m^{1/2}, \tag{4.80}$$

where $\mathbf{F}_i' = \mathbf{F}_i/kT$ and $\mathbf{F}' = \mathbf{F}/kT$. Clearly, $\mathscr{L}_0$ is greater than $\mathscr{L}_1$ by a factor of $(m_0/m)^{1/2}$ and $\mathscr{L}$ goes to $\mathscr{L}_0$ as $m_0/m \to 0$.

We next use an identity similar to that given by Eq. (4.44):

$$e^{i(1-P)\mathscr{L}t} = e^{i\mathscr{L}_0 t} + i\int_0^t d\tau\, e^{i(1-P)\mathscr{L}(t-\tau)}[(1-P)\mathscr{L}_1 - P\mathscr{L}_0]e^{i\mathscr{L}_0\tau}. \tag{4.81}$$

Bringing the operator $e^{i(1-P)\mathscr{L}t}$, represented by the right-hand side of Eq. (4.81), to act on $\dot{\mathbf{p}}$, we find that the random force is given by

$$\mathbf{F}(t) = e^{i(1-P)\mathscr{L}t}\dot{\mathbf{p}} = \mathbf{f}(t) + i\int_0^t d\tau\, e^{i(1-P)\mathscr{L}(t-\tau)}[(1-P)\mathscr{L}_1 - P\mathscr{L}_0]\mathbf{f}(t). \quad (4.82)$$

Since $\mathscr{L}_1$ is smaller than $\mathscr{L}_0$ by a factor of $(m_0/m)^{1/2}$, in the limit of $(m_0/m)^{1/2} \to 0$ the operator $(1 - P)\mathscr{L}_1$ can be neglected. The term involving $P\mathscr{L}_0$ is given by

$$P\mathscr{L}_0\mathbf{f}(\tau) = (1/mkT)\langle[\mathscr{L}_0\mathbf{f}(\tau)]\mathbf{p}^+\rangle\mathbf{p},$$

which can be reduced to

$$P\mathscr{L}_0\mathbf{f}(\tau) = (1/mkT)\{\langle[\mathscr{L}_0\mathbf{f}(\tau)]\mathbf{p}^+\rangle_0\,\mathbf{p} + O(m_0/m)\}, \quad (4.83)$$

where $O(m_0/m)$ represents terms proportional to m_0/m. The subscript 0 to the angular bracket represents the ensemble average taken with respect to the bath Hamiltonian $\mathscr{H}_0$ and not the full Hamiltonian $\mathscr{H}$ given by Eq. (4.66).

However, since $\mathscr{L}_0$ is a Hermitian operator, the first term on the right-hand side of Eq. (4.83) can be written as

$$\langle[\mathscr{L}_0\mathbf{f}(\tau)]\mathbf{p}^+\rangle_0 = \langle\mathbf{f}(\tau)(\mathscr{L}_0\mathbf{p})^+\rangle, \quad (4.84)$$

which vanishes because of Eq. (4.75).

Therefore, in the limit of $m_0/m \to 0$ Eq. (4.82) becomes

$$\mathbf{F}(t) = \mathbf{f}(t) + O(m_0/m). \quad (4.85)$$

The memory-function matrix given in Eq. (4.71) can thus be written as

$$\mathbf{K}(\tau) = (1/mkT)\langle\mathbf{f}(t)\mathbf{f}^+\rangle_0 = (1/3mkT)\langle\mathbf{f}(t)\cdot\mathbf{f}^+\rangle_0\,\mathbf{I}. \quad (4.86)$$

The last equality results from the isotropy of the fluid system. We further note that

$$\mathbf{f} = i\mathscr{L}\mathbf{p} = i\mathscr{L}_1\mathbf{p} = -\sum_i \frac{\partial}{\partial \mathbf{r}_i} U(|\mathbf{r}_i - \mathbf{r}|), \quad (4.87)$$

which is the total force that the fluid particles exert on the Brownian particle.

Substituting Eqs. (4.70), (4.85), and (4.86) into the generalized Langevin equation [Eq. (4.60)], we obtain

$$\frac{d\mathbf{p}(t)}{dt} = -\frac{1}{3mkT}\int_0^t d\tau\,\langle\mathbf{f}(\tau)\cdot\mathbf{f}^+\rangle_0\,\mathbf{p}(t-\tau) + \mathbf{f}(\tau). \quad (4.88)$$

It is interesting to note that in the limit $m_0/m \to 0$ the time-correlation function $\langle\mathbf{f}(\tau)\cdot\mathbf{f}^+\rangle_0$ contains only the microscopically rapid dynamic processes characteristic of the motion of the light fluid particles. Consequently, $\langle \mathbf{f}(\tau) \cdot \mathbf{f}^{+} \rangle_0$ will decay rapidly to zero in the time scale of the motion of the Brownian particle. Thus, to describe the motion of the Brownian particle, it is a good approximation to write

$$(1/3mkT)\langle \mathbf{f}(\tau) \cdot \mathbf{f}^{+} \rangle_0 = (2\xi/m)\delta(\tau), \tag{4.89}$$

where ξ, known as the friction coefficient, is given by

$$\xi = \frac{1}{2kT} \int_0^\infty dt \, \langle \mathbf{f}(t) \cdot \mathbf{f}^{+} \rangle_0. \tag{4.90}$$

[Note that a factor 2 is in Eq. (4.89) because the time integral in Eq. (4.88) extends over only positive times.] This formula was first obtained by Kirkwood (1946) using a rather different technique. It is important to note that the kinetic energy (the particle streaming) of the Brownian particle does not enter the formula of the friction coefficient. The friction coefficient ξ depends only on the area under the time-correlation function of the total force of the fluid particle with respect to a force center $\mathbf{r}$ that represents the position of the Brownian particle. Since the position of the Brownian particle is arbitrary, ξ is independent of $\mathbf{r}$. This result has been tested by a computer molecular-dynamics calculation and is found to be correct.

Substituting Eq. (4.89) into (4.88), we obtain the well-known Langevin equation given by Eq. (4.32).

The time-correlation function for the velocity of the Brownian particle is given by

$$(d/dt)\, \langle \mathbf{u}(t) \cdot \mathbf{u}^{+} \rangle_0 = -(\xi/m)\langle \mathbf{u}(t) \cdot \mathbf{u}^{+} \rangle_0, \tag{4.91}$$

or

$$\langle \mathbf{u}(t) \cdot \mathbf{u}^{+} \rangle = (3kT/m)e^{-\alpha t}, \qquad \alpha = \xi/m, \tag{4.92}$$

because

$$\langle \mathbf{f}(t)\mathbf{u}^{+} \rangle_0 = 0.$$

In conclusion, we note that the Langevin equation is valid only when m_0/m is small and the perturbation of the random force owing to the fluid particle is short so that the memory function can be represented by a δ-function. These assumptions are clearly valid for the motion of a large colloidal particle in a fluid consisting of light particles. Representing the memory function by a δ function as given by Eq. (4.89) is known as a Markov approximation.

The theory of Brownian motion is closely connected to the hydrodynamic theory of particle diffusion. To show this connection, we recall the time-correlation function $F(\mathbf{q}, t)$ [Eq. (2.45)] as

$$F(q, t) = \frac{1}{N}\left\langle \sum_n \exp\{iq \cdot [\mathbf{r}_n(t) - \mathbf{r}_n]\} \right\rangle = \frac{1}{N}\left\langle \sum_n \exp[iq \cdot \Delta\mathbf{r}_n(t)] \right\rangle. \tag{4.93}$$

This time correlation is related to the Van Hove space–time self-correlation function $G(\mathbf{r}, t)$ given as

$$G(\mathbf{r}, t) = \frac{1}{N}\left\langle \sum_n \delta[\mathbf{r} - \Delta\mathbf{r}_n(t)] \right\rangle \tag{4.94}$$

by a spatial Fourier transform:

$$\int_V d^3r\, G(\mathbf{r}, t)e^{i\mathbf{q}\cdot\mathbf{r}} = F(\mathbf{q}, t). \tag{4.95}$$

In Chapter 2, we described the translational diffusion of macromolecules in a dilute solution in terms of the time-correlation function $F(q, t)$. We shall now show, using the memory-function formalism, that under appropriate approximations $G(\mathbf{r}, t)$ is governed by a diffusion equation.

Using Eq. (4.19), we identify $C_A(t)$ to $F(q, t)$ and write

$$\frac{\partial}{\partial t} F(\mathbf{q}, t) = -\int_0^t d\tau\, K(\tau)F(\mathbf{q}, t - \tau), \tag{4.96}$$

where the memory function $K(\tau)$ is given by Eq. (4.30) as

$$K(\tau) = \frac{1}{2\pi i}\oint ds \frac{e^{s\tau}\tilde{\phi}(s)}{[1 - (1/s)\tilde{\phi}(s)]} \tag{4.97}$$

The contour integral, together with the factor $\frac{1}{2}\pi i$ indicates the inverse Laplace transform. Here the function $\tilde{\phi}(s)$ is given by Eqs. (4.7) and (4.18) as

$$\tilde{\phi}(s) = \tilde{\phi}(s)/F(\mathbf{q}, 0) = \tilde{\phi}(s) = \int_0^\infty dt\, e^{-st}\phi(t). \tag{4.98}$$

The function $\phi(t)$, given by Eq. (4.7), is calculated to be

$$\begin{aligned}\phi(t) &= \frac{1}{N}\left\langle \sum_n [\mathbf{q} \cdot \mathbf{u}_n(t)][\mathbf{q} \cdot \mathbf{u}_n^+] \exp[i\mathbf{q} \cdot \Delta\mathbf{r}_n(t)] \right\rangle \\ &\cong \frac{q^2}{3N}\left\langle \sum_n [\mathbf{u}_n(t) \cdot \mathbf{u}_n^+] \exp[iq \cdot \Delta\mathbf{r}_n(t)] \right\rangle \equiv q^2\psi(q, t).\end{aligned} \tag{4.99}$$

Since the time-correlation function $F(\mathbf{q}, t)$ and its time derivatives are local quantities, their values fall off rapidly with increasing $\Delta\mathbf{r}_n(t)$, and thus the time-correlation function $\psi(q, t)$ should be finite as $q \to 0$.

Taking the $q \to 0$ limit, we find that $\psi(\mathbf{q}, t)$ becomes

$$\lim_{q \to 0} \psi(q, t) = \frac{1}{3N} \left\langle \sum_n \mathbf{u}_n(t) \cdot \mathbf{u}_n^+ \right\rangle = \tfrac{1}{3}\langle \mathbf{u}(t) \cdot \mathbf{u}^+ \rangle = C_u(t), \qquad (4.100)$$

which is the velocity–velocity time-correlation function. Here $\mathbf{u}$ is the velocity of a representative particle.

Thus, in the $q \to 0$ limit, $\phi(t)$ is given by

$$\phi(t) = q^2 C_u(t). \qquad (4.101)$$

The memory function $K(\tau)$ is also simplified. Since $\tilde{\phi}(s)$ is proportional to q^2, i.e. to the lowest order in q, $K(\tau)$ is proportional to q^2. This is clearly seen from Eq. (4.97) by expanding the integrand in powers $\tilde{\phi}(s)$. Thus, in the $q \to 0$ limit, we have

$$\lim_{q \to 0} K(\tau) = \phi(t) = q^2 C_u(t). \qquad (4.102)$$

Furthermore, because we are interested in the macroscopic relaxation time, which is longer than the period of the velocity fluctuations, the Markov approximation can be used; i.e., we approximate $K(\tau)$ as

$$K(\tau) = 2Dq^2 \delta(\tau), \qquad (4.103)$$

where D is the diffusion coefficient given by

$$D = \tfrac{1}{3} \int_0^\infty dt \, \langle \mathbf{u}(t) \cdot \mathbf{u}^+ \rangle. \qquad (4.104)$$

Substituting Eq. (4.103) into (4.96) gives

$$\partial F(\mathbf{q}, t)/\partial t = -Dq^2 F(\mathbf{q}, t). \qquad (4.105)$$

Spatial Fourier inversion of $F(\mathbf{q}, t)$ gives $G(\mathbf{r}, t)$. Thus, converting $F(\mathbf{q}, t)$ to $G(\mathbf{r}, t)$ results in the diffusion equation for $G(\mathbf{r}, t)$:

$$\partial G(\mathbf{r}, t)/\partial t = D\nabla^2 G(\mathbf{r}, t) \qquad (4.106)$$

with the initial condition $G(\mathbf{r}, 0) = \delta(\mathbf{r})$.

We have thus established that in the $q \to 0$ limit and under the Markov approximation (or in the hydrodynamic region), the behavior of $G(\mathbf{r}, t)$ is governed by a diffusion equation.

The diffusion coefficient D is related to the friction coefficient ξ by the Einstein relation

$$D = kT/\xi. \qquad (4.107)$$

This can easily be shown by substituting the expression of the velocity–time-correlation function [Eq. (4.92)] into Eq. (4.104) and carrying out the time integration.

We can also calculate the mean distance that the Brownian particle traverses in time t. From the Langevin equation [Eq. (4.32)], we obtain

$$\mathbf{u}(t) - \mathbf{u}(0)e^{-\alpha t} = \int_0^t d\tau\, \mathbf{f}(\tau)e^{-\alpha(t-\tau)}, \tag{4.108}$$

where $\alpha = \xi/m$.

A second integration yields the position $\mathbf{r}(t)$ of the Brownian particle that was initially at $\mathbf{r}(0)$ with velocity $\mathbf{u}(0)$:

$$\mathbf{r}(t) - \mathbf{r}(0) = \mathbf{u}(0)\,\frac{(1 - e^{-\alpha t})}{\alpha} + \frac{1}{\alpha}\int_0^t dt' \int_0^{t'} d\tau\, \mathbf{f}(\tau)e^{-\alpha(t'-\tau)}$$

or

$$\Delta\mathbf{r}(t) = \mathbf{u}(0)\frac{(1 - e^{-\alpha t})}{\alpha} + \frac{1}{\alpha}\int_0^t d\tau\, f(\tau)[1 - e^{-\alpha(t-\tau)}]. \tag{4.109}$$

The mean distance is obtained by squaring $\Delta\mathbf{r}(t)$ and then averaging and calculating the double integrals. We obtain, using the equipartition theorem $(\langle|\mathbf{u}^2(0)|\rangle = 3kT/m)$;

$$\langle|\Delta\mathbf{r}(t)|^2\rangle = (6kT/m\alpha^2)(\alpha t - 1 + e^{-\alpha t}), \tag{4.110}$$

which gives, in the range of large values of time ($t \gg \alpha^{-1}$),

$$\langle|\Delta\mathbf{r}(t)|^2\rangle = 6kTt/\xi = 6Dt. \tag{4.111}$$

On the other hand, for very small values of time ($t \ll \alpha^{-1}$), we get

$$\langle|\Delta\mathbf{r}(t)|^2\rangle = (3kT/m)t^2 = \langle\mathbf{u}^2(0)\rangle t^2, \tag{4.112}$$

which is entirely owing to the uniform kinetic motion of the Brownian particle.

4.6 SPATIAL DEPENDENCE OF THE TIME-CORRELATION FUNCTION

In the microscopic theory of the time-correlation function, such as that given in Chapter 1 and in this chapter, we focused only on the study of time dependence. So far we have not considered the spatial dependence of the dynamic variable. In a continuum system, the dynamic variable varies with space and time. We saw in Chapter 2 that the spatial fluctuation of the dielectric constant causes the scattering of light in the nonforward direction, and the temporal fluctuation causes the shift from the incident-laser frequency. Thus, the spatial variation of the dynamic variable plays just as important a role as the temporal dependence. We shall show how the spatial dependence of the dynamic variable is incorporated into the microscopic theory of the time-correlation function.

Similar to the microscopic expression of the polarizability density given in Chapter 2, a general way of expressing a dynamic variable $A(\mathbf{r}, t)$ is given by

$$A(\mathbf{r}, t) = \sum_i a_i(t)\delta[\mathbf{r} - \mathbf{r}_i(t)], \tag{4.113}$$

where $a_i(t)$ is the molecular quantity associated with particle i at the position $\mathbf{r}_i(t)$ at time t. For example, $A(\mathbf{r}, t)$ is magnetization if a_i is the magnetic moment (spin plus orbital) operator of particle i. On the other hand, $A(\mathbf{r}, t)$ is the polarizability density if a_i is the polarizability of molecule i.

If $A(\mathbf{r}, t)$ is coupled to an external field $F(\mathbf{r}, t)$, then the Hamiltonian in the presence of the external field is given by

$$\mathscr{H}(t) = \mathscr{H} - \int d^3r\, A(\mathbf{r}, t)F(\mathbf{r}, t), \tag{4.114}$$

where $\mathscr{H}$ is the Hamiltonian of the system in the absence of the external field. We shall briefly generalize the linear-response theory as presented in Chapter 1 to include spatial dependence. A detailed discussion of linear response theory including the spatial dependence is given in Appendix. Given the Hamiltonian, we can calculate the linear response of the system associated with the dynamic variable $B(\mathbf{r}, t)$. Following the method of Chapter 1, we find that the induced linear response in the quantal case is given by

$$\begin{aligned}\langle \Delta B(\mathbf{r}, t)\rangle &= \langle B(\mathbf{r}, t)\rangle - \langle B(\mathbf{r}, t)\rangle_e \\ &= \frac{i}{\hbar}\int_0^t dt' \int d^3r' \, \langle [B(\mathbf{r}, t), A(\mathbf{r}', t')]\rangle F(\mathbf{r}', t'),\end{aligned} \tag{4.115a}$$

and in the classical case by

$$\langle \Delta B(\mathbf{r}, t)\rangle = \int_0^t dt' \int d^3r' \, \langle \{B(\mathbf{r}, t), A(\mathbf{r}', t')\}_{\mathrm{PB}}\rangle F(\mathbf{r}', t'). \tag{4.115b}$$

Here the boundary for the spatial integration includes all regions in which the external force has effectively interacted with the ystem.

In a liquid or a homogeneous solid at equilibrium, the ensemble-averaged quantity in Eq. (4.115) is stationary in both space and time, i.e., it depends only on the relative spatial distance $(\mathbf{r} - \mathbf{r}')$ and the time difference $(t - t')$.

Consequently, Eq. (4.115) can be written in terms of the convolution integral in space and time as

$$\langle \Delta B(\mathbf{r}, t) \rangle = \int_0^t dt' \int dr' \, \phi_{BA}(\mathbf{r} - \mathbf{r}', t - t') F(\mathbf{r}', t'), \tag{4.116}$$

where $\phi_{BA}(\mathbf{r}, t)$ is the response-function density and is given, in the quantal case by

$$\phi_{BA}(\mathbf{r}, t) = i/\hbar \langle [B(\mathbf{r}, t), A(\mathbf{0}, 0)] \rangle, \tag{4.117a}$$

and in the classical case by

$$\phi_{BA}(\mathbf{r}, t) = \langle \{B(\mathbf{r}, t), A(\mathbf{0}, 0)\}_{\mathrm{PB}} \rangle. \tag{4.117b}$$

These results are similar to Eqs. (1.97)–(1.99).

With the help of the convolution theorem, we shall now perform a Fourier transform in space and a complex Laplace transform in time to convert Eq. (4.116) into

$$\langle \Delta B(\mathbf{q}, z) \rangle = X_{BA}(\mathbf{q}, z) F(\mathbf{q}, z), \tag{4.118}$$

where

$$\langle \Delta B(\mathbf{q}, z) \rangle = \int_0^\infty dt \int d^3r \, e^{i(\mathbf{q} \cdot \mathbf{r} - zt)} \langle \Delta B(\mathbf{r}, t) \rangle \tag{4.119}$$

$$F(\mathbf{q}, z) = \int_0^\infty dt \int d^3r \, e^{i(\mathbf{q} \cdot \mathbf{r} - zt)} F(\mathbf{r}, t), \tag{4.120}$$

and

$$\chi_{BA}(\mathbf{q}, z) = \int_0^\infty dt \int d^3r \, e^{i(\mathbf{q} \cdot \mathbf{r} - zt)} \phi_{BA}(r, t). \tag{4.121}$$

Note that except for the presence of the additional variable $\mathbf{q}$, these equations are similar to those we obtain in Chapter 1 [compare with Eqs. (1.108)–(1.112)].

Following Chapter 1, we can set the complex variable z identical with a real frequency ω, and $\chi_{BA}(\mathbf{q}, \omega)$ becomes the generalized susceptibility. Further, we can separate $\chi_{BA}(\mathbf{q}, \omega)$ into a real (χ'_{BA}) and an imaginary (χ''_{BA}) part:

$$\chi_{BA}(\mathbf{q}, \omega) = \chi'_{BA}(\mathbf{q}, \omega) - i\chi''_{BA}(\mathbf{q}, \omega). \tag{4.122}$$

In terms of the relaxation function, the real part is given by [compare with Eqs. (1.130) and (1.131)]

$$\chi'_{BA}(\mathbf{q}, \omega) = R_{BA}(\mathbf{q}, 0) - \omega \int_0^\infty dt \, R_{BA}(\mathbf{q}, t) \sin \omega t, \tag{4.123}$$

$$\chi''_{BA}(\mathbf{q}, \omega) = \omega \int_0^\infty dt \, R_{BA}(\mathbf{q}, t) \cos \omega t, \tag{4.124}$$

where $\chi'_{BA}(\mathbf{q}, \omega)$ and $\chi''_{BA}(\mathbf{q}, \omega)$ are related by the Kramers–Kronig relation [Eqs. (1.125) and (1.126)]. From Eqs. (4.122) and (4.123), the generalized susceptibility at zero frequency is equal to the relaxation function $R_{BA}(\mathbf{q}, t)$ at $t = 0$. We set the zero frequency susceptibility to be $\chi_{BA}(\mathbf{q})$.

In most cases, the dynamic variable monitored in an experiment is associated with the same quantity $\mathbf{A}(\mathbf{q}, t)$. To obtain $\langle \Delta A(\mathbf{q}, t) \rangle$, we set B equal to A in the foregoing equations. The imaginary part of the susceptibility $\chi''_{AA}(\mathbf{q}, \omega)$ is of particular importance. It can be written in terms of a Fourier integral [$R_{AA}(\mathbf{q}, t)$ is even in t]:

$$\chi''_{AA}(\mathbf{q}, \omega) = \frac{\omega}{2} \int_{-\infty}^{\infty} dt\, e^{-i\omega t} R_{AA}(\mathbf{q}, t). \tag{4.125}$$

Following the procedure leading to Eq. (1.152), we can reduce these equations to

$$\chi''_{AA}(\mathbf{q}, \omega) = \frac{1}{2\hbar} (e^{\beta\hbar\omega} - 1) \int_{-\infty}^{\infty} dt\, e^{-i\omega t} C_A(\mathbf{q}, t). \tag{4.126}$$

This is the fluctuation–dissipation theory presented in terms of the one-sided time-correlation function $C_A(\mathbf{q}, t) \equiv \langle A(\mathbf{q}, t) A^+(\mathbf{q}) \rangle$. The classical result can be obtained by taking the limit $\beta\hbar\omega \ll 1$ in Eq. (4.126); this results in

$$\chi''_{AA}(\mathbf{q}, \omega) = \frac{\beta\omega}{2} \int_{-\infty}^{\infty} dt\, e^{-i\omega t} C_A(\mathbf{q}, t) = \beta\omega \operatorname{Re} \int_{0}^{\infty} dt\, e^{-i\omega t} C_A(\mathbf{q}, t). \tag{4.127}$$

The sum rule for $\chi'(\mathbf{q}, \omega)$ is related to the moment of the time correlation $C_A(\mathbf{q}, t)$. This can be seen by inverting Eq. (4.126) as

$$C_A(\mathbf{q}, t) = \frac{\hbar}{\pi} \int_{-\infty}^{\infty} d\omega\, e^{i\omega t} \frac{\chi''(\mathbf{q}, \omega)}{(e^{\beta\hbar\omega} - 1)}, \tag{4.128}$$

Since $C_A(\mathbf{q}, t)$ can be expressed in terms of the moment expression as [compare with Eq. (1.184)]

$$C_A(\mathbf{q}, t) = C_A(\mathbf{q}, 0) \left[\sum_{N=0}^{\infty} \frac{(i)^n M_n(\mathbf{q})}{n!} t^n \right], \tag{4.129}$$

we can obtain the sum rule for $\chi''(\mathbf{q}, \omega)$ after expanding the phase function $e^{i\omega t}$ in a Taylor series in time t as

$$\frac{1}{\pi} \int_{-\infty}^{\infty} d\omega \frac{\omega^{2n} \chi''(\mathbf{q}, \omega)}{(e^{\beta\hbar\omega} - 1)} = \frac{1}{\hbar} C_A(\mathbf{q}, 0) M_{2n}(\mathbf{q}). \tag{4.130}$$

In the classical limit, if we replace $(e^{\beta\hbar\omega} - 1)$ by $\beta\hbar\omega$, then Eq. (4.130) is reduced to

$$\frac{1}{\pi} \int_{-\infty}^{\infty} d\omega\, \omega^{2n-1} \chi''_{AA}(\mathbf{q}, \omega) = \chi_{AA}(\mathbf{q}) M_{2n}(\mathbf{q}), \tag{4.131}$$

where we have related the static susceptibility $\chi_{AA}(\mathbf{q})$ to the initial amplitude of the time-correlation function $C_A(\mathbf{q}, 0)$, because in classical systems $R_{AA}(\mathbf{q}, t)$ is equal to $\beta C_A(\mathbf{q}, t)$ at all times [compare with Eq. (1.140)]. Further, in classical systems only the even moments exist, because $\chi''(\mathbf{q}, \omega)$ is an even function of ω. Therefore, for classical systems, correlation functions are even in time.

Since $M_0(\mathbf{q}) = 1$, Eq. (4.131) yields an important sum rule for the classical system as (Martin and Kadanoff, 1963)

$$\frac{1}{\pi}\int_{-\infty}^{\infty} d\omega\, \chi''_{AA}(q, \omega)/\omega = \chi_{AA}(\mathbf{q}). \tag{4.132}$$

It is clear that the inclusion of the spatial dependence in the dynamic variable and in the Hamiltonian does not result in any difficulty in the linear-response theory. The spatial dependence gives a variable $\mathbf{q}$, known as the wave vector (Forster, 1975). The wave vector describes the range of spatial resolution, just as the variable ω determines the range of temporal resolution. All the equations developed in Chapter 1 are preserved if $\mathbf{q}$ is included with ω.

To generalize the memory-function formalism to include the continuum system, we need to choose a dynamic variable $A(\mathbf{q}, t)$ specifically for the "$\mathbf{q}$ mode." According to Eq. (4.113), the dynamic variable $A(\mathbf{q}, t)$ is equal to

$$A(\mathbf{q}, t) = \int d^3r\, A(\mathbf{r}, t)e^{i\mathbf{q}\cdot\mathbf{r}} = \sum_j a_j(t)e^{i\mathbf{q}\cdot\mathbf{r}_j(t)}. \tag{4.133}$$

The equation of motion for $A(\mathbf{q}, t)$ is again given by Eq. (4.34) as

$$\partial A(\mathbf{q}, t)/\partial t = i\mathscr{L}A(\mathbf{q}, t). \tag{4.134}$$

There is no need to transform the Liouville operator because it acts only on the molecular quantities. However, instead of Eq. (4.34), the projection operator is now defined in terms of $A(\mathbf{q}, t)$ as

$$P = A^+(\mathbf{q})\langle A(\mathbf{q})A^+(\mathbf{q})\rangle^{-1}A(\mathbf{q}) = \beta A^+(\mathbf{q})\chi_{AA}^{-1}(\mathbf{q})A(\mathbf{q}), \tag{4.135}$$

which, when acting on a function $G(\mathbf{q}, t)$, gives

$$PG(\mathbf{q}, t) = \beta\langle G(\mathbf{q}, t)A^+(\mathbf{q})\rangle\chi_{AA}^{-1}(\mathbf{q})A(\mathbf{q}). \tag{4.136}$$

The various steps given in Eqs. (4.41)–(4.53) also apply here. Thus, the generalized Langevin equation for $A(\mathbf{q}, t)$ is given by

$$\partial A(\mathbf{q}, t)/\partial t = i\Omega(\mathbf{q})A(\mathbf{q}, t) - \int_0^t d\tau\, K(\mathbf{q}, t - \tau)A(\mathbf{q}, \tau) + F(\mathbf{q}, t) \tag{4.137}$$

where the frequency $\Omega(\mathbf{q})$, memory function $K(\mathbf{q}, t)$, and random force $F(\mathbf{q}, t)$ are given by

$$\Omega(\mathbf{q}) = \langle\{\mathscr{L}A(\mathbf{q})\}A^+(\mathbf{q})\rangle\langle|A(\mathbf{q})|^2\rangle^{-1} = \beta\langle[\mathscr{L}A(\mathbf{q})]A^+(\mathbf{q})\rangle\chi_{AA}^{-1}(\mathbf{q}), \tag{4.138}$$

$$K(\mathbf{q}, t) = \beta\langle\{e^{iQ\mathscr{L}t}F(\mathbf{q})\}F^+(\mathbf{q})\rangle\chi_{AA}^{-1}(\mathbf{q}), \tag{4.139}$$

$$F(\mathbf{q}, t) = e^{iQ\mathscr{L}t}F(\mathbf{q}) = ie^{iQ\mathscr{L}t}Q\mathscr{L}A(\mathbf{q}), \tag{4.140}$$

where $Q = 1 - P$.

The time-correlation function $C_A(\mathbf{q}, t) \equiv \langle A(\mathbf{q}, t)A^+(\mathbf{q})\rangle$ follows the equation

$$\partial C_A(\mathbf{q}, t)/\partial t = i\Omega(\mathbf{q})C_A(\mathbf{q}, t) - \int_0^t d\tau\, K(\mathbf{q}, t - \tau)C_A(\mathbf{q}, \tau). \tag{4.141}$$

Generalization to include several dynamic variables is obvious. In the general case, we obtain results similar to Eqs. (4.60) and (4.65), with the wave vector $\mathbf{q}$ incorporated in the equation together with the time variable t.

4.7 HYDRODYNAMIC APPROXIMATION AND RELAXATION TIMES

In the microscopic theory of Brownian motion, we have shown that in the limit of $(m_0/m) \to 0$, the memory function becomes the time-correlation function of the random force of the fluid particles acting on the Brownian particle. Since the time-correlation function of the random force of fluid particles is expected to decay rapidly, it is approximated as a δ function. As a result, the time-correlation function of the velocity of the Brownian particle relaxes exponentially with a relaxation time given by α^{-1} [Eq. (4.92)].

Another example of the δ-function approximation is the derivation of the diffusion equation for the self part of the probability-correlation function $F(\mathbf{q}, t)$ [the spatial Fourier transform of the van Hove space–time self-correlation function $G(\mathbf{r}, t)$]. There we showed that in the limit $q \to 0$, the memory function $K(\tau) \to q^2C_u(t)$, $C_u(t)$ is the time correlation function of velocity as given in Eq. (4.100). We argued that if we are interested in the relaxation of the probability-correlation function at a time scale longer than the period of the velocity fluctuation of the Brownian particle, we can approximate the time correlation function of velocity by a δ function. Because of this approximation, the function $F(\mathbf{q}, t)$ has a relaxation time equal to $(Dq^2)^{-1}$.

These examples suggest that for problems in which the separation of variables is possible either in the $q \to 0$ limit or in the $(m_0/m) \to 0$ limit, as in Brownian motion, we can select a dynamic variable (or a set of dynamic variables if they are coupled) and consider it to be a slow variable; the other variables not included in the selection are considered to be fast. Analytically, the projection operator P is used to project from the Liouville equation the part associated with the slow variable. All other variables not selected by the projection operator are assumed to fluctuate rapidly. The dynamics of these fast variables is described by the memory function. Thus, for times greater than the largest relaxation time of the memory function, it is permissible to approximate the memory function $K(\mathbf{q}, \tau)$ as

$$K(\mathbf{q}, \tau) = 2\Gamma(\mathbf{q})\,\delta(\tau) \tag{4.142}$$

where $\Gamma(\mathbf{q})$, known as the relaxation-rate constant, is given by

$$\Gamma(\mathbf{q}) = \int_0^\infty dt\, K(\mathbf{q}, t). \tag{4.143}$$

The approximation of the memory function in terms of the δ function is known as the Markov approximation. Substituting Eq. (4.142) into (4.141) yields

$$\partial C_A(\mathbf{q}, t)/\partial t = i\Omega(\mathbf{q})C_A(\mathbf{q}, t) - \Gamma(\mathbf{q})C_A(\mathbf{q}, t), \tag{4.144}$$

where $\Gamma(\mathbf{q})^{-1}$ is called the relaxation time.

In this case the time-correlation function is given by

$$C_A(\mathbf{q}, t) = C_A(\mathbf{q}, 0)\exp[i\Omega(\mathbf{q})t - \Gamma(\mathbf{q})t], \qquad t > 0 \tag{4.145}$$

From Eq. (4.127), we find that the imaginary part of the susceptibility is

$$\chi''_{AA}(q, \omega) = \beta\omega\,\mathrm{Re}\int_0^\infty dt\, e^{i\omega t}C_A(\mathbf{q}, t) = \frac{\omega\Gamma(\mathbf{q})\chi_{AA}(\mathbf{q})}{[\omega - \Omega(\mathbf{q})]^2 + \Gamma(\mathbf{q})^2}. \tag{4.146}$$

This is the well-known Lorentzian line-shape function, with the spectral center located at $\omega = \Omega(\mathbf{q})$.

If $\Omega(\mathbf{q}) = 0$ [e.g., when $A(\mathbf{q})$ and $\dot{A}(\mathbf{q})$ have opposite time-reversal symmetry], then $\Gamma(\mathbf{q}) = q^2 D$ and $\chi_{AA}(\mathrm{q}) = \chi$ as $q \to 0$, and Eq. (4.146) becomes

$$\chi''_{AA}(\mathbf{q}, \omega) = \frac{\omega D q^2 \chi}{\omega^2 + (Dq^2)^2}. \tag{4.147}$$

This is the result for the diffusion model for $F(\mathbf{q}, t)$, previously described. Under this approximation, the relaxation time $\tau(\mathbf{q}) = (Dq^2)^{-1}$ becomes infinite as $q \to 0$. The $q \to 0$ limit together with the Markov approximation is known as the hydrodynamic approximation. The relaxation times of all hydrodynamic modes become infinite as $q \to 0$. Thus, all hydrodynamic modes may be chosen as dynamic variables in the memory-function formalism. From Eq. (4.147), we obtain

$$\chi_A = \lim_{q \to 0} \int \frac{d\omega}{\pi} \chi''_{AA}(\mathbf{q}, \omega)/\omega, \tag{4.148}$$

which is the thermodynamic sum rule obtained previously [Eq. (4.132)]. Further, we note that

$$D\chi_A = \lim_{\omega \to 0} \left[\lim_{q \to 0} \omega \chi''_{AA}(\mathbf{q}, \omega)/q^2 \right], \tag{4.149}$$

which gives a Kubo-type relaxation for the transport coefficient. It is important to note that the $q \to 0$ limit should be taken before the $\omega \to 0$ limit. Note further that in the hydrodynamic approximation, $\chi''(\mathbf{q}, \omega)$, given by Eq. (4.143), fails to yield a correct second moment for $C_A(\mathbf{q}, t)$. In accordance with Eq. (4.131), the second moment can be written as

$$M_2(\mathbf{q}) = \frac{1}{\pi \chi_{AA}(\mathbf{q})} \int_{-\infty}^{\infty} d\omega\, \omega \chi''_{AA}(\mathbf{q}, \omega). \tag{4.150}$$

If $\omega''_{AA}(\mathbf{q}, \omega)$, given by Eq. (4.147), is substituted into this equation, we obtain a divergent $M_2(\mathbf{q})$. Evidently, the Lorentzian function for $\chi''_{AA}(\mathbf{q}, \omega)$ obtained from the Markov approximation does not fall rapidly at large ω. We can correct this by abandoning the Markov approximation in the memory function. We assume for simplicity that the memory function decays by a single relaxation time τ_0, i.e., that in the $q \to 0$ limit

$$K(\mathbf{q}, \tau) = (Dq^2/\tau_0)e^{-\tau/\tau_0}, \tag{4.151}$$

where Dq^2/τ_0 is the initial value of $K(\mathbf{q}, \tau)$. This equation reduces to the Markov approximation for $\tau_0 \to 0$. We substitute this into Eq. (4.141) and obtain (assuming $\Omega = 0$)

$$\frac{\partial}{\partial t} C_A(\mathbf{q}, t) = -\frac{Dq^2}{\tau_0} \int_0^t d\tau\, e^{-(t-\tau)/\tau_0} C_A(\mathbf{q}, \tau). \tag{4.152}$$

This equation can be solved by Laplace transformation:

$$\tilde{C}_A(\mathbf{q}, s) = \frac{C_A(\mathbf{q}, 0)(s + \tau_0^{-1})}{s(s + \tau_0^{-1}) + Dq^2\tau_0^{-1}}, \tag{4.153}$$

where s is the Laplace variable and

$$\tilde{C}_A(\mathbf{q}, s) = \int_0^{\infty} dt\, e^{-st} C_A(\mathbf{q}, t). \tag{4.154}$$

The susceptibility $\chi''_{AA}(\mathbf{q}, \omega)$ is calculated according to Eq. (4.127) as

$$\chi''_{AA}(\mathbf{q}, \omega) = \beta\omega \operatorname{Re} \tilde{C}_A(\mathbf{q}, s)|_{s=i\omega} = \frac{\omega D q^2 \chi}{\omega^2 + (Dq^2 - \omega^2\tau_0)^2}. \quad (4.155)$$

This is the Drude–Maxwell line-shape function frequently used to fit the dielectric response. For small frequencies such that $\tau_0\omega^2 \ll Dq^2$, it reduces to the Lorentzian function obtained previously under the Markov approximation. The interesting feature of this line-shape function is that $\chi''(q, \omega)$ now falls off faster in the wings; therefore, the second moment is finite. According to Eq. (4.127), we find

$$M_2(\mathbf{q}) = \frac{Dq^2}{\pi} \int_0^\infty d\omega \frac{\omega^2}{\omega^2 + (Dq^2 - \omega^2\tau_0)^2} = \frac{Dq^2}{\sqrt{1 - 4Dq^2\tau_0}} (\sqrt{\lambda_+} - \sqrt{\lambda_-}), \quad (4.156)$$

where $\lambda_\pm$ are given by

$$\lambda_\pm = 1/\lambda\tau_0^2[(1 - 2Dq^2\tau_0) \pm \sqrt{1 - 4Dq^2\tau_0}]. \quad (4.157)$$

In the limit $Dq^2\tau_0 \ll 1$, the second moment reduces to

$$M_2(q) = (Dq^2/\tau_0)[1 + O(Dq^2\tau_0)]. \quad (4.158)$$

Equations (4.158) or (4.156) make an interesting prediction about the diffusion coefficient near a critical point, say in the liquid–gas transition in molecular liquids. As the phase transition is approached, the static susceptibility $\chi_{AA}(\mathbf{q})$ increases without bound because of the increase long-range spatial correlation. In accordance with Eq. (4.150), the second moment $M_2(\mathbf{q})$ should decrease rapidly to zero. Since we do not expect the microscopic time τ_0 to be much affected, Eq. (4.158) thus shows that the diffusion coefficient is expected to decrease to zero as the critical point is approached. This result is experimentally verified.

Although the result for the time-correlation function or for $\chi''_{AA}(\mathbf{q}, \omega)$ is somewhat improved by introducing Eq. (4.151) for the memory function, the line-shape function $\chi''_{AA}(\mathbf{q}, \omega)$ obtained in this approximation [Eq. (4.155)] still yields a divergent fourth moment.

We now generalize the calculation so that more complex situations can be treated using the memory-function approach. The generalized Langevin equation is an integrodifferential equation. It has a convolution form and can be solved by the Laplace transform technique. From Eq. (4.141) we can always write the rigorous expression for the time-correlation function as

$$\tilde{C}_A(\mathbf{q}, s) = C_A(\mathbf{q})/[s - i\Omega(\mathbf{q}) + K(\mathbf{q}, s)], \quad (4.159)$$

which is analogous to Eq. (4.16), with a finite frequency $\Omega(\mathbf{q})$ and inclusion of the wave vector $\mathbf{q}$.

Equation (4.159) can be regarded as the generalization of the hydrodynamic results for one single-dynamic variable. As pointed out previously, if the dynamic variable A has pure time-reversal symmetry, as is often the case, then $\Omega(\mathbf{q}) = 0$. The quantity $K(\mathbf{q}, s)$ is the Laplace transform of the memory function given by

$$K(\mathbf{q}, s) = \int_0^\infty dt\, e^{-st} K(\mathbf{q}, t). \tag{4.160}$$

The Markov approximation is obtained if the Laplace variable s in this equation is set equal to zero. Under this approximation, $K(\mathbf{q}, s)$ is independent of s. However, as shown here, the Markov approximation leads to an infinite second moment for the time-correlation function $C_A(\mathbf{q}, t)$. On the other hand, if $K(\mathbf{q}, t)$ is approximated as a single exponential function, the result is a finite second moment; however, the fourth moment is infinite.

It is reasonable to guess (a correct guess) that the assumption of a two-exponential function for $K(\mathbf{q}, t)$ will lead to a finite fourth moment and an n-exponential function for $K(\mathbf{q}, t)$ will lead to a finite $2n$th moment; higher moments than the $2n$th are still divergent. Thus, to insure the convergence of all moments we need, in principle, to have an infinite-exponential function representating the memory function:

$$K(\mathbf{q}, t) = \int_0^\infty d\tau\, f(\mathbf{q}, \tau) e^{-t/\tau}, \tag{4.161}$$

where $f(\mathbf{q}, \tau)$ corresponds to the distribution function of the relaxation times τ. However, since the distribution function $f(\mathbf{q}, \tau)$ is difficult to obtain, Eq. (4.161) has little practical application. A useful mathematical model for the memory function with finite moments of any order is the Gaussian function

$$K(\mathbf{q}, t) = K(\mathbf{q}) e^{-t^2/\tau_0^2}, \tag{4.162}$$

where $K(\mathbf{q})$ is the initial value of the memory function.

From Eq. (4.127), we find that

$$\begin{aligned} \chi_A''(\mathbf{q}, \omega) &= \beta \operatorname{Re} \frac{\omega \chi_A(\mathbf{q})}{(i\omega + K(\mathbf{q}) \int_0^\infty dt\, e^{-i\omega t} e^{-t^2/\tau_0^2})} \\ &= 2\sqrt{\pi}\, \beta \chi \tau_0 K) \frac{e^{-\omega^2 \tau_0^2/4}}{[(\sqrt{\pi} K \tau_0)^2 + (\omega - K \int_0^\infty dt\, e^{-t^2/\tau_0^2} \sin \omega t)^2]}. \end{aligned} \tag{4.163}$$

In the second line of Eq. (4.163) the q dependence in $\chi(\mathbf{q})$ and $K(\mathbf{q})$ is not explicitly written out. This equation is very useful even though it contains only two parameters, $K(\mathbf{q})$ and τ_0; it permits a variety of spectroscopic line-shape problems to be fitted. A specific application of this equation was given in solid-state NMR problems (Demco *et al.*, 1975) and Raman spectroscopy (Steele, 1981).

In this section we considered the correlation function involving only one single dynamic variable. The calculation can be extended to several coupled variables. In the several-variable problem, Eq. (4.141) becomes a matrix equation with C_A, Ω, and K represented by a square matrix. In this case, we begin again by converting the equation into its Laplace transform. Then, by rearranging terms in the transformed equation, we obtain a matrix equation for the correlation-function matrix as

$$\mathbf{C}(\mathbf{q}, s) = [s\mathbf{I} - i\mathbf{\Omega}(\mathbf{q}) + \mathbf{K}(\mathbf{q}, s)]^{-1} \cdot \mathbf{C}(\mathbf{q}) \equiv \mathbf{Z}(\mathbf{q}, s) \cdot \mathbf{C}(\mathbf{q}), \quad (4.164)$$

where $\mathbf{Z}(\mathbf{q}, s)$ is referred to as the generalized transport matrix. We shall now apply this equation to analyze the dynamic motion of the transverse and longitudinal components of the linear-momentum density without making the Markov approximation.

4.8 GENERALIZED HYDRODYNAMICS OF THE MOMENTUM DENSITY

The linear-momentum density for a system of particles is given by

$$\mathbf{g}(\mathbf{r}) = \sum_l \boldsymbol{\pi}_l \, \delta(\mathbf{r} - \mathbf{r}_l), \quad (4.165)$$

where $\boldsymbol{\pi}_l$ is the linear momentum of molecule l. When we convert the position $\mathbf{r}$ in Eq. (4.165) into a Fourier variable, this equation at time t takes the form

$$\mathbf{g}(\mathbf{q}, t) = \int d^3r \, e^{i\mathbf{q}\cdot\mathbf{r}} \mathbf{g}(\mathbf{r}, t) = \sum_l \boldsymbol{\pi}_l(t) e^{i\mathbf{q}\cdot\mathbf{r}_l(t)}. \quad (4.166)$$

There are three Cartesian components for $\mathbf{g}$. It is useful to choose a coordinate system with $\mathbf{q}$ parallel to the z axis. The component g_z represents the momentum density along $\mathbf{q}$ and is called the longitudinal-momentum density. The components $g_x(\mathbf{q})$ and $g_y(\mathbf{q})$, perpendicular to $\mathbf{q}$, are called the transverse-momentum densities. In an isotropic system, the longitudinal and transverse components are uncoupled. This is well known in hydrodynamics.

In the presence of an external symmetry-breaking field, the longitudinal and transverse components are coupled and cross-correlation effects in the two vertical components may be present. However, we shall not consider that case here.

Furthermore, the transverse-momentum density can couple to molecular orientation. This coupling is in fact a very important mechanism that makes it possible to study shear waves in molecular liquids by light scattering (see Chapter 2). Coupling of the transverse-shear wave to molecular reorientation can also become a mechanism for phase transitions in molecular solids. We shall provide microscopic theories to account for these effects in some later sections of this chapter. Here, we shall consider only a generalized hydrodynamic theory for the transverse-momentum density.

We begin by choosing a dynamic variable, which for this case is

$$A(\mathbf{q}) = g_{\mathrm{T}}(\mathbf{q}) = \frac{1}{\sqrt{2}} \sum_l (\pi_{lx}\hat{x} + \pi_{ly}\hat{y})e^{iqz_l}. \tag{4.167}$$

The two transverse components, g_x and g_y, have different reflection symmetry. That is, if g_x is transformed by a symmetry element chosen to be a reflection plane containing both the x and z axes, then it remains unchanged, but g_y changes sign. Thus, g_x and g_y are not mixed. Considering the symmetry operation before actual calculations are performed can save a lot of work, especially for problems dealing with several dynamic variables.

Since g_x and g_y do not mix, we can do the calculation by using a scalar quantity $g_{\mathrm{T}}(\mathbf{q})$. Analysis of this problem can be carried out by using the generalized Langevin equation for one single dynamic variable. Since $g_{\mathrm{T}}(\mathbf{q})$ has definite time-reversal symmetry, the frequency $\Omega(\mathbf{q})$ vanishes.

Next, we note from the equipartition principle that

$$C_{\mathrm{T}}(\mathbf{q}) = \langle |g_{\mathrm{T}}(\mathbf{q})|^2 \rangle = \sum_l \langle (\pi_{lx}^2 + \pi_{ly}^2)/2 \rangle = NkTm. \tag{4.168}$$

The cross terms with $l \neq l'$ in this equation vanish because $\langle \pi_{li}\pi_{l'g} \rangle = \langle \pi_{li} \rangle \langle \pi_{l'j} \rangle = 0$, where $i, j = x, y$. Calculating the memory function is more involved. We note that the random force associated with $g_{\mathrm{T}}(\mathbf{q})$ is given by

$$F(\mathbf{q}) = iQ\mathscr{L}g_{\mathrm{T}}(\mathbf{q}) = i(1 - P)\mathscr{L} \sum_l \pi_{l\mathrm{T}}^{iqz_l} = \sum_l (\dot{\pi}_{l\mathrm{T}} + iq\pi_{l\mathrm{T}}\pi_{lz}/m)e^{iqz_l}, \tag{4.169}$$

where $\pi_{l\mathrm{T}}$ is the transverse component of π_l. For small q, we expand e^{iqz_l} in a power series of q and keep only the lowest-order term in q. In the $q \to 0$ limit, $F(\mathbf{q})$ becomes

$$F(q) = iq \sum_l (\dot{\pi}_{l\mathrm{T}} z_l + \pi_{l\mathrm{T}}\pi_{lz}/m) = iq\sigma_{\mathrm{T}z}. \tag{4.170}$$

In obtaining this equation, we used the fact that $\sum_l \dot{\pi}_l = 0$, owing to conservation of linear momentum. For a molecular system in which the interaction potential depends on the orientation as well as on the relative distance, the total central force for all molecules in the system $\sum_l \dot{\pi}_l$ still adds up to zero if the particles constituting the molecule interact through central forces, such as the atom–atom-interaction model for intermolecular potential. However, in this case, the total angular momentum acting with respect to the center of mass of each molecule is not conserved, i.e.,

$$\sum_l \mathbf{r}_l \times \dot{\pi}_l \neq 0. \tag{4.171a}$$

But we have

$$\sum_l (\mathbf{r}_l \times \dot{\pi}_l + \mathbf{L}_l) = 0, \tag{4.171b}$$

where $\mathbf{L}_l$ is the body torque owing to the angle-dependent force exerted on molecule l. Equation (4.171b) has an important consequence in the translation–rotation-coupled system. Research to clarify problems in this area is in progress.

To provide a generalized hydrodynamics theory for the transverse-momentum density, we shall now return to $F(q)$ as given by Eq. (4.170). The memory function is given by

$$K(\mathbf{q}, t) = \langle [e^{iQ\mathscr{L}} F(\mathbf{q})] F^+(\mathbf{q}) \rangle \langle |g_T(\mathbf{q})|^2 \rangle^{-1} = (q^2/NkTm) \langle (e^{iQ\mathscr{L}t} \sigma_{Tz}) \sigma_{Tz}^+ \rangle. \tag{4.172}$$

We now show that for $q \to 0$,

$$e^{iQ\mathscr{L}t} \sigma_{Tz} = e^{i\mathscr{L}t} \sigma_{Tz}. \tag{4.173}$$

From the operator identity

$$e^{iQ\mathscr{L}t} = e^{i\mathscr{L}t} - i \int_0^t d\tau\, e^{iQ\mathscr{L}(t-\tau)} P \mathscr{L} e^{i\mathscr{L}\tau}, \tag{4.174}$$

we have

$$e^{iQ\mathscr{L}t} \sigma_{Tz} = \sigma_{Tz}(t) - i \int_0^t d\tau\, e^{iQ\mathscr{L}(t-\tau)} P \mathscr{L} \sigma_{Tz}(\tau), \tag{4.175a}$$

where

$$\sigma_{Tz}(t) = e^{i\mathscr{L}t} \sigma_{Tz}. \tag{4.175b}$$

But the term in the integrand involving the projection operator can be written as

$$\begin{aligned} P\mathscr{L}\sigma_{Tz}(\tau) &= \langle [\mathscr{L}\sigma_{Tz}(\tau)] g_T^+(\mathbf{q}) \rangle \langle |g_T(\mathbf{q})|^2 \rangle^{-1} g_T(\mathbf{q}) \\ &= (1/NkTm) \langle \sigma_{Tz}(\tau) [\mathscr{L} g_T(\mathbf{q})]^+ \rangle g_T(\mathbf{q}) \\ &= (iq/NkTm) \langle \sigma_{Tz}(\tau) \sigma_{Tz}^+ \rangle g_T(\mathbf{q}) + O(q^2). \end{aligned} \tag{4.176}$$

To obtain the second equality on the right-hand side, we used the Hermitian property of $\mathscr{L}$. The third equality is the result of the q expansion [Eq. (4.170)]. Thus, in the limit $q \to 0$, we obtain Eq. (4.173). The memory function given in Eq. (4.172) now reduces to

$$K(\mathbf{q}, t) = (q^2/NkTm)\langle \sigma_{\mathrm{T}z}(t)\sigma_{\mathrm{T}z}^{+} \rangle. \tag{4.177}$$

Therefore, the correlation function $C_{\mathrm{T}}(\mathbf{q}, s)$ can now be written as

$$\tilde{C}_{\mathrm{T}}(\mathbf{q}, s) = \frac{NmkT}{s + q^2 \eta_{\mathrm{T}z}(s)/\rho_m}, \tag{4.178}$$

where $\rho_m = Nm/V$, the mass density, and

$$\begin{aligned} \eta_{\mathrm{T}z}(s) &= \frac{1}{VkT} \int_0^\infty dt\, e^{-st} \langle \sigma_{\mathrm{T}z}(t)\sigma_{\mathrm{T}z}^{+} \rangle \\ &= \frac{1}{2VkT} \int_0^\infty dt\, e^{-st} [\langle \sigma_{xz}(t)\sigma_{xz}^{+} \rangle + \langle \sigma_{yz}(t)\sigma_{yz}^{+} \rangle]. \end{aligned} \tag{4.179}$$

Equation (4.178) is the basic equation of generalized hydrodynamics for transverse-momentum density. This provides the derivation of Eq. (2.128) obtained phenomenologically.

By setting $s = i\omega$, we obtain an expression for frequency-dependent viscosity, known as dynamic shear viscosity, as

$$\eta_{\mathrm{T}z}(\omega) = \frac{1}{2VkT} \int_0^\infty dt\, e^{-i\omega t} [\langle \sigma_{xz}(t)\sigma_{xz}^{+} + \langle \sigma_{yz}(t)_{yz}^{+} \rangle]. \tag{4.180}$$

For very low frequency, $\eta_{\mathrm{T}z}(\omega)$ reduces to the usual shear viscosity

$$\eta_{\mathrm{S}} = \lim_{\omega \to 0} \eta_{\mathrm{T}z}(\omega) = \frac{1}{2VkT} \int_0^\infty dt\, [\langle \sigma_{xz}(t)\sigma_{xz}^{+} \rangle + \langle \sigma_{yz}(t)\sigma_{yz}^{+} \rangle]. \tag{4.181}$$

If the frequency dependence can be neglected as in the Markov approximation, then Eq. (4.178) becomes

$$\tilde{C}_{\mathrm{T}}(\mathbf{q}, s) = \frac{NmkT}{s + q^2 \eta_{\mathrm{S}}/\rho_m}, \tag{4.182}$$

which is equivalent to Eq. (2.88), where z is used for s. In terms of the transverse velocity in the time domain, we have

$$\langle u_{\mathrm{T}}(\mathbf{q}, t) u_{\mathrm{T}}(\mathbf{q}) \rangle = \langle |u_{\mathrm{T}}(\mathbf{q})|^2 \rangle \exp(-q^2 \eta_{\mathrm{s}} t/\rho_m), \tag{4.183}$$

where $\langle |u_{\mathrm{T}}(q)|^2 \rangle = (NkT/m)$. The relaxation time given in Eq. (4.183) is identical to that given in Eq. (2.96) from the classical hydrodynamic theory.

We have considered extremely low frequencies. We shall now consider high frequencies, which we can expand Eq. (4.179) for $\eta_{Tz}(s)$ in a power series in $1/s$ as

$$\eta_{Tz}(s) = (1/s)(\eta^{(0)} + \eta^{(1)}/s + \cdots + \eta^{(n)}/s^n + \cdots), \tag{4.184}$$

where

$$\eta^{(0)} = (1/VkT)\langle|\sigma_{Tz}|^2\rangle, \tag{4.185}$$

$$\eta^{(1)} = (i/VkT)\langle[\mathscr{L}\sigma_{Tz}]\sigma_{Tz}^{+}\rangle = 0, \tag{4.186}$$

$$\eta^{(2)} = (1/VkT)\langle|\mathscr{L}\sigma_{Tz}|^2\rangle. \tag{4.187}$$

For asymptotically high frequencies, we can drop all terms except $\eta^{(0)}$, the first term on the right-hand side of Eq. (4.184). Thus,

$$\lim_{s\to\infty} \eta_{Ts}(s) = (1/sVkT)\langle|\sigma_{Tz}|^2\rangle = G_\infty/s, \tag{4.188}$$

where G_∞ is the shear modulus at very high frequencies.

Substituting Eq. (4.188) into (4.182) gives

$$\tilde{C}_g(\mathbf{q}, s) = \frac{NkTs}{s^2 + q^2 G_\infty/\rho_m}. \tag{4.189}$$

This equation has two poles on the imaginary axis, at $s = \pm iq\sqrt{G_\infty/\rho_m}$. Thus, at very high frequency the transverse-momentum density oscillates with frequencies given by $\pm q\sqrt{G_\infty/\rho_m}$. These are the frequencies of shear waves. The velocities of shear waves are equal to

$$V_\infty = \pm\sqrt{G_\infty/\rho_m}. \tag{4.190}$$

This result was obtained in Eq. (2.128) for viscoelastic systems. For the intermolecular potential, which depends only on the relative distance, Mountain and Zwanzig (1965) have identified the high-frequency shear modulus as

$$G_\infty = \rho kT\left\{1 + \frac{2\pi}{15}\rho\beta\int_0^\infty dr\, g(r)\frac{\partial}{\partial r}\left[r^2\frac{\partial U(r)}{\partial r}\right]\right\}, \tag{4.191}$$

where $g(r)$ is the radial distribution and $U(r)$ the intermolecular potential of two particles separated by a distance r.

We shall now consider the response at intermediate frequencies. We first assume that

$$(1/VkT)\langle\sigma_{Tz}(t)\sigma_{Tz}^{+}\rangle = G_\infty e^{-t/\tau_0}, \tag{4.192}$$

where

$$\tau_0 = \left(\frac{1}{G_\infty}\right)\frac{1}{VkT}\int_0^\infty dt\,\langle\sigma_{Tz}(t)\sigma_{Tz}^{+}\rangle = \frac{\eta_s}{G_\infty}, \tag{4.193}$$

which is the Maxwell relaxation time. Then, using Eq. (4.179), we find

$$\eta_{\mathrm{T}z}(s) = G_\infty/(s + \tau_0^{-1}). \tag{4.194}$$

Thus, $\tilde{C}_g(\mathbf{q}, s)$ as given by Eq. (4.178) becomes

$$\tilde{C}_g(\mathbf{q}, s) = \frac{NkT(s + \tau_0^{-1})}{s(s + \tau_0^{-1}) + q^2 V_\infty^2}, \tag{4.195}$$

where $V_\infty^2 = G_\infty/\rho_m$. Using partial fractions, we can write

$$\tilde{C}_q(\mathbf{q}, s) = (NkT)[A_+/(s - \lambda_+) + A_-/(s - \lambda_-)], \tag{4.196}$$

where $\lambda_\pm$ are the roots of the equation

$$s^2 + s\tau_0^{-1} + q^2 V_\infty^2 = 0, \tag{4.197}$$

i.e.,

$$\lambda_\pm = \frac{-1 \pm (1 - 4q^2 V_\infty^2 \tau_0^2)^{1/2}}{2\tau_0}, \tag{4.198}$$

$$A_+ = \frac{\lambda_+ + \tau_0^{-1}}{\lambda_+ - \lambda_-}, \quad \text{and} \quad A_- = \frac{\lambda_- + \tau_0^{-1}}{\lambda_- - \lambda_+}. \tag{4.199}$$

At high viscosity such that $4q^2 V_\infty^2 \tau_0 = 4q^2 \eta_s/\rho \gg 1, \lambda_\pm$ becomes complex, and oscillation of damped shear waves occurs. The frequencies of the oscillation are given by

$$\omega_\pm = (1/\tau_0)(\sqrt{q^2 V_\infty^2 \tau_0^2 - \tfrac{1}{4}}). \tag{4.200}$$

Thus, at high viscosity the expression for $\lambda_\pm$ can be written as

$$\lambda_\pm = \pm i\omega_\pm - 1/2\tau_0, \tag{4.201}$$

which shows that the shear waves oscillating with frequencies $\omega_\pm$ are damped with the time constant $2\tau_0$ rather than τ_0. Interestingly, according to Eq. (4.200), the shear-wave frequency is not precisely proportional to q. It is only when the Maxwell relaxation time becomes very long ,e.g.,

$$4q^2 V_\infty^2 \tau_0 \ll 1, \tag{4.202}$$

where we can drop the factor $\frac{1}{4}$ in Eq. (4.200), that $\omega_\pm$ becomes $\pm qV_\infty/\sqrt{\tau_0}$. This result has not yet been tested experimentally.

We shall next analyze the dynamic behavior of the longitudinal-momentum density. The fluctuation dynamics of the longitudinal-momentum density is complicated because of coupling to other longitudinal variables such as fluctuations in number density and energy density. To facilitate the calculation involving several variables, examining the symmetry properties of the Hamiltonian and the dynamic variables is very useful. Understanding the symmetry properties can reduce the amount of calculation considerably, because the terms that vanish owing to symmetry can be excluded before starting. Moreover, the frequency and the wave-vector dependences of the various terms in the transport matrix can also be elucidated from symmetry consideration alone. For these reasons, we shall consider some of the more important symmetry properties before calculating the dynamics of the longitudinal-momentum density.

4.9 SYMMETRY CONSIDERATION§

TIME-REVERSAL SYMMETRY

On time reversal, the coordinate **r** remains unchanged and the conjugate momentum **p** changes to $-\mathbf{p}$. If a dynamic variable $A_i(\mathbf{r}, \mathbf{p})$ has definite time-reversal symmetry, then under the phase-space transformation $(\mathbf{r}, \mathbf{p}) \to (\mathbf{r}, -\mathbf{p})$ we have

$$A_i \to \gamma_i A_i, \tag{4.203}$$

where $\gamma_i = +1$ for even time-reversal symmetry and $\gamma_i = -1$ for odd time-reversal symmetry.

Since the Hamiltonian $\mathscr{H}$ of a conservative system is a quadratic function of the momenta, $\mathscr{H}$ is invariant ($\gamma_{\mathscr{H}} = +1$) under time reversal. The Liouville operator $\mathscr{L}$ is odd in **p**, and thus $\mathscr{L} \to -\mathscr{L}$. Therefore, the elements of the time-correlation matrix under time reversal transforms as

$$C_{ij}(t) = \langle A_i(t)A_j^+\rangle \to \gamma_i\gamma_j C_{ij}(-t) = \gamma_i\gamma_j C_{ji}^*(t). \tag{4.204}$$

Thus, if $i = j$, then the autocorrelation function in $C_{ii}(t)$ is real and is also an even function of time. On the other hand, if A_i and A_j have opposite time reversal symmetry,

$$C_{ij}(t) = -C_{ij}(-t) = -C_{ji}^*(t). \tag{4.205}$$

The first equality shows that $C_{ij}(t)$ is an odd function of time, and the second equality shows it to be a complex function. Clearly, in this case the static elements $C_{ij}(t = 0) = 0$.

The element of the frequency matrix Ω_{ij} vanishes unless A_i and A_j have opposite time-reversal symmetry, i.e., under time reversal we have

$$\Omega_{ij} \to -\gamma_i\gamma_j\Omega_{ij}. \tag{4.206}$$

§ See also chapter 10 of Berne and Pecora (1976).

The matrix elements of the random-force–correlation function are transformed under time reversal in a way similar to the matrix elements of the time-correlation function $C_{ij}(t)$, i.e.,

$$R_{ij}(t) \equiv \langle (e^{iQ\mathscr{L}t} F_i)(F_k)^* \rangle \to \gamma_i \gamma_j R_{ij}(-t) = \gamma_i \gamma_j R^*_{ji}(t). \tag{4.207}$$

Thus, the transport coefficients Z_{ij} transform as

$$Z_{ij} = \beta \int_0^\infty dt\, R_{ij}(t) \to \gamma_i \gamma_j Z^*_{ji}. \tag{4.208}$$

Equation (4.208) forms the basis of the Onsager reciprocal relations.

The elements of the relaxation matrix $\mathbf{\Gamma}$ are related to the elements of the transport matrix $\mathbf{Z}$ by $\mathbf{\Gamma} = \mathbf{Z} \cdot \boldsymbol{\chi}^{-1}$, or $Z_{ij} = \sum_k \Gamma_{ik} \chi_{kj}$. If the susceptibility matrix $\boldsymbol{\chi}$ is diagonal, we obtain $Z_{ij} = \Gamma_{ij} \chi_{jj}$. Likewise, $Z^*_{ji} = \Gamma^*_{ji} \chi_{ii}$ (χ_{ii} is real). But according to Eq. (4.208), $Z_{ij} = \gamma_i \gamma_j Z^*_{ji}$; it follows that

$$\Gamma_{ij} \chi_{jj} = \gamma_i \gamma_j \Gamma^*_{ji} \chi_{ii} \tag{4.209a}$$

or

$$\Gamma_{ij} = \gamma_i \gamma_j (\chi_{ii}/\chi_{jj}) \Gamma^*_{ji} = \gamma_i \gamma_j (\langle |A_i|^2 \rangle / \langle |A_j|^2 \rangle) \Gamma^*_{ji}. \tag{4.209b}$$

This is a very useful expression that relates different elements of the relaxation matrix. If $i = j$, then we have $\Gamma_{ii} = \Gamma^*_{ii}$, and thus the diagonal elements of the relaxation matrix is real. Since this property follows from a diagonal $\boldsymbol{\chi}$, it is useful to begin the calculation using a set of orthogonal dynamic variables.

Inversion Symmetry

Under inversion, both $\mathbf{r}$ and $\mathbf{p}$ change sign, i.e., $(\mathbf{r}, \mathbf{p}) \to (-\mathbf{r}, -\mathbf{p})$. Under inversion A_i transforms as

$$A_i \to \varepsilon_i A_i, \tag{4.210}$$

where $\varepsilon_i = +1$ for even parity and $\varepsilon_i = -1$ for odd parity.

For a system of molecules having centers of inversion symmetry and with no symmetry-breaking force acting on the system, the Hamiltonian is invariant to inversion. The Liouville operator is also invariant to inversion. Thus, under inversion, the matrix elements of the time-correlation function transform as

$$C_{ij}(t) \to \varepsilon_i \varepsilon_j C_{ij}(t). \tag{4.211}$$

It follows that dynamic variables with opposite parity are uncorrelated for all values of time. The same is true for the memory-function matrix:

$$K_{ij}(t) \to \varepsilon_i \varepsilon_j K_{ij}(t). \tag{4.212}$$

Obviously, this applies to the elements of the frequency matrix, to the elements of the transport matrix, and to the relaxation matrix.

Finally, the dynamic variables $A_i(\mathbf{q})$ transform under inversion as

$$A_i(\mathbf{q}) \to \varepsilon_i A_i(-\mathbf{q}). \tag{4.213}$$

Under inversion, the elements of all the matrices discussed here will be odd functions of $\mathbf{q}$ if they have opposite parity and even functions of $\mathbf{q}$ if they have the same parity.

Reflection Symmetry

Under the phase-space transformation

$$(x_j y_j z_j, P_{jx}, P_{jy}, P_{jz}) \to (x_j, -y_j, z_j, P_{jx}, -P_{jy}, P_{jz}).$$

Dynamic variables A_i, which transform as

$$A_i \to \delta_i A_i, \qquad \delta_i = \pm 1, \tag{4.214}$$

are said to possess reflection symmetry in the xz plane. For such properties, the elements of $\mathbf{K}(t)$, $\mathbf{C}(t)$, and Ω are zero unless $\delta_i = \delta_j$, because the Hamiltonian and the Liouville operator are invariant to reflection. Thus, for a set of dynamic variables $\mathbf{A}$ with definite reflection symmetry, the variable where $\delta_i = 1$ is totally uncorrelated with that where $\delta_i = -1$.

4.10 GENERALIZED HYDRODYNAMICS THEORY OF THE ISOTROPIC LIGHT-SCATTERING SPECTRUM

We are now ready to carry out an analysis of the dynamics of the three longitudinal variables. These variables are longitudinal-momentum density g_z, number density ρ, and energy density ε. They are given by

$$g_z(\mathbf{q}) = \sum_l \pi_{lz} e^{iqz_l}, \tag{4.215}$$

$$\rho(\mathbf{q}) = \sum_l e^{iqz_l}, \tag{4.216}$$

$$\varepsilon(\mathbf{q}) = \sum_l \left(\frac{\pi_l^2}{2m_l} + \frac{1}{2}\sum_{l \neq l'} u_{ll'}\right) e^{iqz_l} \equiv \sum_l \varepsilon_l e^{iqz_l}, \tag{4.217}$$

where ε_l represents the energy of molecule l, and $u_{ll'}$ is the interaction potential energy between molecules l and l'.

We shall first develop the dynamic equation for the correlation-function matrix involving the variables $g_z(\mathbf{q})$, $\delta\rho(\mathbf{q})$, and $\delta\varepsilon(\mathbf{q})$ and then calculate in detail the spectrum associated with the density fluctuation. The spectrum of the momentum density is closely related to the spectrum of the density fluctuation. The spectrum of the energy fluctuation, if needed, can be obtained from the matrix equation for the correlation function. We shall write the dynamic variables $\delta\rho(\mathbf{q})$ and $\delta\varepsilon(\mathbf{q})$ for $\rho(\mathbf{q})$ and $\varepsilon(\mathbf{q})$, because for $q \neq 0$, $\rho(\mathbf{q}) = \delta\rho(\mathbf{q})$ and $\varepsilon(\mathbf{q}) = \delta\varepsilon(\mathbf{q})$. The term $g_z(\mathbf{q})$ is equal to the momentum-density fluctuation because there is no new flow for a system at equilibrium.

We note that ρ and g_z have opposite time-reversal symmetry; thus, $\langle \delta\rho\, g_z^+ \rangle = 0$. Similarly, $\langle g_z\, \delta\varepsilon^+ \rangle = 0$. However, $\langle \delta\rho\, \delta\varepsilon^+ \rangle \neq 0$. To facilitate the calculation and to keep the algebra to a minimum, instead of using $\delta\rho$, g_z, and $\delta\varepsilon$ as coupled dynamic variables, we shall use a set of normalized orthogonal variables (Lin and Wang, 1979) given by

$$A = \begin{bmatrix} \xi_1 \\ \xi_2 \\ \xi_3 \end{bmatrix}, \tag{4.218}$$

where

$$\xi_1 = \delta\rho / \langle |\delta\rho|^2 \rangle^{1/2}, \tag{4.219}$$

$$\xi_2 = g_z / \langle |g_z|^2 \rangle^{1/2}, \tag{4.220}$$

$$\xi_3 = [\delta\varepsilon / \langle |\delta\varepsilon|^2 \rangle^{1/2} - S\xi_1](1 - S^2)^{-1/2}, \tag{4.221}$$

and

$$S = \langle \delta\varepsilon\, \delta\rho^* \rangle / \langle |\delta\varepsilon|^2 \rangle^{1/2} \langle |\delta\rho|^2 \rangle^{1/2}. \tag{4.222}$$

Using this set of variables, we have, $\langle AA^+ \rangle = \mathbf{I}$, where $\mathbf{I}$ is a 3×3 unit matrix. It can be shown that ξ_3 is proportional to the entropy fluctuation (see Chapter 2).

The frequency matrix is found to be

$$\mathbf{\Omega} = \langle (\mathscr{L}\mathbf{A})\mathbf{A}^+ \rangle = \begin{bmatrix} 0 & \omega_{\mathrm{T}} & 0 \\ \omega_{\mathrm{T}} & 0 & R \\ 0 & R^* & 0 \end{bmatrix} \tag{4.223}$$

where ω_{T} is the isothermal sound frequency given by

$$\omega_{\mathrm{T}} = q(NkT/m)^{1/2} \langle |\delta\rho|^2 \rangle^{-1/2}, \tag{4.224}$$

which reduces to $q(m\rho\chi_{\mathrm{T}})^{-1/2}$ in the small-q limit. Here χ_{T} is isothermal compressibility [compare with Eq. (2.142)].

The parameter R is given by

$$R = \langle(\mathscr{L}\xi_2)\xi_2^*\rangle = (1 - S^2)^{-1/2} \frac{\langle(\mathscr{L}g_z)\,\delta\varepsilon^*\rangle}{\langle|g_z|^2\rangle^{1/2}\langle|\delta\varepsilon|^2\rangle^{1/2}} - iS\omega_{\mathrm{T}}, \quad (4.225)$$

which is proportional to q in the $q \to 0$ limit. We shall show later that S is related to the difference between the isothermal and the adiabatic sound frequency.

The random force is given by

$$\mathbf{F} = iQ\mathscr{L}\mathbf{A} = i\mathscr{L}\mathbf{A} - i\mathbf{\Omega}\mathbf{A} = \begin{bmatrix} 0 \\ f_2 \\ f_3 \end{bmatrix}, \quad (4.226)$$

where

$$f_2 = i\mathscr{L}\xi_2 - i\omega_{\mathrm{T}}\xi_1 - iR\xi_3, \quad (4.227)$$

$$f_3 = i\mathscr{L}\xi_3 - iR^*\xi_2. \quad (4.228)$$

We shall now consider the functional forms of f_2 and f_3 in the small-q limit. For small q, we find

$$i\mathscr{L}\xi_2 = iq\sigma_{zz}/(NmkT)^{1/2}, \quad (4.229)$$

where

$$\sigma_{zz} = \sum_l (F_{lz}z_l + \pi_{lz}^2/m) \quad (4.230)$$

and

$$i\mathscr{L}\xi_3 = iq\sum_i \varepsilon_l\dot{z}_l/(kT^2C_V)^{1/2} + \cdots = iqJ_z/(kT^2C_V)^{1/2} + \cdots, \quad (4.231)$$

where the terms not written out explicitly are partially cancelled by the $iR^*\xi_2$ term in Eq. (4.228). The quantity R is proportional to q and is given by

$$R = -iq\{S(\rho_m\chi_{\mathrm{T}})^{-1/2} - \langle\sigma_{zz}\delta\varepsilon^*\rangle/[(1 - S^2)(NmkT)(\langle|\delta\varepsilon|^2\rangle)]^{1/2}\} = qR', \quad (4.232)$$

where $\delta\varepsilon$ is given by $\sum_l(\varepsilon_l - \langle\varepsilon_l\rangle)$, which is independent of q. The quantity S is also independent of q for small q. Thus, in the limit $q \to 0$, both f_2 and f_3 are proportional to q.

Therefore, for small q, the memory-function matrix $\mathbf{K}(\tau)$ will have the general form

$$\mathbf{K}(\tau) = \begin{bmatrix} 0 & 0 & 0 \\ 0 & K_{22}(\tau) & 0 \\ 0 & 0 & K_{33}(\tau) \end{bmatrix}, \quad (4.233)$$

where

$$K_{22}(\tau) = [q^2/(NmkT)]\langle\sigma_{zz}(\tau)\sigma_{zz}\rangle + q^2V_l^2\langle\xi_1(\tau)\xi_1^+\rangle + q^2R'^2\langle\xi_3(\tau)\xi_3^+\rangle \tag{4.234}$$

with $V_l = (\rho_m\chi_T)^{-1/2}$, and

$$K_{33}(\tau) \equiv [q^2/kT^2C_V]\langle J_z(\tau)J_z\rangle + q^2R'^2\langle\xi_3(\tau)\xi_3^+\rangle. \tag{4.235}$$

The $K_{23}(\tau)$ and $K_{32}(\tau)$ terms that result from dynamic coupling are strictly nonvanishing at finite τ, although they vanish at $\tau = 0$. However, in the $q \to 0$ limit these terms are proportional to q^4 and can be neglected in the small-q calculation.

Having obtained the matrices $\boldsymbol{\Omega}$ and $\mathbf{K}(\tau)$, we can substitute these into the correlation-function equation [Eq. (4.71)] or into the generalized form of Eq. (4.147). In terms of the Laplace transform, Eq. (4.147) takes the form

$$[s\mathbf{I} - i\boldsymbol{\Omega} + \mathbf{K}(s)]\mathbf{C}(s) \equiv \mathbf{X}(s)\mathbf{C}(s) = \mathbf{C}(0). \tag{4.236}$$

Because the dynamic variables are orthonormal, we have $\mathbf{C}(0) = \langle AA^+\rangle = \mathbf{I}$. Writing out Eq. (4.236) explicitly, we have

$$\begin{bmatrix} s & -i\omega_T & 0 \\ -i\omega_T & s + K_{22}(s) & -iR \\ 0 & -iR^* & s + K_{33}(s) \end{bmatrix} \times \begin{bmatrix} C_{\rho\rho}(s) & C_{\rho P}(s) & C_{\rho\varepsilon}(s) \\ C_{P\rho}(s) & C_{PP}(s) & C_{P\varepsilon}(s) \\ C_{\varepsilon\rho}(s) & C_{\varepsilon P}(s) & C_{\varepsilon\varepsilon}(s) \end{bmatrix}$$
$$= \begin{bmatrix} 1 & 0 & 0 \\ 0 & 1 & 0 \\ 0 & 0 & 1 \end{bmatrix}, \tag{4.237}$$

where

$$\begin{aligned} C_{\rho\rho}(s) &= \langle\xi_1(s)\xi_1^+\rangle, \\ C_{PP}(s) &= \langle\xi_2(s)\xi_2^+\rangle, \\ C_{P\rho}(s) &= \langle\xi_2(s)\xi_1^+\rangle, \\ C_{\varepsilon\varepsilon}(s) &= \langle\xi_3(s)\xi_3^+\rangle. \end{aligned} \tag{4.238}$$

Equation (4.237) contains complete information about the correlation functions of the conserved variables in the $q \to 0$ limit. However, the $\langle\xi_1(\tau)\xi_1^+\rangle$ and $\langle\xi_3(\tau)\xi_3^+\rangle$ terms in the $K_{22}(\tau)$ expression and the $\langle\xi_3(\tau)\xi_3^+\rangle$ term in the $K_{33}(\tau)$ expression should be excluded from the memory functions [see Eqs. (4.234) and (4.235)], because when Eq. (4.235) is explicitly multiplied out, these terms combine with the quantities in the correlation-function matrix to give higher order terms. Therefore, rather than using the expression given by Eq. (4.233), the memory-function matrix for the linearized, generalized hydrodynamic theory should be

$$\mathbf{K}(\tau) = \begin{bmatrix} 0 & 0 & 0 \\ 0 & q^2 a(\tau) & 0 \\ 0 & 0 & q^2 b(\tau) \end{bmatrix}, \tag{4.239}$$

where

$$a(\tau) = (1/NmkT)\langle \sigma_{zz}(\tau)\sigma_{zz} \rangle, \tag{4.240}$$

$$b(\tau) = (1/kT^2 C_V)\langle J_z(\tau) J_z \rangle. \tag{4.241}$$

With this modification, the memory-function matrix to be used for subsequent calculations will correspond to the Laplace transform of Eq. (4.239).

The correlation functions $\langle \xi_1(s)\xi_1^+ \rangle$ and $\langle \xi_2(s)\xi_2^+ \rangle$ are readily found to be

$$C_{\rho\rho}(s) = \langle \xi_1(s)\xi_1^+ \rangle = \{[s + q^2 a(s)][s + q^2 b(s)] + R^2\}/|X| \tag{4.242}$$

and

$$C_{pp}(s) = \langle \xi_2(s)\xi_2^+ \rangle = s(s + q^2 b(s))/|X|, \tag{4.243}$$

where $a(s)$ and $b(s)$ are the Laplace transform of $a(t)$ and $b(t)$, respectively; $|X|$, the determinant of $X(s)$, is given by

$$|X| = \{s[s + q^2 a(s)] + \omega_T^2\}\{s + q^2 b(s)\} + sR^2. \tag{4.244}$$

The Rayleigh–Brillouin spectral density is thus equal to

$$\begin{aligned} I(\mathbf{q}, \omega) &= \frac{1}{\pi} \operatorname{Re}\langle \xi_1(s)\xi_1^+ \rangle \Big|_{s=i\omega} \\ &= \frac{1}{\pi} \operatorname{Re}\left(\frac{[s + q^2 a(s)][s + q^2 b(s)] + R^2}{\{s[s + q^2 a(s)] + \omega_T^2\}\{s + q^2 b(s)\} + sR^2} \right) \Bigg|_{s=i\omega} \end{aligned} \tag{4.245}$$

The resonance behavior of $C_{\rho\rho}(s)$ and $C_{pp}(s)$ are related by

$$C_{\rho\rho}(s) = C_{pp}(s)[(1 + q^2 a(s)/s) + R^2/s(s + q^2 b(s))]. \tag{4.246}$$

Thus, in the limit of high frequency (large s), $C_{\rho\rho}(s)$ and $C_{pp}(s)$ have an identical spectral feature. To analyze the resonance behavior of $C_{\rho\rho}(s)$ or $C_{pp}(s)$, it is necessary to investigate the structure of $|X|$. We shall analyze the spectral feature of $I(\mathbf{q}, \omega)$ in detail.

Before studying $I(\mathbf{q}, \omega)$ further, we shall associate the quantity R with the well-known physical quantity. We can relate this quantity to the adiabatic sound frequency and show this by diagonalizing the frequency matrix (Eq. 4.223). This yields eigenvalues

$$\lambda_+ = (\omega_T^2 + R^2)^{1/2} = \omega_s, \qquad \lambda_- = -(\omega_T^2 + R^2)^{1/2} = \omega_s, \qquad \lambda_0 = 0. \tag{4.247}$$

The corresponding normalized eigenvectors are found to be

$$\phi_{\pm} = (1/\sqrt{2})\begin{bmatrix} \omega_T/\omega_s \\ \pm 1 \\ R/\omega_s \end{bmatrix}, \tag{4.248}$$

$$\phi_0 = \begin{bmatrix} R/\omega_s \\ 0 \\ \omega_T/\omega_s \end{bmatrix}. \tag{4.249}$$

In the absence of relaxation effects, ξ_1 can be expressed in terms of $\phi_{\pm}$ and ϕ_0 as

$$\xi_1 = \delta\rho/\langle|\delta\rho|^2\rangle^{1/2} = (1/\sqrt{2})(\omega_T\phi_+/\omega_s + \omega_T\phi_-/\omega_s + \sqrt{2R}\,\phi_0/\omega_s). \tag{4.250}$$

The eigenvectors ϕ_+ and ϕ_- are associated with the acoustic waves that propagate with frequencies ω_s and $-\omega_s$. The third eigenvector is associated with the unshifted Rayleigh central peak.

These results can also be obtained from Eq. (4.245) by neglecting relaxation, i.e., by setting $a(s)$ and $b(s)$ to zero. In this case, we obtain

$$I(\mathbf{q}, \omega) = (1/\pi)\,\mathrm{Re}\{(R^2/\omega_s^2)(1/s) + (\omega_T^2/2\omega_s^2)[1/(s + i\omega_s) + 1/(s - i\omega_s)]_{s=i\omega}\},$$

or

$$I(\mathbf{q}, \omega) = (R^2/\omega_s^2)\,\delta(\omega) + (\omega_T^2/2\omega_s^2)[\delta(\omega + \omega_s) + \delta(\omega - \omega_s)]. \tag{4.251}$$

The Landau–Placzek ratio in the absence of relaxation in thus equal to

$$\mathscr{L}P = I_C/2I_B = R^2/\omega_T^2 = \omega_s^2/\omega_T^2 - 1 = \gamma - 1, \tag{4.252}$$

where γ is C_P/C_V.

We shall now examine the effect of relaxation. To do this, we recall Eq. (4.242). The resonance behavior of $I(\mathbf{q}, \omega)$ depends significantly on the dynamic response of the dissipating functions $a(i\omega)$ and $b(i\omega)$. From this, we see that in the absence of dissipation $I(\mathbf{q}, \omega)$ gives three infinitely sharp resonance lines at $\omega = 0, \pm\omega_s$. The effect of dissipation is to broaden these resonances and to introduce new spectral features, depending on the frequency dependence of the functions $a(i\omega)$ and $b(i\omega)$.

In the case of fast relaxation, the frequency dependence in $a(i\omega)$ and $b(i\omega)$ can be neglected, and it is expected that the usual hydrodynamic result given in Chapter 2 will be recovered. To show this, we set

$$a(i\omega) \simeq D_V = \int_0^\infty dt\, a(t) = \frac{1}{NmkT}\int_0^\infty dt\,\langle\sigma_{zz}(t)\sigma_{zz}\rangle, \tag{4.253}$$

$$b(i\omega) \simeq D_T = \int_0^\infty dt\, b(t) = \frac{1}{KT^2C_V}\int_0^\infty dt\,\langle J_z(k)J_z\rangle. \tag{4.254}$$

The resonance behavior can be found by setting $s = i\omega$ and also setting the quantity $|X|$ in Eq. (4.244) to zero:

$$(i\omega)^3 + (i\omega)^2(\lambda_2 + \lambda_3) + (i\omega)(\lambda_2\lambda_3 + \omega_s^2) + \omega_T^2\lambda_3 = 0, \qquad (4.255)$$

where $\lambda_2 = q^2 D_V$ and $\lambda_3 = q^2 D_T$.

Equation (4.255) corresponds exactly to the results obtained from the linearized equation of hydrodynamics described in Chapter 2 [Eq. (2.246)]. Equation (4.255), a cubic equation, has three roots: two correspond to the Brillouin sidebands and one to the unshifted Rayleigh peak. For most fluids, the condition $\omega_s > \lambda_i$, $(i = 2, 3)$, applies; following the technique of Chapter 2, the perturbative solution of Eq. (4.255) gives three roots as

$$\omega_0 = i(1/\gamma)\lambda_3 = iq^2 D_T, \qquad (4.256)$$

$$\begin{aligned}\omega_\pm &= \pm\omega_s + (i/2)\{(1 - 1/\gamma)\lambda_3 + \lambda_2\} \\ &= \pm\omega_s + (i/2)\{(1 - 1/\gamma)q^2 D_T + q^2 D_V\}.\end{aligned} \qquad (4.257)$$

This calculation provides a derivation of the thermal diffusivity D_T and the longitudinal kinematic viscosity D_V. The Kubo formulas for D_V and D_T are given by Eqs. (4.253) and (4.254), respectively. We can write $\sigma_{zz}(t)$ as

$$\sigma_{zz}(t) = \sigma(t) + \sigma_a(t), \qquad (4.258)$$

where $\sigma = \frac{1}{3}(\sigma_{xx} + \sigma_{yy} + \sigma_{zz})$, corresponding to the isotropic average of the $\boldsymbol{\sigma}$ tensor; and $\sigma_a = (\sigma_{zz} - \sigma)$, corresponding to the anisotropy of $\boldsymbol{\sigma}$. For a rotationally isotropic fluid, using the technique developed in Chapter 3 it is easy to show that Eq. (4.254) reduces to

$$D_V = (1/\rho_m)(\eta_V + 4\eta_s/3), \qquad (4.259)$$

where η_s is the shear viscosity [the Kubo expression of η_s was given previously by Eq. (4.181)], and η_V, the bulk viscosity, is given by

$$\eta_V = \frac{1}{VkT}\int_0^\infty dt\,\langle\sigma(t)\sigma\rangle. \qquad (4.260)$$

As given in Chapter 2, the thermal diffusivity D_T is related to the thermal conductivity λ by $D_T = \lambda/\rho_m C_V$.

When the relaxation rate of the system is not fast compared with ω_T or ω_s, the frequency dependence of $a(i\omega)$ and $b(i\omega)$ cannot be neglected. To study this general case, we first separate $a(i\omega)$ and $b(i\omega)$ into real and imaginary parts:

$$\lambda_2(\omega) = q^2 \int_0^\infty dt\, e^{-i\omega t} a(t) = \alpha_2'(\omega) + i\alpha_2''(\omega), \tag{4.261}$$

$$\lambda_3(\omega) = q^2 \int_0^\infty dt\, e^{-i\omega t} b(t) = \alpha_3'(\omega) + i\alpha_3''(\omega), \tag{4.262}$$

Substituting Eqs. (4.261) and (4.262) into (4.245) and then carrying out some algebraic manipulations, we obtain

$$I(\mathbf{q}, \omega) = (N_1 D_1 + N_2 D_2)/(D_1^2 + D_2^2), \tag{4.263}$$

where

$$\begin{aligned}
N_1 &= -\omega^2\{1 + [\alpha_2''(\omega)/\omega]\}\{1 + [\alpha_3''(\omega)/\omega]\} + \alpha_2'(\omega)\alpha_3'(\omega) + (1 - 1/\gamma)\omega_s^2, \\
N_2 &= \omega\{\alpha_2'(\omega) + \alpha_3'(\omega) + [\alpha_2''(\omega)/\omega]\alpha_3'(\omega) + [\alpha_3''(\omega)/\omega]\alpha_2'(\omega)\}, \\
D_1 &= -\omega^2\{\alpha_2'(\omega) + \alpha_3'(\omega) + [\alpha_2''(\omega)/\omega]\alpha_3'(\omega) \\
&\quad + [\alpha_3''(\omega)/\omega]\alpha_2'(\omega)\} + [\alpha_3'(\omega)/\omega]\omega_T^2, \\
D_2 &= \omega(-\omega^2\{1 + [\alpha_2''(\omega)/\omega]\}\{1 + [\alpha_3''(\omega)/\omega]\} \\
&\quad + \omega_s^2\{1 + [\alpha_3''(\omega)\gamma/\omega]\} + \alpha_2'(\omega)\alpha_3'(\omega)).
\end{aligned} \tag{4.264}$$

Clearly, in this general case the resonance behavior of $I(q, \omega)$ is very complicated. The functional forms of $\alpha_2'(\omega)$, $\alpha_2''(\omega)$, $\alpha_3'(\omega)$ and $\alpha_3''(\omega)$ depend on the behavior of the dissipating functions $a(t)$ and $b(t)$. Approximations are needed to simplify these expressions. Information so far obtained shows that the molecular dynamics involved in affecting the thermal diffusivity (or thermal conductivity) are mainly the result of intramolecular processes. The intramolecular processes are generally faster than $q^2 D_T$, which is on the order of several megahertz at ordinary temperature; thus, for analyzing the Rayleigh–Brillouin spectrum, a very good approximation that involves setting $\alpha_3' = q^2 D_T$ and $\alpha_3'' = 0$ can be used to simplify the foregoing expressions.

The frequency dependence in the $\lambda_2(\omega)$ term has a very important effect on the line shape of $I(q, \omega)$. This frequency dependence has two origins: one is thermal relaxation owing to the coupling of *internal* degrees of freedom of a molecule to translation. Mountain (1968) showed that coupling of translation to low-lying vibrational or librational modes will give a relaxation central peak. A second but more important mechanism is structural relaxation. Reorganization of the structure is very sensitive to temperature. This process is closely associated with the density fluctuation in polymeric-chain molecules, even if the overall center-of-mass motion does not take place. A good example of this is the segmental motion of polymer chains in bulk polymer liquids or in amorphous solids. The segmental motion contributes significantly to the density fluctuation in the hypersonic frequency region, even if the overall center-of-mass motion occurs very slowly. Coupling of the structural-relaxation process to density fluctuation is known to yield new features in the Rayleigh–Brillouin spectrum. To analyze the effect of the structural relaxation in $I(q, \omega)$, we return to Eq. (4.245). We set $\lambda_3 = q^2 b(s) = q^2 D_T$ and also assume that $a(t)$ is relaxed by a single relaxation time τ_2. Thus

$$\lambda_2(s) = q^2 a(s) = q^2 \tau_2 a/(1 + s\tau_2), \tag{4.265}$$

where a is the high frequency modulus given by

$$a = \langle \sigma_{zz}^2 \rangle / NmKT = (K_\infty + \tfrac{4}{3} G_\infty), \tag{4.266}$$

with K_∞ and G_∞ being the bulk and shear rigidity modulus, respectively. For viscoelastic liquids above T_g, G_∞ is less than K_∞ by one to two orders of magnitude (Ferry, 1980). Thus the bulk modulus K_∞ dominates the value of a. The quantity τ^2, the relaxation time of the longitudinal modulus, is given by

$$\tau_2 = \frac{1}{\langle \sigma_{zz}^2 \rangle} \int_0^\infty dt \, \langle \sigma_{zz}(t) \sigma_{zz} \rangle. \tag{4.267}$$

Since the longitudinal modulus is sensitive to structural variations, τ_2 is also designated as the structural relaxation time. As a result of these approximations, Eq. (4.244) becomes

$$s^3(1 + s\tau_2) + s^2[aq^2\tau_2 + D_T q^2(1 + s\tau_2)] + s[(aq^2\tau_2)(D_T q^2) + \omega_s^2(1 + s\tau_2)] + D_T q^2 \omega_T^2 (1 + s\tau_2) = 0, \tag{4.268}$$

which is a quartic equation and hence has four roots. To find these roots, we note that $(q^2 D_T)^2$ and $q^2 a$ are in general only a fraction of ω_s^2. Thus Eq. (4.268) can be solved by the perturbation expansion of $q^2 D_T/\omega_s$ and $q^2 a/\omega_s^2$. Using the technique given in Chapter 2, we find that the perturbation solutions of Eq. (4.268) to the lowest order in $q^2 D_T/\omega_s$ and $q^2 a/\omega_s^2$ are

$$s_1 = D_T q^2, \tag{4.269}$$

$$s_{2,3} = i\omega_\pm = \pm i\left\{\omega_s + \frac{q^2 a \omega_s \tau_2^2}{2[1 + (\omega_s \tau_2)^2]}\right\} + \frac{1}{2}\left\{\frac{q^2 a \tau_2}{[1 + (\omega_s \tau_2)^2]} + \left(1 - \frac{1}{\gamma}\right) q^2 D_T\right\}, \tag{4.270}$$

$$s_4 = 1/\tau_2 \tag{4.271}$$

The root s_1 is identical to the unshifted Rayleigh peak given previously in Eq. (4.256), indicating that if the frequency dependence of λ_3 is ignored, this peak will be dissociated from the structural relaxation effect. The roots $s_{2,3}$ correspond to the Brillouin doublet. Each consists of a real part and an imaginary part. The real part corresponds to the frequency shift and the imaginary part to the linewidth. Both parts are affected by structural relaxation. Note that in the limit of $\omega_s \tau_2 \ll 1$, $s_{2,3}$ reduces to the classical result given in Eq. (4.257).

A new feature is associated with the s_4 root. Like the s_1 peak, the s_4 peak is unshifted from the laser frequency and has a linewidth determined by the structural-relaxation time τ_2.

It is important to note that structural relaxation has a significant effect on the line shape of the Rayleigh–Brillouin spectrum. At high temperature, when the structural relaxation rate is fast compared with the frequency of the hypersonic wave, the intensity associated with the structural relaxation is smeared over a wide range of frequency and contributes as a uniform, weak

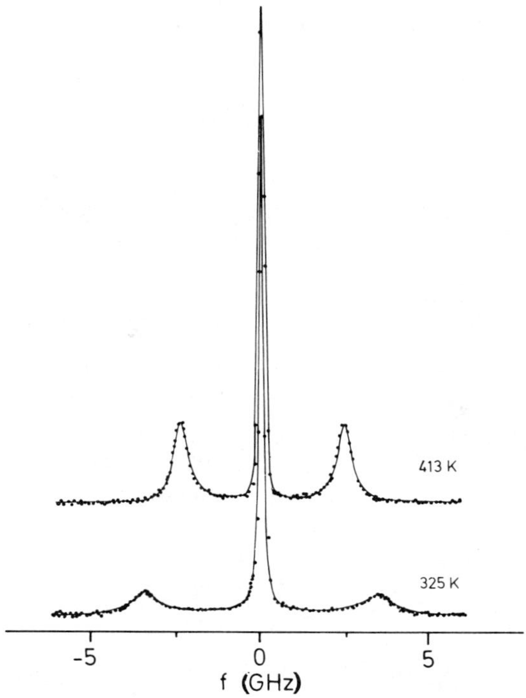

FIG. 4.1 Comparison of experimental spectra (dots) of diphenyl siloxane (DPS) at two temperatures with those calculated (lines) using Eq. (4.245). $\lambda_3 = 0.2$ GHz, $\tau_2 = 0.3e^{1890/T}$ psec, $\omega_s = (10.45\text{–}1.35 \times 10^{-3}T)$ GHz, and $aq^2/\omega_s^2 = (3.2\text{–}2.5 \times 10^{-3}T)$, where T is temperature in Kelvins.

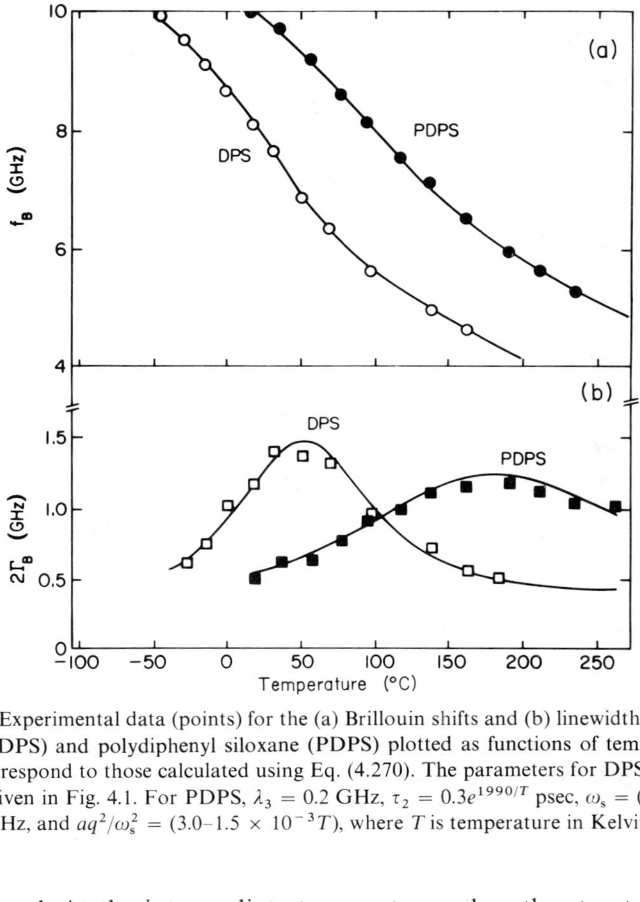

FIG. 4.2 Experimental data (points) for the (a) Brillouin shifts and (b) linewidths for diphenyl siloxane (DPS) and polydiphenyl siloxane (PDPS) plotted as functions of temperature. The curves correspond to those calculated using Eq. (4.270). The parameters for DPS are the same as those given in Fig. 4.1. For PDPS, $\lambda_3 = 0.2$ GHz, $\tau_2 = 0.3e^{1990/T}$ psec, $\omega_s = (11.70\text{–}1.28 \times 10^{-3}T)$ GHz, and $aq^2/\omega_s^2 = (3.0\text{–}1.5 \times 10^{-3}T)$, where T is temperature in Kelvins.

background. At the intermediate temperature, when the structural relaxation rate $1/\tau_2$ is of the order of ω_s, the linewidth of the Brillouin doublet displays a maximum, and the hypersonic velocity displays considerable dispersion. In this case, the intensity of the structural relaxation mode contributes to the intensity of the Brillouin doublet as well as to the intensity in the region between the unshifted Rayleigh peak and the Brillouin doublet. At low temperature, when the structural relaxation is slow compared with the hypersonic frequency, the structural relaxation manifests itself as a broad peak centered at the laser frequency. Comparison of the experimental and theoretical spectra calculated using Eq. (4.263) is illustrated in Fig. 4.1 for diphenyl disiloxane at two temperatures. The effects of structural relaxation on the frequency and line width of the Rayleigh–Brillouin spectrum at various temperatures in diphenyl disiloxane and polydiphenyl siloxane are shown in

Fig. 4.2. Both experimental and calculated results are shown. Good agreement between experiment and theory is also found in other viscoelastic fluids.

4.11 GENERALIZED HYDRODYNAMICS THEORY OF THE DEPOLARIZED LIGHT-SCATTERING SPECTRUM

We analyzed the shear-wave motion in a liquid in Section 4.8 and showed that at high temperature the shear wave relaxes with a relaxation rate equal to $\eta_s q^2/\rho_m$. At low temperature, the shear wave becomes oscillatory and the medium becomes viscoelastic. Observation of the shear-wave motion by light scattering is made possible through its coupling to the density of the polarizability anisotropy. The macroscopic theory of the depolarized spectrum of scattered light, which takes into account coupling of the shear wave to the orientation density, was given in Chapter 2. We present here the microscopic molecular theory using the technique of the generalized Langevin equation.

We begin by considering a system of rigid, symmetric-top molecules, since this simplifies the theory considerably and still yields all the salient features of the effect of translation–rotation coupling observed in the experimental spectra.

Using the scattering geometry shown in Fig. 2.5, the fluctuation in the polarizability anisotropy $\alpha_{\rm VH}(\mathbf{q}, t)$, which gives the VH spectrum, has two components:

$$\alpha_{\rm VH}(\mathbf{q}, t) = \alpha_{yz}(\mathbf{q}, t) \sin \theta/2 + \alpha_{xy}(\mathbf{q}, t) \cos \theta/2. \tag{4.272}$$

For symmetric-top molecules, $\alpha_{yz}(\mathbf{q}, t)$ and $\alpha_{xy}(\mathbf{q}, t)$ are given by

$$\alpha_{yz}(q, t) = \beta \sum_j n_y^{(j)} n_z^{(j)} e^{iqz_j}, \tag{4.273a}$$

$$\alpha_{xy}(\mathbf{q}, t) = \beta \sum_j n_x^{(j)} n_y^{(j)} e^{iqz_j}, \tag{4.273b}$$

where $n_\alpha^{(j)}$ is the projection of the symmetry axis of molecule j onto the α axis of the laboratory coordinate system.

The first step in applying the memory-function formalism to the calculation of the VH spectrum is to examine the symmetry properties of the secondary variables that can couple to the primary variables. We shall designate the polarizability density $\alpha_{xy}(\mathbf{q}, t)$ and $\alpha_{yz}(\mathbf{q}, t)$ as primary variables, because the VH spectra are directly determined by them. The other slow variables that couple to the polarizability density are designated as secondary. Since α_{yz} and α_{xy} have respectively odd and even reflection symmetry with respect to the yz plane ($n_x^{(j)} \to -n_x^{(j)}$), they are uncoupled at all times. Thus, the depolarized spectrum reduces to

$$I_{\mathrm{VH}}(\mathbf{q}, \omega) = (1/2\pi) \int_{-\infty}^{\infty} dt\, e^{-i\omega t}\{\langle \alpha_{yz}(\mathbf{q}, t)\alpha_{yz}^*(\mathbf{q})\rangle \cos^2 \theta/2 + \langle \alpha_{xy}(\mathbf{q}, t)\, \delta\alpha_{xy}^*(\mathbf{q})\rangle \sin^2 \theta/2\}, \quad (4.274)$$

as given previously.

The secondary variables to consider for coupling to α_{yz} and α_{xy} are, of course, the conserved hydrodynamic variables; number density $\delta\rho(q, t)$, momentum density $\mathbf{g}(\mathbf{q}, t)$, and energy density $\delta\varepsilon(\mathbf{q}, t)$. Symmetry consideration indicates that only the y component of the momentum density $g_y(\mathbf{q}, t)$ can couple to α_{yz}, whereas α_{xy} does not couple to any of the foregoing hydrodynamic variables. Thus, the calculation of $\langle \alpha_{yz}(\mathbf{q}, t)\alpha_{yz}^*(\mathbf{q})\rangle$ will involve a two-variable theory with A given by (Andersen and Pecora, 1971a, b)

$$\mathbf{A}(\mathbf{q}, t) = \begin{bmatrix} \alpha_{yz}(\mathbf{q}, t) \\ P_y(\mathbf{q}, t) \end{bmatrix} \quad (4.275)$$

Consider now the frequency matrix $\mathbf{\Omega}$, given by

$$\mathbf{\Omega} = \begin{bmatrix} \langle(\mathscr{L}\alpha_{yz})\alpha_{yz}^*\rangle & \langle(\mathscr{L}\alpha_{yz})P_y^*\rangle \\ \langle(\mathscr{L}P_y)\alpha_{yz}^*\rangle & \langle(\mathscr{L}P_y)P_y^*\rangle \end{bmatrix} \begin{bmatrix} \langle|\alpha_{yz}|^2\rangle & \langle\alpha_{yz}P_y^*\rangle \\ \langle P_y\alpha_{yz}^*\rangle & \langle|P_y|^2\rangle \end{bmatrix}^{-1}. \quad (4.276)$$

Since α_{yz} and P_y have opposite time-reversal symmetry, $\langle\alpha_{yz}P_y^*\rangle$ and $\langle P_y\alpha_{yz}^*\rangle$ vanish. Similarly, $\langle(\mathscr{L}\alpha_{yz})\alpha_{yz}^*\rangle$ and $\langle(\mathscr{L}P_y)P_y^*\rangle$ also vanish because the Liouville operator is odd under time reversal.

We now define the susceptibility for polarizability as

$$\chi_\alpha(\mathbf{q}) = \beta\langle|\alpha_{yz}(q)|^2\rangle = \beta\langle|\alpha_{yz}(\mathbf{q})|^2\rangle. \quad (4.277)$$

The last identity follows because of rotational isotropy of the fluid. Furthermore, since

$$\langle|P_y|^2\rangle = \sum_{l,l'} \langle\pi_{ly}\pi_{l'y}\exp[iq(z_l - z_{l'})]\rangle = NmkT, \quad (4.278)$$

the frequency matrix can be written as

$$\mathbf{\Omega} = \begin{bmatrix} 0 & \dfrac{1}{NmkT}\langle(\mathscr{L}\alpha_{yz})P_y^*\rangle \\ \dfrac{\langle(\mathscr{L}P_y)\alpha_{yz}^*\rangle}{kT\chi_\alpha} & 0 \end{bmatrix}. \quad (4.279)$$

Thus, from the symmetry properties of the dynamic variables A, given in Eq. (4.276), the frequency matrix $\mathbf{\Omega}$ has the form of Eq. (4.279). We shall next show by explicit calculation that the term $\langle(\mathscr{L}\alpha_{yz})P_y^*\rangle$ also vanishes.

Since

$$\mathscr{L}P_y = i\sum_i (F_{ly} + iq\pi_{ly}\pi_{lz}/m)\exp(iqz_l), \tag{4.280}$$

we have

$$\langle(\mathscr{L}P_y)\alpha_{yz}^*\rangle = -i\sum_l\sum_{l'}\langle(F_{ly} + iq\pi_{ly}\pi_{lz}/m)\alpha_{yz}^{(l')}\exp[iq(z_l - z_{l'})]\rangle, \tag{4.281}$$

where F_{ly} is the y component of the force acting on molecule l. In Eq. (4.281), the ensemble average over the quantity $\pi_{ly}\pi_{lz}$ vanishes, and there is no contribution from the second term on the right-hand side of Eq. (4.281). The first term also vanishes because

$$\begin{aligned}\langle F_{ly}\alpha_{yz}^{(l')}\exp[iq(z_l - z_{l'})]\rangle &= \frac{-1}{Z}\int d^N\mathbf{r}\exp(-\beta U)\left(\frac{\partial U}{\partial y_l}\right)\alpha_{yz}^{(l')}\exp[iq(z_l - z_{l'})] \\ &= \frac{kT}{Z}\int d^N r\left(\frac{\partial}{\partial y_l}e^{-\beta U}\right)\alpha_{yz}^{(l')}\exp[iq(z_l - z_{l'})] \\ &= \frac{-kT}{Z}\int d^N r\exp(-\beta U)\frac{\partial}{\partial y_l} \\ &\quad\times\{\alpha_{yz}^{(l')}\exp[iq(z_l - z_{l'})]\},\end{aligned} \tag{4.282}$$

where the last step is obtained by partial integration. In Eq. (4.282), U is the potential energy of the system and Z the configuration integral given by $Z = \int d^N r\exp(-\beta U)$.

Equation (4.282) vanishes identically because $\alpha_{yz}^{(l')}$ is independent of the position coordinate y_l [see Eq. (4.273a)]. As a result, we have

$$\langle(\mathscr{L}P_y)\alpha_{yz}^*\rangle = \langle(\mathscr{L}\alpha_{yz})P_y^*\rangle^* = 0, \tag{4.283}$$

and the frequency matrix $\mathbf{\Omega}$ is a null matrix.

The memory-function matrix is more complicated. Consider first the random force $\mathbf{F}(t)$. By the definition of the random force [Eq. (4.140)], we have

$$\mathbf{F}(\mathbf{q}, t) = e^{i(1-P)\mathscr{L}t}(1 - P)\begin{bmatrix} i\mathscr{L}\alpha_{yz} \\ i\mathscr{L}P_y\end{bmatrix} = e^{i(1-P)\mathscr{L}t}\begin{bmatrix} i\mathscr{L}\alpha_{yz} \\ i\mathscr{L}P_y\end{bmatrix} \equiv \begin{bmatrix}\bar{F}_\alpha(\mathbf{q}, t) \\ \bar{F}_{\mathrm{P}}(\mathbf{q}, t)\end{bmatrix}. \tag{4.284}$$

At $t = 0$, we have $\bar{F}_\alpha(\mathbf{q}) = i\mathscr{L}\alpha_{yz} \equiv F_\alpha(\mathbf{q})$ and $\bar{F}_{\mathrm{P}}(q) = i\mathscr{L}P_y \equiv F_{\mathrm{P}}(\mathbf{q})$. The overbar on $\bar{F}_\alpha(\mathbf{q}, t)$ and $\bar{F}_{\mathrm{P}}(\mathbf{q}, t)$ emphasizes the fact that time dependence is defined with respect to projected space. The operator P in Eq. (4.284) is the projection operator that, when acting on any function $G(t)$, gives

$$PG(t) = \langle G(t)\mathbf{A}^+\rangle\langle \mathbf{A}\mathbf{A}^+\rangle^{-1}\mathbf{A}. \tag{4.285}$$

The memory-function matrix is given by

$$\begin{aligned}\mathbf{K}(\mathbf{q}, t) &= \begin{bmatrix} \langle \bar{F}_\alpha(\mathbf{q}, t)F_\alpha^*(\mathbf{q})\rangle/kT\chi_\alpha(\mathbf{q}) & \langle \bar{F}_\alpha(\mathbf{q}, t)F_{\mathrm{P}}^*(\mathbf{q})\rangle/NmkT \\ \langle \bar{F}_{\mathrm{P}}(\mathbf{q}, t)F_\alpha^*(\mathbf{q})\rangle/kT\chi_\alpha(\mathbf{q}) & \langle \bar{F}_{\mathrm{P}}(\mathbf{q}, t)F_{\mathrm{P}}^*(\mathbf{q})\rangle/NmkT \end{bmatrix} \\ &\equiv \begin{bmatrix} K_{11}(\mathbf{q}, t) & K_{12}(\mathbf{q}, t) \\ K_{21}(\mathbf{q}, t) & K_{22}(\mathbf{q}, t) \end{bmatrix}. \end{aligned} \tag{4.286}$$

Most of the elements $K_{ij}(t)$ in the memory-function matrix cannot be eliminated by symmetry consideration. It should be noted that $K_{ij}(t)$ depends on q, but for brevity of notation the q dependence is not explicitly written out. Inversion symmetry shows that the diagonal elements are an even function of q and the off-diagonal elements are odd in q.

In the limit of small q, we can carry out the q expansion as

$$\begin{aligned} F_\alpha(\mathbf{q}) &= i\mathscr{L}\dot{\alpha}_{yz}(q) = \sum_l i\mathscr{L}(\alpha_{yz}^{(l)} e^{iqz_l}) \\ &= \sum_l (\dot{\alpha}_{yz}^{(l)} + iq\alpha_{yz}^{(l)}\pi_{lz}/m)e^{iqz_l} = \dot{\alpha}_{yz} + O(\mathbf{q}), \end{aligned} \tag{4.287}$$

where

$$\alpha_{yz} = \sum_l \alpha_{yz}^{(l)}. \tag{4.288a}$$

This gives

$$\dot{\alpha}_{yz} = \sum_l \dot{\alpha}_{yz}^{(l)}. \tag{4.288b}$$

Furthermore,

$$\begin{aligned} F_{\mathrm{P}}(q) &= i\mathscr{L}P_y(\mathbf{q}) = \sum_l i\mathscr{L}(\pi_{ly} e^{iqz_l}) \\ &= \sum_l (\dot{\pi}_{ly} + iq\pi_{ly}\pi_{lz}/m)e^{iqz_l} = iq\sigma_{yz} + O(q^2), \end{aligned} \tag{4.289}$$

where σ_{yz} is the yz term of the stress tensor given by [see Eq. (4.169)]

$$\sigma_{yz} \equiv \sum_l (\dot{\pi}_{lz} z_l + \pi_{ly}\pi_{lz}/m). \tag{4.290}$$

In obtaining Eq. (4.289), we used the fact that $\sum_l \dot{\pi}_l = 0$ because the total force acting on all molecules in the system vanishes.

Consider further the susceptibility $\chi_\alpha(\mathbf{q})$:

$$kT\chi_\alpha(\mathbf{q}) = \sum_{ll'} \langle \alpha_{yz}^{(l)}\alpha_{yz}^{(l')} \exp[iq(z_l - z_{l'})]\rangle$$
$$= N\langle|\alpha_{yz}^{(1)}|^2\rangle + N(N-1)\langle\alpha_{yz}^{(1)}\alpha_{yz}^{(2)} \exp[iq(z_1 - z_2)]\rangle, \tag{4.291}$$

where $\alpha_{yz}^{(1)}$ and $\alpha_{yz}^{(2)}$ are the molecular polarizabilities of the representative molecules 1 and 2. For small q, Eq. (4.291) can be written as

$$\lim_{q\to 0} \chi_\alpha(\mathbf{q}) = N/kT\langle|\alpha_{yz}^{(1)}|^2\rangle g_2 \equiv \chi_1 g_2, \tag{4.292}$$

where g_2, a static orientational pair-correlation factor, is given by

$$g_2 = 1 + (N-1)\left\{\frac{\langle\alpha_{yz}^{(1)}\alpha_{yz}^{(2)}\rangle}{\langle|\alpha_{yz}^{(1)}|^2\rangle}\right\} \simeq 1 + Nf. \tag{4.293}$$

Thus, in the $q \to 0$ limit the explicit expressions for the elements of the memory function are given by

$$K_{11}(\mathbf{q}, t) = \langle\dot{\alpha}_{yz}(t)\dot{\alpha}_{yz}\rangle_Q/kT\chi_1 g_2, \tag{4.294a}$$

$$K_{12}(\mathbf{q}, t) = -iq\langle\dot{\alpha}_{yz}(t)\sigma_{yz}\rangle_Q/NmkT = qK_{12}(t), \tag{4.294b}$$

$$K_{21}(\mathbf{q}, t) = iq\langle\sigma_{yz}(t)\dot{\alpha}_{yz}\rangle_Q/kT\chi_1 g_2 = qK_{21}(t), \tag{4.294c}$$

$$K_{22}(\mathbf{q}, t) = q^2\langle\sigma_{yz}(t)\sigma_{yz}\rangle/NmkT, \tag{4.294d}$$

where the subscript Q indicates that the time dependence is expressed in projected space according to Eq. (4.160), the Laplace transform of the correlation-function matrix is given by

$$\begin{bmatrix} \langle\alpha_{yz}(\mathbf{q}, s)\alpha_{yz}^*(\mathbf{q})\rangle & \langle\alpha_{yz}(\mathbf{q}, s)P_y^*(\mathbf{q})\rangle \\ \langle P_y(\mathbf{q}, s)\alpha_{yz}^*(\mathbf{q})\rangle & \langle P_y(\mathbf{q}, s)P_y^*(\mathbf{q})\rangle \end{bmatrix}$$
$$= \begin{bmatrix} \dfrac{s + q^2\eta(s)}{D(\mathbf{q}, s)} & \dfrac{-qK'_{12}(s)}{D(\mathbf{q}, s)} \\ \dfrac{-qK'_{21}(s)}{D(\mathbf{q}, s)} & \dfrac{s + \Gamma(s)}{D(\mathbf{q}, s)} \end{bmatrix} \begin{bmatrix} kT\chi_1 g_2 & 0 \\ 0 & NmkT \end{bmatrix}, \tag{4.295}$$

where $D(\mathbf{q}, s)$ is the determinant of the $[s\mathbf{I} + \mathbf{K}(\mathbf{q}, s)]$ matrix and is given by

$$D(\mathbf{q}, s) = s^2 + [\Gamma(s) + q^2\eta(s)]s + q^2[\eta(s)\Gamma(s) + \delta(s)], \tag{4.296}$$

where $\Gamma(s)$, $\eta(s)$, and $\delta(s)$ are

$$\Gamma(s) = \frac{1}{kT\chi_1 g_2} \int_0^\infty dt\, e^{-st} \langle \dot{\alpha}_{yz}(t)\dot{\alpha}_{yz} \rangle_Q, \tag{4.297}$$

$$\eta(s) = \frac{1}{\rho_m VkT} \int_0^\infty dt\, e^{-st} \langle \sigma_{yz}(t)\sigma_{yz} \rangle, \tag{4.298}$$

$$\delta(s) = K'_{12}(s)K'_{21}(s) = \frac{1}{Nm(kT)^2\chi_1 g_2} \left| \int_0^\infty dt\, e^{-st} \langle \dot{\alpha}_{yz}(t)\sigma_{yz} \rangle_Q \right|^2. \tag{4.299}$$

The spectral power density is found to be (Anderson and Pecora, 1971a, b)

$$\begin{aligned} I_{yz}(\mathbf{q}, \omega) &= (1/2\pi) \int_{-\infty}^{\infty} dt\, e^{-i\omega t} \langle \alpha_{yz}(\mathbf{q}, t)\alpha_{yz}^*(q) \rangle \\ &= (1/\pi)\, \mathrm{Re} \langle \alpha_{yz}(\mathbf{q}, s)\alpha_{yz}^*(q) \rangle |_{s=i\omega} \\ &= (kT\chi_1 g_2/\pi)\, \mathrm{Re}\{[i\omega + q^2\eta(i\omega)]/D(\mathbf{q}, i\omega)\}. \end{aligned} \tag{4.300}$$

Assuming for a moment that the Markov approximation can be used and the frequency dependence in $\Gamma(i\omega)$, $\eta(i\omega)$, and $\delta(i\omega)$ can be neglected, we then obtain from Eq. (4.300) an expression for the spectral power density $I_{yz}(\mathbf{q}, \omega)$ as

$$\begin{aligned} I_{yz}(\mathbf{q}, \omega) &= \frac{kT\chi_1 g_2}{\pi} \left\{ \frac{\omega^2\Gamma_S(\mathbf{q}) + q^2\eta(q^2\omega_S^2 - \omega^2)}{(q^2\omega_S^2 - \omega^2)^2 + \omega^2\Gamma_S^2(\mathbf{q})} \right\} \\ &= \frac{kT\chi_1 g_2}{\pi} \frac{\omega^2\Gamma + q^2\eta(q^2\omega_S^2)}{(q^2\omega_S^2 - \omega^2)^2 + \omega^2\Gamma_S^2(\mathbf{q})} \end{aligned} \tag{4.301}$$

In this equation, we have introduced $\Gamma_S(\mathbf{q}) = \Gamma + q^2\eta$ and $\omega_S^2 = q^2(\eta\Gamma + \delta)$.

To complete the calculation of the depolarized spectrum, the terms involving $\delta\alpha_{xy}$ in Eq. (4.275) must be accounted for. The spectrum resulting from the fluctuation of $\delta\alpha_{xy}$ is given by

$$\begin{aligned} I_{xy}(\mathbf{q}, t) &= (1/2\pi) \int_{-\infty}^{\infty} dt\, e^{-i\omega t} \langle \delta\alpha_{xy}(\mathbf{q}, t)\, \delta\alpha_{xy}^*(\mathbf{q}) \rangle \\ &= (1/\pi)\, \mathrm{Re} \langle \delta\alpha_{xy}(\mathbf{q}, s)\, \delta\alpha_{xy}^*(\mathbf{q}) \rangle |_{s=i\omega}. \end{aligned} \tag{4.302}$$

As previously mentioned, $\delta\alpha_{xy}$ does not couple to any of the hydrodynamic modes. Thus, for small $\mathbf{q}$ the autocorrelation function of $\delta\alpha_{xy}(\mathbf{q}, t)$ in the Markov approximation is given by

$$(\partial/\partial t)\langle \alpha_{xy}(\mathbf{q}, t)\alpha_{xy}^*(\mathbf{q}) \rangle = -\Gamma \langle \alpha_{xy}(\mathbf{q}, t)\alpha_{xy}^*(\mathbf{q}) \rangle \tag{4.303}$$

or

$$\langle \alpha_{xy}(\mathbf{q}, s)\alpha_{xy}^*(\mathbf{q}) \rangle = \langle |\alpha_{xy}(\mathbf{q})|^2 \rangle [1/(s + \Gamma)]. \tag{4.304}$$

According to Eq. (4.302), the spectral power density is thus

$$I_{xy}(\mathbf{q}, \omega) = (kT\chi_1 g_2/\pi)[\Gamma/(\Gamma^2 + \omega^2)]. \tag{4.305}$$

In Eq. (4.303), the orientational relaxation rate associated with $\delta\alpha_{xy}$ is taken to be identical to that associated with $\delta\alpha_{yz}$. This is justified because of the rotational isotropy of the fluid.

Using the results of Eqs. (4.301) and (4.305), we find that the VH spectral density $I_{\mathrm{VH}}(\mathbf{q}, \omega)$ [Eq.(4.275)] depends on the orientational relaxation rate Γ and the other two parameters η and ω_s. However, in the $q \to 0$ limit and under the Markov approximation, a relationship between the parameter ω_s and the ordinary shear viscosity can be found by comparing the $q \to 0$ limit of the autocorrelation function of the transverse-momentum density $\langle P_y(\mathbf{q}, t)P_y(\mathbf{q})^*\rangle$. From Eq. (4.295), we obtain (in the Markov approximation)

$$\langle P_y(\mathbf{q}, i\omega)P_y(\mathbf{q})^*\rangle = (NmkT)\{i\omega + q^2[\eta + \delta/(i\omega + \Gamma)]\}^{-1}. \tag{4.306}$$

Since in the limits of small q and small ω the shear viscosity η_s is related to $\langle P_y(\mathbf{q}, i\omega)P_y^*(\mathbf{q})\rangle$ [see Eqs. (4.178) and (4.181)] as

$$\lim_{\omega\to 0}\lim_{q\to 0}\langle P_y(\mathbf{q}, i\omega)P_y^*(\mathbf{q})\rangle = \frac{NmkT}{q^2\eta_s/\rho_m}. \tag{4.307}$$

By taking the $q \to 0$ and $\omega \to 0$ limits with respect to the right-hand side of Eq. (4.306), we expect to have

$$\frac{NmkT}{q^2\eta_s/\rho_m} = \frac{(NmkT)}{q^2(\eta + \delta/\Gamma)} \tag{4.308}$$

or

$$\eta_s/\rho_m = (\Gamma\eta + \delta)/\Gamma = \omega_s^2/\Gamma. \tag{4.309}$$

From Eq. (4.309) we now introduce a coupling parameter R:

$$\eta = (\eta_s/\rho_m)(1 - R) = v_s(1 - R), \tag{4.310}$$

where v_s is the kinematic viscosity and

$$R = \frac{\rho_m}{\eta_s}\frac{\delta}{\Gamma} = [V(kT)^2\chi_1 g_2\eta_s\Gamma]^{-1}\left|\int_0^\infty dt\,\langle\dot{\alpha}_{yz}(t)\sigma_{yz}\rangle_Q\right|^2. \tag{4.311}$$

If $R = 0$ (no coupling between orientation and transverse velocity), then η will be equal to v_s. In the presence of coupling, η acquires a new component Rv_s. Thus, the parameter R corresponds to the measure of that fraction of the shear viscosity resulting from rotational motion.

Note that Eq. (4.301) is identical to Eq. (2.193), in which the effect of rotation–translation on the depolarized spectrum was analyzed using a macroscopic theory. This section thus provides a microscopic derivation of the various relaxation coefficients and the coupling parameter R. Note that the molecular theory gives the orientational relaxation rate Γ and the coupling parameter R in terms of the projected space. The parameter R is determined by the dynamic coupling between shear stress and the angular momentum of the molecules through the term $\dot{\alpha}_{yz}$.

4.12 LIGHT SCATTERING FROM VISCOELASTIC LIQUIDS

The analysis in the preceding section was carried out for the case in which the correlation time of $\langle\sigma_{yz}(t)\sigma_{yz}\rangle$ is so short that the frequency dependence of $\eta(s)$ [Eq. (4.298)] can be neglected. In this limit, η becomes the renormalized shear viscosity (i.e., the shear viscosity corrected by translation–rotation coupling). For supercooled liquids or liquids at low temperature, $\langle\sigma_{yz}(t)\sigma_{yz}\rangle$ does not relax rapidly, and the frequency dependence of $\eta(s)$ is important. The system is said to be viscoelastic if $\eta(s)$ displays a pronounced frequency dependence. The viscoelastic system displays shear-wave sidebands, as discussed in Chapter 2. The shape of the central component is determined by $\Gamma(s)$ [Eq. (4.297)]. The frequency dependence of $\Gamma(s)$ is closely related to the dynamic response of the molecular reorientation process that leads to the decay of the polarizability anisotropy. Studies of the time-correlation function associated with the central component as a function of temperature, using the photon-correlation technique, indicate that the time-correlation function has the shape of a William–Watts function: $\exp[-(t/\tau_0)^\beta]$, with $\beta < 1$ (Wang *et al.*, 1983). This form shows that there is a continuous distribution of the relaxation times, and the central component is reflected strongly by the effects of the dynamic-relaxation process. However, the mean relaxation time $\bar{\tau} = \int_0^\infty dt \exp[-(t/\tau_0)^\beta] = \tau_0 \Gamma(\beta^1)/\beta$ associated with the central component is a large value; thus, the mean relation rate is slow compared with the peak position and the width of the shear-wave doublet; as far as the description of the spectral distribution is concerned, it suffices to assume that Γ is independent of frequency.

The dynamic-coupling coefficient $\delta(s)$ can be assumed to be frequency independent because dynamic coupling is effective only during molecular reorientation. The assumption of a frequency-independent Γ also implies a frequency-independent $\delta(s)$.

From the foregoing discussion, for the study of the dynamics of the shear-wave doublet in the viscoelastic liquid, it suffices to keep only the frequency dependence in $\eta(s)$ and to rewrite Eq. (4.300) as

$$I_{yz}(\mathbf{q}, \omega) = \left(\frac{kT\chi_1 g_2}{\pi}\right)\mathrm{Re}\left\{\frac{i\omega + q^2\eta(i\omega)}{-\omega^2 + [\Gamma + q^2\eta(i\omega)]i\omega + q^2[\eta(i\omega)\Gamma + \delta]}\right\}, \tag{4.312}$$

where $\eta(i\omega)$ is given by Eq. (4.298), with $s = i\omega$,

$$\eta(i\omega) = \frac{1}{\rho_m VkT}\int_0^\infty dt\, e^{-i\omega t}\langle\sigma_{yz}(t)\sigma_{yz}\rangle. \tag{4.313}$$

Further development of the theory will be associated with the calculation of $\eta(i\omega)$. Assuming that $\sigma_{yz}(t)$ is subject to rapid modulation of the random force with very short memory and $\sigma_{yz}(t)$ does not couple back to the primary variable $\alpha_{yz}(q, t)$ or to the secondary variable $P_y(q, t)$, then the correlation function $\langle\sigma_{yz}(t)\sigma_{yz}\rangle$ will be given by the first-order kinetic equation (Wang and Liu, 1983)

$$(\partial/\partial t)\langle\sigma_{yz}(t)\sigma_{yz}\rangle = -\gamma\langle\sigma_{yz}(t)\sigma_{yz}\rangle \tag{4.314}$$

or

$$\langle\sigma_{yz}(t)\sigma_{yz}\rangle = \langle\sigma_{yz}^2\rangle e^{-\gamma t}, \tag{4.315}$$

where γ is a q-independent rate constant given by

$$\gamma = \int_0^\infty dt\,\langle\dot{\sigma}_{yz}(t)\dot{\sigma}_{yz}\rangle/\langle\sigma_{yz}^2\rangle. \tag{4.316}$$

According to Eq. (4.315), we write γ^{-1} as

$$\gamma^{-1} = \frac{1}{\langle\sigma_{yz}^2\rangle}\int_0^\infty dt\,\langle\sigma_{yz}(t)\sigma_{yz}\rangle = \frac{P_m VkT}{\langle\sigma_{yz}^2\rangle}\eta_s = \frac{\eta_s}{G_\infty}, \tag{4.317}$$

where $G_\infty = \langle\sigma_{yz}^2\rangle/\rho_m VkT$ is the infinite-frequency shear modulus. With reference to Eq. (2.206), γ^{-1} corresponds to the Maxwell relaxation time. Thus, Eq. (4.314) corresponds to the Maxwell model. The expression of $I_{yz}(\mathbf{q}, \omega)$ for a Maxwell liquid was given in Chapter 2.

Clearly, the Maxwell liquid is rather restricted because the validity of this model relies on the assumption that $\sigma_{yz}(t)$ relaxes exponentially and further, that it does not couple back to the primary or the secondary variable. The assumption of exponential decay for the stress tensor is not necessarily valid because, as, $t \to \infty$, $\sigma_{yz}(t)$ approaches a finite value rather than zero in some systems. However, for systems with an exceedingly long relaxation time, the Maxwell model can realistically represent the dynamic response over some frequency range.

If the reaction of $\sigma_{yz}(t)$ to $P_y(q, t)$ is included in the theory, it would be necessary to include $\sigma_{yz}(t)$ [or more appropriately, $\dot{P}_y(q, t)$] as another member of the set of secondary variables. Since $\dot{P}_y(q, t)$ is not a conserved variable, the choice of $\dot{P}_y$ as the third variable is not unique. Other possibilities of choosing the third variable also exist, and there is no agreement on the nature of the third variable (Wang *et al.*, 1984). The choice of $\dot{P}_y(q, t)$ as the third dynamical variable results in an expression for $I_{yz}(\mathbf{q}, t)$ similar to that of the Voigt model discussed in Chapter 2 (Chappell and Kivelson, 1981).

4.13 SINGLE-PARTICLE AND COLLECTIVE ORIENTATIONAL-RELAXATION TIMES

As a final section of this chapter, we shall derive an expression that relates the single-particle orientational-relaxation time to the collective orientational-relaxation time.

The collective orientational-relaxation rate in the $q \to 0$ limit is given by Eq. (4.297) as

$$\Gamma = \frac{1}{kT\chi_1 g_2}\int_0^\infty dt\ \langle \dot{\alpha}_{yz}(t)\dot{\alpha}_{yz}\rangle_Q. \tag{4.318}$$

(The frequency dependency is neglected; this corresponds to setting $s = 0$). Here Γ is referred to as collective because the quantity $\dot{\alpha}_{yz}$ is defined as the sum of all molecules within the scattering volume [see Eq. (4.288)]. Thus, Γ can be regarded as the collective tumbling rate of all molecules in the system. Substituting Eq. (4.288) into (4.318) and also into (4.293) for g_2 results in N terms involving a one-particle index and $N(N - 1)$ terms involving different particle indices. The N terms involving only a one-particle index are called single-particle terms, and the $N(N - 1)$ terms involving the indices of two different particles are called pair terms. The pair terms represent interference and do not vanish in Rayleigh scattering, because Rayleigh scattering is coherent.

The single-particle terms (with appropriate modification of the projected time dependence) are associated with the single-particle relaxation rate and represent the tumbling rate of one single molecule influenced by the random motion of the rest of the molecules.

Since the rotation–translation coupling mechanism does not affect the orientational-relaxation rate, we shall neglect the coupling in the derivation of the relationship between the single particle and collective orientational-relaxation rates. We shall use the generalized Langevin equation for $C_\alpha(t) = \langle \alpha_{yz}(t)\alpha_{yz} \rangle$ as a starting point and carry out the calculation by extensive use of the matrix-multiplication technique.

Since $\alpha_{yz}(t)$ is equal to the sum of the single molecular-polarizability anisotropy of all molecules in the scattering volume, i.e.,

$$\alpha_{yz}(t) = \sum_l \alpha_{yz}^{(l)}(t), \tag{4.319}$$

we can write $C_\alpha(t)$ as

$$C_\alpha(t) = \sum_l \sum_{l'} \langle \alpha_{yz}^{(l)}(t)\alpha_{yz}^{(l')} \rangle. \tag{4.320}$$

It is convenient to construct a square time-correlation matrix $\mathbf{C}(t)$, in which the trace of $\mathbf{C}(t)$ gives the single-particle time-correlation functions, and the sum of all off-diagonal elements gives the time correlations involving different molecules, i.e.,

$$C_\alpha(t) = \sum_l \langle \alpha_{yz}^{(l)}(t)\alpha_{yz}^{(l)} \rangle + \sum_{l \neq l'} \sum_{l'} \langle \alpha_{yz}^{(l)}(t)\alpha_{yz}^{(l')} \rangle = C_{\text{self}}(t) + C_{\text{pair}}(t). \tag{4.321}$$

Thus, we choose a dynamic variable $\mathbf{A}$ given by a column vector (Wang, 1982)

$$\mathbf{A} = \begin{bmatrix} \alpha_{yz}^{(1)} \\ \alpha_{yz}^{(2)} \\ \vdots \\ \alpha_{yz}^{(N)} \end{bmatrix}. \tag{4.322}$$

The time-correlation matrix $\mathbf{C}(t)$ is given by

$$\mathbf{C}(t) = \langle \mathbf{A}(t)\mathbf{A}^+ \rangle. \tag{4.323}$$

The generalized Langevin equation for $\mathbf{C}(t)$ (the frequency matrix vanishes) is given by

$$d\mathbf{C}(t)/dt = -\int_0^\infty dt\, \mathbf{K}(\tau)\mathbf{C}(t-\tau), \tag{4.324}$$

where the memory function $\mathbf{K}(\tau)$ is

$$\mathbf{K}(\tau) = \langle (e^{iQ\mathscr{L}\tau} i\mathscr{L}\mathbf{A})(i\mathscr{L}\mathbf{A})^+ \rangle \langle \mathbf{A}\mathbf{A}^+ \rangle^{-1} \equiv \langle \mathbf{F}(\tau)\mathbf{F} \rangle \mathbf{C}(0)^{-1}. \tag{4.325}$$

It is more convenient to carry out the calculation in terms of the Laplace transform of $\mathbf{C}(t)$. Taking the Laplace transform of Eq. (4.324) and solving for $\mathbf{C}(s)$, we obtain

$$\mathbf{C}(s) = [s\mathbf{I} + \mathbf{K}(s)]^{-1}\mathbf{C}(0), \tag{4.326}$$

where $\mathbf{C}(s)$ and $\mathbf{K}(s)$ are the Laplace transforms of $\mathbf{C}(t)$ and $\mathbf{K}(t)$, respectively.

We now calculate $\mathbf{K}(s)$. According to Eq. (4.325), the calculation involves multiplication of the diagonal elements of $\langle \mathbf{F}(\tau)\mathbf{F}^{+}\rangle$ and $\mathbf{C}(0)^{-1}$. Since the diagonal elements of $\mathbf{C}(0)$ are identical, we set each diagonal element as

$$\langle \alpha_{yz}^{(1)}\alpha_{yz}^{(1)}\rangle = \langle \alpha_{yz}^{(2)}\alpha_{yz}^{(2)}\rangle = \cdots = \langle \alpha_{yz}^{(N)}\alpha_{yz}^{(N)}\rangle \equiv a = \beta\chi_1 \tag{4.327a}$$

All the off-diagonal elements of $\mathbf{C}(0)$ are also identical and are related to the static pair-orientation factor f ($f = \langle \alpha_{yz}^{(1)}\alpha_{yz}^{(2)}\rangle / \langle |\alpha_{yz}^{(1)}|^2\rangle$) as

$$\langle \alpha_{yz}^{(1)}\alpha_{yz}^{(2)}\rangle = \langle \alpha_{yz}^{(2)}\alpha_{yz}^{(3)}\rangle = \cdots = \langle \alpha_{yz}^{(1)}\alpha_{yz}^{(N)}\rangle \equiv b = \beta\chi_1 f. \tag{4.327b}$$

Thus, the $\mathbf{C}(0)$ matrix has all the diagonal elements equal to a and all the off-diagonal elements equal to b:

$$\mathbf{C}(0) = \begin{bmatrix} a & b & & \cdots & b \\ b & a & b & \cdots & b \\ \vdots & & & \ddots & \vdots \\ & & & & b \\ b & \cdots & & b & a \end{bmatrix} = (a - b)\mathbf{I} + b\mathbf{U}, \tag{4.328}$$

where $\mathbf{I}$ is a unit matrix and $\mathbf{U}$ is a matrix with all its elements equal to one. Using the identity

$$\mathbf{U}^n = N^{n-1}\mathbf{U}, \tag{4.329}$$

we find the inverse of $\mathbf{C}(0)$ as

$$\begin{aligned} \mathbf{C}(0)^{-1} &= (a - b)^{-1}\left(\mathbf{I} + \frac{b\mathbf{U}}{a - b}\right)^{-1} = (a - b)^{-1}\left\{\mathbf{I} + \sum_{n=1}^{\infty}\left(\frac{b}{b - a}\right)^n \mathbf{U}^n\right\} \\ &= (a - b)^{-1}\left\{\mathbf{I} - \frac{b\mathbf{U}}{a + (N - 1)b}\right\}. \end{aligned} \tag{4.330}$$

Similarly, the $\langle \mathbf{F}(t)\mathbf{F}^{+}\rangle$ matrix has all the diagonal elements equal to each other, and all the off-diagonal elements are also equal to each other:

$$\langle F_i(s)F_i^*\rangle = \int_0^{\infty} dt\, e^{-st}\langle \alpha_{yz}^{(i)}(t)\alpha_{yz}^{(i)}\rangle_Q = d(s), \qquad i = 1, 2, \ldots, N \tag{4.331}$$

and

$$\langle F_i(s)F_j^*\rangle = \int_0^\infty dt\, e^{-st}\langle \alpha_{yz}^{(i)}(t)\alpha_{yz}^{(j)}\rangle_Q = h(s), \qquad i, j = 1, \ldots, N, \quad i \neq j. \tag{4.332}$$

Thus, the $\langle \mathbf{F}(\tau)\mathbf{F}^+\rangle$ matrix can be written as

$$\langle \mathbf{F}(\tau)\mathbf{F}^+\rangle = (d - h)\mathbf{I} + h\mathbf{U}. \tag{4.333}$$

Substituting Eqs. (4.330) and (4.333) into the Laplace transform of Eq. (4.325) and carrying out the matrix multiplication, we find the memory-function matrix $\mathbf{K}(s)$ to be

$$\begin{aligned}\mathbf{K}(s) &= [(d - h)\mathbf{I} + h\mathbf{U}](a - b)^{-1}\{\mathbf{I} - b\mathbf{U}/[a + (N - 1)b]\} \\ &= [D/(1 - f)][(1 - g)\mathbf{I} + (g - f)\mathbf{U}/[1 + (N - 1)f]],\end{aligned} \tag{4.334}$$

where D, the single-particle relaxation rate, is given by

$$D = \frac{d}{a} = \frac{1}{\langle |\alpha_{yz}^{(1)}|^2\rangle}\int_0^\infty dt\, e^{-st}\langle \dot{\alpha}_{yz}^{(1)}(t)\dot{\alpha}_{yz}^{(1)}\rangle_Q, \tag{4.335}$$

where g is the dynamic orientational pair-correlation factor given by

$$g = \int_0^\infty dt\, e^{-st}\langle \dot{\alpha}_{yz}^{(1)}(t)\dot{\alpha}_{yz}^{(2)}\rangle_Q \bigg/ \int_0^\infty dt\, e^{-st}\langle \alpha_{yz}^{(1)}(t)\alpha_{yz}^{(1)}\rangle_Q. \tag{4.336}$$

Substituting Eq. (4.334) into (4.326) and carrying out several matrix-multiplication steps, we obtain

$$\mathbf{C}(s) = \frac{a}{s + (1 - g)D/(1 - f)}\left[(1 - f)\mathbf{I} + \frac{(1 - T)f + T}{1 - TN}\mathbf{U}\right], \tag{4.337}$$

where

$$T = \frac{D(f - g)}{(1 - f)[1 + (N - 1)f][s + D(1 - g)/(1 - f)]}. \tag{4.338}$$

The self terms of $\mathbf{C}(s)$ are equal to $Tr\ \mathbf{C}(s)$ and are found as

$$\begin{aligned}C_{\text{self}}(s) = \text{Tr } \mathbf{C}(s) &= \frac{Na}{s + (1-g)/(1-f)} \\ &\quad \times \{(1-f) + [(1-T)f + T]/(1-TN)\} \\ &= \left[\frac{Na}{s + (1-g)D/(1-f)}\right] \\ &\quad \times \left(\frac{1 + (N-1)D(g-f)/[s + D(1-g)/(1-f)][1 + (N-1)f]}{1 + ND(g-f)/\{(1-g)[s + D(1-g)/(1-f)][1 + (N-1)f]\}}\right).\end{aligned} \tag{4.339}$$

For large N and large V, we approximate $1 - g \simeq 1$, $1 - f \simeq 1$, and $N - 1 \simeq N$; under this approximation, Eq. (4.339) reduces to

$$C_{\text{self}}(s) \simeq Na/(s + D) \tag{4.340}$$

The pair term of $\mathbf{C}(s)$, the sum of all off-diagonal matrix elements of the matrix $\mathbf{C}(s)$, is equal to

$$C_{\text{pair}}(s) = \frac{\{N(N-1)[(1-T)f + T]a\}/(1-NT)}{s + D(1-g)/(q-f)}. \tag{4.341}$$

In the large V and large N limit, the pair term reduces to

$$C_{\text{pair}}(s) \simeq \left[\frac{N^2 a}{s + D}\right]\left[\frac{f - D(g-f)/(s+D)(1+Nf)}{1 + ND(g-f)/(s+D)(1+Nf)}\right]. \tag{4.342}$$

The time-correlation function $C_\alpha(s)$ is equal to

$$\begin{aligned}C_\alpha(s) &= C_{\text{self}}(s) + C_{\text{pair}}(s) \\ &= \frac{Na}{(s+D)}\left[1 + \frac{Nf - ND(g-f)/(s+D)(1+Nf)}{1 + ND(g-f)/(s+D)(1+Nf)}\right] \\ &= \frac{Na(1+Nf)}{s + D + ND(g-f)/(1+Nf)},\end{aligned}$$

or

$$C_\alpha(s) = \frac{Na(1+Nf)}{s + D(1+Ng)/(1+Nf)}. \tag{4.343}$$

If the Markov approximation is assumed for the single-particle relaxation rate D and for the dynamic pair-orientation factor g, then Eq. (4.343) implies that $C_\alpha(t)$ will decay with a relaxation rate Γ given by (Keyes and Kivelson, 1972)

$$\Gamma = D(1 + Ng)/(1 + Nf), \tag{4.344}$$

or, equivalently, in terms of relaxation times $\tau_s = D^{-1}$ and $\tau_{\mathscr{L}s} = \Gamma^{-1}$,

$$\tau_{\mathscr{L}s} = \tau_s(1 + Nf)/(1 + Ng). \tag{4.345}$$

Thus, the collective light-scattering relaxation time $\tau_{\mathscr{L}s}$ is proportional to the single-particle relaxation time τ_s. In most cases, $\tau_{\mathscr{L}s}$ is found to be longer than τ_s, indicating that the pair orientational-correlation factors are non-vanishing.

For a system of symmetric-top molecules, we can show that

$$Nf = g_2 - 1 = \sum_{i \neq j} \left\langle \frac{P_2(\hat{u}_i \cdot \hat{u}_j)}{N - 1} \right\rangle; \tag{4.346}$$

Nf is given by a more complicated expression for asymmetric-top molecules. Experimentally, it is found that Nf is positive and varies with the concentration of anisotropic molecules. In principle, Nf can be negative, but this is seldom found.

Although there is no a priori reason to expect that the dynamic-pair orientation factor g will vanish, it is expected to be small because g is associated with the correlation of the angular momenta of different molecules. This result was confirmed experimentally by the dynamic light-scattering study of optically anisotropic molecules dissolved in optically isotropic solvents (Kivelson and Madden, 1980).

5 Nuclear Magnetic Resonance Spectroscopy

5.1 INTRODUCTION

The principle of nuclear magnetic resonance (NMR) spectroscopy is based on the application of a strong magnetic field to a system of nuclear spins. This leads to a lifting of the degeneracy of nuclear spin energy levels. Transitions between the energy levels are induced by the application of a radio-frequency (rf) field. Since the nuclear spins are coupled to other degrees of freedom, the line shape of the resonance spectrum and the relaxation effect of the nuclear spins are closely connected to the dynamics of molecular motion. In this chapter, we specifically discuss NMR, although a similar formation can also be obtained for electron spin resonance (ESR).

Since the pioneering work of Bloembergen, Purcell, and Pound (BPP) (1948), NMR has proved to be an extremely powerful tool in the study of molecular motion of materials in the condensed phase. Not only is this technique useful in the study of the dynamics of "mass flow," or the translational diffusion of atoms and molecules, but also it is useful in the study of molecular-reorientation processes when mass flow is absent. Furthermore, because of much activity in the development of this technique, most of the theoretical problems concerning NMR spectroscopy have been clarified.

This enables the researcher to interpret experimental results on a firm theoretical ground and permits unambiguous conclusions about the nature of molecular motion in condensed matter. In this aspect, the field of magnetic-resonance spectroscopy enjoys a greater advantage than other techniques.

Relaxation of the nuclear spin energy to its environment typically takes place on a time scale of 0.1 msec to tens of seconds. This scale is easily accessible to measurement and provides the opportunity to externally influence a spin system in various useful ways. No other field of molecular spectroscopy has been as successful as NMR in designing external means for manipulating the system under study. This is because it is possible to apply both static and time-dependent magnetic fields to interact more strongly with the spin system than with either the interaction between spins or the interaction of the spins with other nonspin degrees of freedom, which in the parlance of NMR is known as the lattice.

The use of the external magnetic field to influence the dynamics of the spin system has resulted in sophisticated multiple-pulse experiments developed primarily by Waugh and his collaborators (Waugh *et al.*, 1968). In the multiple-pulse experiments, the use of the external magnetic field to dominate the motion of the spin system in a controlled way is particularly important. The theory of coherent averaging forms the basis for the analysis of many of these experiments. In this chapter, we shall present basic concepts, e.g., local field, spin temperature, and effective Hamiltonians, that are central to interpretation of NMR experiments. Theories that have been developed to describe nuclear spin relaxations will also be presented.

5.2 MOTION OF FREE SPINS

A fundamental problem in NMR is describing the motion of a free spin in an external magnetic field. A system of noninteracting spins can also be considered a free spin with an angular momentum $\hbar\mathbf{I}$ and a magnetic moment $\mathbf{M} = \gamma\hbar\mathbf{I}$, where $\hbar$ is the Planck constant divided by 2π and γ the gyromagnetic ratio.

The equation of motion of the angular-momentum operator $\hbar\mathbf{I}$ is governed by the Heisenberg equation of motion

$$\hbar\, \partial\mathbf{I}/\partial t = i[\mathscr{H}, \mathbf{I}], \tag{5.1}$$

where the Hamiltonian $\mathscr{H}$ for the free spin in the external field $\mathbf{H}$ is given by

$$\mathscr{H} = -\gamma\hbar\mathbf{H}\cdot\mathbf{I}. \tag{5.2}$$

For brevity, we shall write the Hamiltonian in the frequency unit. This is equivalent to setting $\hbar = 1$. (In this chapter we shall write the Hamiltonian in the frequency unit except for situations where confusion may arise in doing so.)

Evaluation of the commutation on the right-hand side of Eq. (5.1) gives

$$\partial \mathbf{I}/\partial t = \gamma(\mathbf{H} \times \mathbf{I}) \tag{5.3}$$

or, in terms of the magnetic moment **M**, we have

$$\partial \mathbf{M}/\partial t = -\gamma(\mathbf{M} \times \mathbf{H}), \tag{5.4}$$

which is identical to the classical equation.

Consider next that a sample containing a large number of identical spins is placed in a time-dependent magnetic field consisting of a very large static field along the laboratory z axis and a somewhat weaker rotating magnetic field in the xy plane. The amplitude of the static field is H_0, and the rotating magnetic field is given as

$$\mathbf{H}_1(t) = H_1(\hat{x} \cos \omega t + \hat{y} \sin \omega t). \tag{5.5}$$

We want to know the magnetization at time t after application of the rotating field at time 0. The calculation of the magnetization can be carried out using the equation of motion for the density operator $\rho(t)$, given by

$$\partial \rho(t)/\partial t = -i[\mathscr{H}(t), \rho(t)]. \tag{5.6}$$

The magnetization at time t is proportional to $\langle \mathbf{I}(t) \rangle$, given by

$$\langle \mathbf{I}(t) \rangle = \mathrm{Tr}[\mathbf{I}\rho(t)]. \tag{5.7}$$

It is convenient to express the Hamiltonian in Eq. (5.6) by

$$\begin{aligned} \mathscr{H}(t) &= -\gamma[H_0 I_z + H_1(I_x \cos \omega t + I_y \sin \omega t)] \\ &= -\omega_0 I_z - \tfrac{1}{2}\omega_1(I_+ e^{-i\omega t} + I_- e^{i\omega t}), \end{aligned} \tag{5.8}$$

where $I_\pm = I_x \pm iI_y$, $\omega_0 = \gamma H_0$, and $\omega_1 = \gamma H_1$.

The solution of Eq. (5.6) with the Hamiltonian given by Eq. (5.8) can be found by first making a transformation to the rotating coordinate frame, which rotates at frequency ω; i.e.,

$$\rho(t) = e^{i\omega I_z t}\sigma(t)e^{-i\omega I_z t}, \tag{5.9}$$

where $\sigma(t)$ is the density operator in the rotating frame.

Substituting Eq. (5.9) into (5.6) gives the equation of motion for the density operator in the rotating frame as

$$\partial \sigma(t)/\partial t = -i[\Delta I_z + \omega_1 I_x, \sigma(t)], \tag{5.10}$$

where $\Delta = \omega_0 - \omega$ is the offset from the resonance frequency.

The solution of Eq. (5.10) is

$$\sigma(t) = \exp[-i(\Delta I_z + \omega_1 I_x)t]\sigma(0)\exp[i(\Delta I_z + \omega_1 I_x)t]$$
$$= \exp[-ia(\hat{n}\cdot\mathbf{I})^x t]\sigma(0), \tag{5.11}$$

where $a = (\Delta^2 + \omega_1^2)^{1/2}$, $\hat{n} = \hat{x}\sin\theta + \hat{z}\cos\theta$, $\theta = \tan^{-1}(\omega_1/\Delta)$. In Eq. (5.11), we have used the superoperator notation discussed in Chapter 1.

Since at $t = 0$, $\rho(0) = \sigma(0)$, the density operator at time t is

$$\rho(t) = \exp(-i\omega I_z^x t)\exp[-ia(\hat{n}\cdot\mathbf{I})^x t]\rho(0). \tag{5.12}$$

Knowing $\rho(t)$, we can calculate the expectation value of any observable Q for the system by the operation

$$\langle Q(t)\rangle = \mathrm{Tr}[\rho(t)Q]. \tag{5.13}$$

Thus

$$\langle I_z(t)\rangle = \mathrm{Tr}(\{\exp(-i\omega I_z^x t)\exp[-ia(\hat{n}\cdot\mathbf{I})^x t]\rho(0)\}I_z)$$
$$= \mathrm{Tr}(\{\exp[-ia(\hat{n}\cdot\mathbf{I})^x t]\rho(0)\}I_z) = \mathrm{Tr}\{\rho(0)\exp[ia(\hat{n}\cdot\mathbf{I})^x t]I_z\}. \tag{5.14}$$

To obtain the second step of Eq. (5.14), we used the fact that the trace operation is invariant to cyclic permutation. Further, using the identity that

$$\exp[-ia(\hat{n}\cdot\mathbf{I})t] = \exp[-ia(I_x\sin\theta + I_z\cos\theta)t]$$
$$= \exp(-i\theta I_y)\exp(-iaI_z t)\exp(i\theta I_y), \tag{5.15}$$

it is easy to show that

$$\exp[ia(\hat{n}\cdot\mathbf{I})^x t]I_z = (I_z\cos\theta + I_x\sin\theta)\cos\theta$$
$$-(I_y\cos\theta - I_z\sin\theta)\sin\theta\cos at - I_y\sin\theta\sin at \tag{5.16}$$

Substituting Eq. (5.16) into (5.14), we obtain

$$\langle I_z(t)\rangle = \langle I_z\rangle_0(\cos^2\theta + \sin^2\theta\cos at)$$
$$+ \langle I_x\rangle_0\sin\theta\cos\theta(1 - \cos at) - \langle I_y\rangle_0\sin\theta\sin at, \tag{5.17a}$$

where $\langle I_i\rangle_0 = \mathrm{Tr}[\rho(0)I_i]$, $i = x, y, z$.

The quantities $\langle I_x(t)\rangle$ and $\langle I_y(t)\rangle$ can be similarly calculated. They are found to be

$$\begin{aligned}\langle I_x(t)\rangle = \langle I_x\rangle_0[(\cos^2\theta\cos at + \sin^2\theta)\cos\omega t + \cos\theta\sin at\sin\omega t]\\ + \langle I_y\rangle_0(\cos\theta\sin at\cos\omega t - \cos at\sin\omega t)\\ + \langle I_z\rangle_0[\sin\theta\cos\theta\,(1-\cos at)\cos\omega t - \sin at\sin\theta\sin\omega t];\end{aligned} \tag{5.17b}$$

$$\begin{aligned}\langle I_y(t)\rangle = \langle I_x\rangle_0[\cos\theta\sin at\cos\omega t + (\cos^2\theta\cos at + \sin^2\theta)\sin\omega t]\\ + \langle I_y\rangle_0(\cos at\cos\omega t - \cos\theta\sin at\sin\omega t)\\ + \langle I_z\rangle_0[\sin\theta\cos\theta\,(1-\cos at)\sin\omega t - \sin\theta\sin at\cos\omega t].\end{aligned} \tag{5.17c}$$

If the rotating magnetic field is applied at time $t = 0$, then $\rho(0)$ is the equilibrium-density operator, given by

$$\rho(0) = e^{\beta\hbar\omega_0 I_z}/Z \tag{5.18}$$

where Z is the partition function equal to Tr $e^{\beta\hbar\omega_0 I_z}$ and $\beta = 1/kT$. In this case, $\langle I_x\rangle_0 = \langle I_y\rangle_0 = 0$, because the equilibrium magnetization has only a z component, given by $M_z = \gamma\langle I_z\rangle_0 = \chi_0 H_0$, where χ_0 is the static nuclear susceptibility. As a result, Eqs. (5.17) reduce to

$$\begin{aligned}\langle I_x(t)\rangle &= \langle I_z\rangle_0[\sin\theta\cos\theta\,(1-\cos at)\cos\omega t - \sin\theta\sin at\sin\omega t];\\ \langle I_y(t)\rangle &= \langle I_z\rangle_0[\sin\theta\cos\theta\,(1-\cos at)\sin\omega t - \sin\theta\sin at\cos\omega t];\\ \langle I_z(t)\rangle &= \langle I_z\rangle_0(\cos^2\theta + \sin^2\theta\cos at).\end{aligned} \tag{5.19}$$

Equations (5.19) show that at resonance ($\Delta = 0$ or $\omega = \omega_0$), $\theta = \pi/2$ and $a = \omega_1$. As a result, Eq. (5.19) reduces to

$$\begin{aligned}\langle I_x(t)\rangle &= -\langle I_z\rangle_0\sin\omega_1 t\sin\omega_0 t,\\ \langle I_y(t)\rangle &= -\langle I_z\rangle_0\sin\omega_1 t\cos\omega_0 t,\\ \langle I_z(t)\rangle &= \langle I_z\rangle_0\cos\omega_1 t.\end{aligned} \tag{5.20}$$

On the other hand, when the frequency ω is far from ω_0 so that $\Delta \gg \omega_1$, then $\theta \simeq 0$. In this case, the oscillating field has a negligible effect on the spin system because $\langle I_z(t)\rangle \simeq \langle I_z\rangle_0$ and $\langle I_x(t)\rangle = \langle I_y(t)\rangle \simeq 0$. The foregoing calculation for a system of free spins illustrates that a rotating field with an amplitude H_1, which may be small compared with the static field H_0, can appreciably orient a magnetic moment when the rotating frequency ω is near the Larmor frequency ω_0. The width of the resonance is the value $|\Delta| = |\omega - \omega_0|$, below which the effect becomes negligible. Clearly, from this calculation, the width of the resonance is on the order of ω_1.

5.3 THE BLOCH EQUATIONS

Note from Eq. (5.20) that the amplitude of the magnetization is conserved, i.e., $[\langle I_x(t)\rangle^2 + \langle I_y(t)\rangle^2 + \langle I_z(t)\rangle^2]^{1/2} = \langle I_z\rangle_0$, as it should be, because for a system of noninteracting spins, no mechanism is present to relax or to cause decay of the magnetization.

However, without a relaxation mechanism it would be impossible to observe a resonance spectrum, because the population of nuclear magnetic-energy levels would be quickly saturated and no transition would be possible afterward. The fact that the resonance signal can be readily detected using an appropriate apparatus indicates that the relaxation mechanism plays an important role in NMR spectroscopy.

In 1946, Felix Bloch proposed a set of simple equations to describe the behavior of an ensemble of nuclei subject to an externally applied magnetic field. He wrote, on the basis of phenomenological arguments, a set of three equations to describe the dynamic behavior of the components of the magnetization along and perpendicular to the direction of the strong, static magnetic field. The Bloch equations have proven exceedingly useful for liquids in which molecular motion occurs very rapidly.

As described in the previous section, the dynamic state of the nuclear magnetization subject to an external field in the absence of the relaxation mechanism is given by Eq. (5.4). Magnetization is then brought by some means, such as the application of a radio frequency at or near a resonance field to a nonequilibrium state. As a result of the interaction of spins with each other and with the surroundings to which the spins are coupled, the component of the magnetization along the field is expected to relax toward the equilibrium value (M_0) with a time constant T_1. The quantity T_1 is known as the longitudinal (or spin–lattice) relaxation time. The component of the magnetization perpendicular to the static field will relax and decay to zero with a time constant T_2. The quantity T_2 is known as the transverse (or spin–spin) relaxation time. In general, the two relaxation times are not equal, although experimentally we find that in ordinary liquids at high temperature $T_1 = T_2$.

Encompassing the relaxation effect, Bloch argued that nuclear magnetization in the presence of the external magnetic field is described by

$$\partial \mathbf{M}/\partial t = -\gamma(\mathbf{M} \times \mathbf{H}) - M_x\hat{x} + M_y\hat{y}/T_2 - (M_z - M_0)\hat{z}/T_1 \quad (5.21)$$

where $\hat{x}$ and $\hat{y}$ represent the unit vectors perpendicular to the static magnetic field; $\hat{z}$ is parallel to the static field.

Equation (5.21) is a vector equation. In terms of the Cartesian components, it becomes three operations:

$$\begin{aligned} \partial M_x/\partial t &= (M_y H_z - M_z H_y) - M_x/T_2, \\ \partial M_y/\partial t &= (M_z H_x - M_x H_z) - M_y/T_2, \\ \partial M_z/\partial t &= (M_x H_y - M_y H_x) - (M_z - M_0)/T_1. \end{aligned} \tag{5.22}$$

For very viscous liquids or solids in which spins are strongly interacting with each other, the Bloch equations become invalid. In the next section, we shall examine the consequence of Block equations.

5.4 STEADY-STATE SOLUTION OF THE BLOCH EQUATIONS AND SATURATION

In a magnetic-resonance experiment, the external magnetic field is a sum of a static field H_0 applied in the z direction and a linearly polarized rf field oscillating with the frequency ω applied in the direction perpendicular to the z axis. The linearly polarized oscillating field can be considered a superposition of two oppositely rotating fields, as

$$\begin{aligned} \mathbf{H}_1(t) = 2H_1\hat{x}\cos\omega t &= H_1(\hat{x}\cos\omega t + \hat{y}\sin\omega t) \\ &+ H_1(\hat{x}\cos\omega t - \hat{y}\sin\omega t) \end{aligned} \tag{5.23}$$

We see that the oscillating magnetic field contains a component identical to the rotating field given in Eq. (5.5). The second component represents the counter-rotating component. In the rotating frame, the effect of the counter-rotating component can be neglected because it is rotating at the frequency 2ω with respect to the rotating frame.

Thus, in the coordinate frame rotating about the static field, the time-dependent rotating field becomes static, and the magnetic field along the z axis is reduced from H_0 to $H_0 + \omega/\gamma$. That is, the effective magnetic field in the rotating frame is

$$\mathbf{H}_{\text{eff}} = (H_0 + \omega/\gamma)\hat{z} + H_1\hat{x} = (\Delta\hat{z} - \omega_1\hat{x})/\gamma, \tag{5.24}$$

where, as before, $\Delta = \omega - \omega_0$, $\omega_0 = \gamma H_0$, and $\omega_1 = \gamma H_1$.

The Bloch equations in the rotating frame become

$$\begin{aligned}\partial\tilde{M}_x/\partial t &= \Delta\tilde{M}_y - \tilde{M}_z/T_2,\\ \partial\tilde{M}_y/\partial t &= -\Delta\tilde{M}_x - \omega_1\tilde{M}_z - \tilde{M}_y/T_2,\\ \partial\tilde{M}_z/\partial t &= \omega_1\tilde{M}_y - (\tilde{M}_z - M_0)/T_1,\end{aligned} \tag{5.25}$$

where $\tilde{M}_x$, $\tilde{M}_y$, and $\tilde{M}_z$ are three components of the magnetization in the rotating frame. They are related to the components in the laboratory frame by

$$M_x = \tilde{M}_x \cos\omega t - \tilde{M}_y \sin\omega t, \qquad M_y = \tilde{M}_x \sin\omega t + \tilde{M}_y \cos\omega t,$$
$$M_z = \tilde{M}_z. \tag{5.26}$$

Equation (5.25) is a set of three coupled, first-order, linear, differential equations. These can be solved exactly. However, we are interested in the steady-state solution of this set of equations. The steady-state solution is applicable in the case wherein the spin system is placed in the external magnetic field (static and oscillating) for a time sufficiently long that all the transient effects have decayed away. The steady-state solution can be obtained by setting $\partial\tilde{M}_x/\partial t = \partial\tilde{M}_y/\partial t = \partial\tilde{M}_z/\partial t = 0$ in Eq. (5.25). As a result, we obtain

$$\begin{aligned}\tilde{M}_x &= \frac{\Delta\gamma H_1 T_2^2}{1 + (\Delta T_2)^2 + T_1 T_2 \omega_1^2} M_0,\\ \tilde{M}_y &= \frac{\gamma H_1 T_2}{1 + (\Delta T_2)^2 + T_1 T_2 \omega_1^2} M_0,\\ \tilde{M}_z &= \frac{1 + (\Delta T_2)^2}{1 + (\Delta T_2)^2 + T_1 T_2 \omega_1^2} M_0,\end{aligned} \tag{5.27}$$

Since the three components of the magnetization in the laboratory frame are related to $\tilde{M}_x$, $\tilde{M}_y$, and $\tilde{M}_z$ by Eq. (5.26), the components M_x and M_y in the laboratory frame vary sinusoidally with time at a frequency ω. The time variation induces in the coil a detectable signal at the frequency ω.

We shall now use the information discussed in Chapter 1 to show that the absorption amplitude is proportional to $\tilde{M}_y$. We proceed by assuming that the oscillating field $2H_1 \cos\omega t$ is sufficiently small and that the linear-response theory previously developed can be used to calculate the response $M_x(t)$ for the rf excitation given by $2H_1 \cos\omega t$. As shown in Chapter 1, linear-response theory gives [compare with Eq. (1.97)]

$$M_x(t) = 2H_1[\chi'(\omega)\cos\omega t + \chi''(\omega)\sin\omega t], \tag{5.28}$$

where $\chi'(\omega)$ and $\chi''(\omega)$ are the real and the imaginary parts, respectively, of the magnetic susceptibility induced by the oscillating rf field. Comparing Eq.

(5.28) with (5.26), we find that $\tilde{M}_x$ is associated with $\chi'(\omega)$ and $\tilde{M}_y$ with $\chi''(\omega)$. Furthermore, we recall that power dissipation is given by Eq. (1.162) as

$$p = -\partial\langle E(t)\rangle/\partial t = -2\omega H_1^2\chi''(\omega). \tag{5.29}$$

Therefore, $\tilde{M}_y$ is proportional to power absorption. Dispersion and absorption are related by the Kramers–Kronig relation [Eq. (1.125) or (1.126)], and $\tilde{M}_x$ is related to the dispersion. Clearly, the dispersion vanishes at resonance, i.e., $\Delta = 0$ [see Eq. (5.27)].

Returning now to Eq. (5.27) and considering the case when a low level rf field is applied, i.e., $\gamma^2 H_1 T_1 T_2 \ll 1$, we can rewrite the expression for $\tilde{M}_y$ as

$$\tilde{M}_y \simeq \frac{\gamma H_1 T_2 M_0}{1 + (\Delta T_2)^2} = \pi\gamma H_1 M_0 f(\Delta), \tag{5.30}$$

where $f(\Delta)$ is the normalized Lorentzian line-shape function having half-width at the half-maximum intensity equal to T_2^{-1}. In other words, in the case of a weak rf field (or negligible saturation), the Bloch equation predicts a Lorentzian absorption line shape.

In the case of appreciable saturation, we rewrite the $\tilde{M}_y$ expression as

$$\tilde{M}_y = \frac{\pi\gamma H_1 M_0}{(1 + T_1 T_2 \omega_1^2)^{1/2}} f_{T_2'}(\Delta), \tag{5.31}$$

where

$$f_{T_2'}(\Delta) = \frac{1}{\pi}\frac{T_2'}{1 + (\Delta T_2')^2}, \tag{5.32}$$

and

$$T_2' = \frac{T_2}{(1 + T_1 T_2 \omega_1^2)^{1/2}}. \tag{5.33}$$

Equation (5.31) shows that in the presence of appreciable saturation the resonance absorption curve still has the Lorentzian shape. However, in this case the half-width at half-maximum is given by $(T_2')^{-1}$ rather than T_2^{-1}. From Eq. (5.33), the ratio of the two resonance half-widths is given by

$$(T_2')^{-1}/T_2^{-1} = (1 + \gamma^2 H_1^2 T_1 T_2)^{1/2}. \tag{5.34}$$

Thus, the Bloch equations predict that the linewidth of the absorption spectrum at large rf field strength is wider than that at weak rf field strength. Although this result is approximately valid in liquids, we shall show in subsequent sections that in solids the opposite result is observed.

Equation (5.31) shows that the absorption maximum is given by

$$(\tilde{M}_y)_{\max} = \frac{\gamma H_1 T_2 M_0}{1 + \gamma^2 T_1 T_2 H_1^2}. \tag{5.35}$$

This result is to be contrasted with the small-H_1 case in which the absorption maximum is proportional to H_1. Equation (5.35) shows that maximum absorption is obtained when $\gamma^2 H_1 T_1 T_2 = 1$, at which the absorption amplitude is equal to

$$(\tilde{M}_y)_{\max} = \tfrac{1}{2} M_0 (T_2/T_1)^{1/2}. \tag{5.36}$$

Clearly, the absorption amplitude decreases with increasing value of H_1 and becomes vanishingly small at very large H_1.

Consider finally the behavior of M_z. From Eq. (5.27), at resonance,

$$M_z = M_0/(1 + \gamma^2 H_1^2 T_1 T_2). \tag{5.37}$$

Thus, at small H_1, M_z differs from M_0 only to second order in H_1; at large H_1, M_z is inversely proportional to H_1^2 and vanishes at very large H_1. This corresponds to the situation of complete saturation.

The Bloch approach previously outlined describes the relationships between the macroscopic relaxation times and the absorption spectrum in a liquid. To extract information about the dynamics of molecular motion from the measured macroscopic relaxation times, a microscopic theory is required to provide a detailed understanding of the manner in which molecular motion affects the absorption (or dispersion) spectrum. To accomplish this objective, we shall first consider the effect of dipolar linewidth in a rigid lattice.

5.5 DIPOLAR INTERACTION IN A RIGID LATTICE

The linewidth of the resonance spectrum of a spin system in an inhomogeneous magnetic field has a certain contribution owing to the spread of the Larmor frequencies of individual spins. The linewidth owing to the spread of the resonance frequencies of individual spins rather than that owing to the interaction among them is known as inhomogeneous broadening.

The other contribution to the spectral linewidth is owing either to the interaction between two spins or to the interaction between the spin and other nonspin degrees of freedom. In the absence of molecular motion, we need only to consider the spin–spin interaction. This is the rigid-lattice situation. The strength of the spin–spin interaction depends on the magnitude and orientation of the two interacting spins, as well as on the distance and orientation of the vector connecting them. The linewidth resulting from this mechanism contributes to what is known as homogeneous broadening.

The interaction Hamiltonian between two nuclear spins with the magnetic moments $\mathbf{M}_1 = \gamma_1 \hbar \mathbf{I}_1$ and $\mathbf{M}_2 = \gamma_2 \hbar \mathbf{I}_2$ is given (in frequency units) by

$$\begin{aligned} \mathscr{H}_{\mathrm{dd}}(1, 2) &= -\mathbf{M}_2 \cdot \mathbf{H}_{12} = -\gamma_2 \hbar \mathbf{I}_2 \cdot \mathbf{H}_{12} \\ &= (\gamma_1 \gamma_2 \hbar / r_{12}^3)[\mathbf{I}_1 \cdot \mathbf{I}_2 - 3(\mathbf{I}_1 \cdot \hat{r}_{12})(\mathbf{I}_2 \cdot \hat{r}_{12})], \end{aligned} \tag{5.38}$$

where $\hat{r}_{12}$ is the unit vector connecting the two spins and $\mathbf{H}_{12}$ the dipolar field produced by spin 1 at the site of spin 2. In a rigid lattice, the local dipolar field owing to a proton in an organic solid is on the order of a few gauss and makes a dominant contribution to the spectral linewidth of a resonance spin in the solid state.

The local dipolar field $\mathbf{H}_{12}$ resulting from spin 1 will have two effects on spin 2. The effects can be understood by decomposing $\mathbf{H}_{12}$ into two components, one parallel and one perpendicular to the direction of the external static field to which spin 2 is subject. The parallel component supplements the applied static field $\mathbf{H}_0$ and shifts the resonance frequency from the free-spin value. For a system with many spins, there is a multiplicity of contributions to the dipolar field from spins at different sites. The superposition of the dipolar fields at a particular spin site is known as the local field. The local field results in a smooth distribution of the parallel components about $\mathbf{H}_0$ and gives a width spread in the Larmor frequencies on the order of several kilohertz. On the other hand, the perpendicular component will rotate about the external static field at the Larmor frequency of spin 1. Just like the externally applied oscillating rf field, this rotating component can flip or induce transition of spin 1 if spin 1 and spin 2 are alike. This provides a mechanism of spin diffusion. However, the effect of the perpendicular component is unimportant if spins 1 and 2 differ significantly in gyromagnetic ratios.

Consider first the effect of the dipole–dipole interaction of like spins on the spectral line shape. The Hamiltonian of a system of like spins in a large static-magnetic field is

$$\mathscr{H} = \mathscr{H}_0 + \mathscr{H}_{\mathrm{dd}}, \tag{5.39}$$

where $\mathscr{H}_0$ is the main Hamiltonian in the frequency unit owing to the external field and is given by

$$\mathscr{H}_0 = -\sum_i \mathbf{M}_i \cdot \mathbf{H}_0 = -\gamma \sum_i \mathbf{I}_i \cdot \mathbf{H}_0 = -\omega_0 I_z, \tag{5.40}$$

where the direction of $\mathbf{H}_0$ is taken to be along the z axis of the laboratory coordinate system and I_z is the z component of the total spins in the system.

The interaction Hamiltonian $\mathscr{H}_{\mathrm{dd}}$ is the sum of the dipole–dipole Hamiltonians of all interaction pairs:

$$\mathscr{H}_{\mathrm{dd}} = \sum_{i<j} \mathscr{H}_{\mathrm{dd}}(ij) = \gamma^2 \hbar \sum_{i<j} r_{ij}^{-3}[\mathbf{I}_i \cdot \mathbf{I}_j - 3(\mathbf{I}_i \cdot \hat{r}_{ij})(\mathbf{I}_j \cdot \hat{r}_{ij})]. \tag{5.41}$$

To further the calculation, it is more convenient to express $\mathscr{H}_{\mathrm{dd}}$ in terms of components of irreducible spherical tensor of rank 2. This can be accomplished by first setting $\hat{r}_{ij}$ to be along the z axis of the molecular frame. In this frame, $\mathscr{H}_{\mathrm{dd}}$ can be written as

$$\mathscr{H}_{\mathrm{dd}} = \gamma^2 \hbar \sum_{i<j} r_{ij}^{-3} (\mathbf{I}_i \cdot \mathbf{I}_j - 3 I_{iz} I_{jz}). \tag{5.42}$$

According to Chapter 3 [Eq. (3.53)], the spin operator function $(\mathbf{I}_i \cdot \mathbf{I}_j - 3I_{iz}I_{jz})$ transforms as T_{20} (i.e., an irreducible tensor of rank 2, component 0). To express $\mathscr{H}_{\mathrm{dd}}$ in terms of the irreducible spherical-tensor components in the coordinate system with $\mathbf{H}_0$ lying along the z axis of the laboratory, we define θ_{ij} and ϕ_{ij} to be the polar and azimuthal angles, respectively, of the vector r_{ij} with respect to the direction of the external magnetic field $\mathbf{H}_0$.

Writing $\mathscr{H}_{\mathrm{dd}}$ in terms of the irreducible spherical tensor as

$$\mathscr{H}_{\mathrm{dd}} = -\gamma^2 \hbar \sum_{i<j} r_{ij}^{-3} T_{20}^{(\mathrm{M})}(ij), \tag{5.43}$$

we can now perform a transformation of $T_{20}^{(\mathrm{M})}(ij)$ to the laboratory coordinate frame using the property of irreducible spherical tensor discussed in Chapter 3 [see Eq. (3.70)]:

$$\begin{aligned} T_{20}^{(\mathrm{M})}(ij) &= \sum_m D_{m0}^{(2)}(\theta_{ij}, \phi_{ij}, 0) T_{2m}^{(\mathrm{L})}(ij) \\ &= \left(\frac{4\pi}{5}\right)^{1/2} \sum_m Y_{2m}^*(\theta_{ij}, \phi_{ij}) T_{2m}^{(\mathrm{L})}(ij) \end{aligned} \tag{5.44}$$

Thus

$$\mathscr{H}_{\mathrm{dd}} = -\gamma^2 \hbar \left(\frac{4\pi}{5}\right)^{1/2} \sum_{i<j} r_{ij}^{-3} \sum_m Y_{2m}^*(\theta_{ij} \phi_{ij}) T_{2m}^{(\mathrm{L})}(ij), \tag{5.45}$$

where $Y_{2m}(\theta_{ij}, \phi_{ij})$ are spherical harmonics, and $T_{2m}^{(\mathrm{L})}$ are irreducible spherical tensors constructed from the spin operators defined in the laboratory frame. The various components of $T_{2m}^{(\mathrm{L})}(ij)$ are given by [see Eq. (3.59)]

$$\begin{aligned} T_{20}^{(\mathrm{L})}(ij) &= (1/\sqrt{6})(3I_{iz}I_{jz} - \mathbf{I}_i \cdot \mathbf{I}_j), \\ T_{2\pm1}^{(\mathrm{L})}(ij) &= \mp \tfrac{1}{2}(I_{iz}I_{j\pm} + I_{jz}I_{i\pm}), \\ T_{2\pm2}^{(\mathrm{L})}(ij) &= \tfrac{1}{2} I_{i\pm} I_{j\pm}. \end{aligned} \tag{5.46}$$

If the spin state is quantized along the laboratory z axis, then we have

$$\mathscr{H}_0 |m\rangle = -\omega_0 I_z |m\rangle = -m\omega_0 |m\rangle, \tag{5.47}$$

where $|m\rangle$ is a short-hand notation denoting the spin state $|m_1 m_2 \cdots m_N\rangle$ of the system.

Since $T_{20}^{(L)}$ commutes with I_z, it follows that $|m\rangle$ is also the eigenstate of $T_{20}^{(L)}$. The part of the dipole-interaction Hamiltonion associated with the $T_{20}^{(L)}$ term is called the secular part. The nonsecular part is associated with the $T_{2\pm1}^{(L)}$ or the $T_{2\pm2}^{(L)}$ term; the nonsecular terms have the effect of admixing a small amount of the state $|m \pm 1\rangle$ or $|m \pm 2\rangle$ into the state $|m\rangle$.

In Eq. (5.47), the unperturbed energy owing to the static field associated with the state $|m\rangle$ is highly degenerate. This is because there are many ways in which the energy of the individual spins can be added to give a total value of $m(m = \sum_i m_i$, where m_i is the quantum number of an individual spin). However, in the presence of the dipole–dipole interaction, the degeneracy is partially lifted. Only the secular part of the dipole Hamiltonian can contribute in first order to the splitting of the energy level given by $-m\omega_0$; the nonsecular part connects spin states differing either by one or two units of spin energy and thus contributes only to the second order. In a solid, the thermal motion is slow compared with the Larmor frequency, and it is only necessary to use the secular part of the dipole Hamiltonian to calculate the line shape resulting from the dipolar interaction. In other words, an appropriate dipolar Hamiltonian to calculate the dipolar linewidth is

$$\begin{aligned}\mathscr{H}'_{\rm dd} &= -\gamma^2\left(\frac{4\pi}{5}\right)^{1/2} \sum_{i<j} r_{ij}^{-3} Y_{20}(\theta_{ij}, \phi_{ij}) T_{20}^{(L)}(ij) \\ &= \gamma^2 \sum_{i<j} r_{ij}^{-3} P_2(\cos\theta_{ij})(\mathbf{I}_i \cdot \mathbf{I}_j - 3I_{iz}I_{jz}),\end{aligned} \tag{5.48}$$

which is the rigid-lattice dipolar Hamiltonian for like spins.

For unlike spins, terms of the type $(I_{xi}I_{xj} + I_{yi}I_{yj})$ also become nonsecular and should be rejected. Thus, the rigid-lattice dipolar Hamiltonian for a system of unlike spins is

$$\mathscr{H}'_{\rm dd} = -2\hbar \sum_{i<j} \gamma_i \gamma_j r_{ij}^{-3} p_2(\cos\theta_{ij}) I_{iz} I_{jz}. \tag{5.49}$$

5.6 FREE-INDUCTION DECAY AND ABSORPTION LINE SHAPE

We previously discussed the absorption line shape in a steady-state experiment in terms of the Bloch equations. A general steady-state experiment is carried out by sweeping the external magnetic field, keeping the frequency of the applied rf field fixed. As the external magnetic field is swept through resonance, an absorption envelope for a solid reflects the distribution of the local fields. The power absorption is proportional to $\chi''(\omega)$, which, according to the fluctuation–dissipation theorem [see Eq. (1.167)], is related to the Fourier transform of the time-correlation function of the magnetization by

$$\begin{aligned}\chi''_{MM}(\omega) &= (2\pi/\hbar)\tanh(\beta\hbar\omega/2)\,\hat{C}_{MM}(\omega)\\ &\simeq (\omega/4kT)\int_{-\infty}^{\infty} dt\, e^{-i\omega t}\langle M_x(t)M_x + M_xM_x(t)\rangle,\end{aligned} \tag{5.50}$$

where the second step is owing to the high-temperature approximation ($\beta\hbar\omega \ll 1$) and $\tanh(\beta\hbar\omega/2)$ is replaced by $\beta\hbar\omega/2$.

In a solid in which the rigid-lattice approximation is valid, the time-dependent operator $M_x(t)$ is given by

$$M_x(t) = e^{i\mathscr{H}t/\hbar}M_xe^{-i\mathscr{H}t/\hbar}, \tag{5.51}$$

where $\mathscr{H}$ is given by Eq. (5.39), with the simplification of replacing the total $\mathscr{H}_{\mathrm{dd}}$ by its secular part.

In addition, the ensemble average represented by the angular brackets in Eq. (5.49) is evaluated strictly with respect to the equilibrium-density operator given by

$$\rho(0) = (1/Z)\exp[-\beta(\mathscr{H}_0 + \mathscr{H}'_{\mathrm{dd}})]. \tag{5.52}$$

Since the local field is about a few gauss and is generally small compared with the Zeeman Hamiltonian given by $\mathscr{H}_0$, $\mathscr{H}'_{\mathrm{dd}}$ can be neglected in Eq. (5.51). Furthermore, under the high-temperature approximation, we can approximate $e^{-\beta\mathscr{H}_0} = \boldsymbol{\varepsilon} - \beta\mathscr{H}_0$, $\boldsymbol{\varepsilon}$ being a unit operator, and to the first order in $\beta\mathscr{H}_0$ we write

$$\rho(0) \simeq (\boldsymbol{\varepsilon} - \beta\mathscr{H}_0)/\mathrm{Tr}\,\boldsymbol{\varepsilon} = (\boldsymbol{\varepsilon} - \beta\mathscr{H}_0)/(2I+1)^N, \tag{5.53}$$

where N is the number of nuclear spins in the system $2I + 1$, the total number of states of an individual spin.

By the approximation of Eq. (5.52), we obtain

$$\begin{aligned}\langle M_x(t)M_x\rangle &= \mathrm{Tr}[\rho(0)M_x(t)M_x]\\ &\simeq [1/(2I+1)^N]\,\mathrm{Tr}[M_x(t)M_x] = \langle M_xM_x(t)\rangle.\end{aligned} \tag{5.54}$$

That is, the left-sided time correlation is equal to the right-sided one. Thus, Eq. (5.50) becomes

$$\chi''_{\mathrm{MM}}(\omega) = \frac{\omega}{2kT(2I+1)^N}\int_{-\infty}^{\infty} dt\, e^{-i\omega t}G(t), \tag{5.55}$$

where $G(t)$ is a time-correlation function, given by

$$G(t) = \mathrm{Tr}[M_x(t)M_x]. \tag{5.56}$$

The time-correlation function can be simplified. Since $\mathscr{H}_0$ and $\mathscr{H}'_{\mathrm{dd}}$ commute, we can write Eq. (5.55) as

$$G(t) = \mathrm{Tr}[\exp(i\mathscr{H}'_{\mathrm{dd}}t/\hbar)\exp(i\mathscr{H}_0 t/\hbar)M_x \exp(-i\mathscr{H}_0 t/\hbar)\exp(-i\mathscr{H}'_{\mathrm{dd}}t/\hbar)M_x]. \tag{5.57}$$

Since

$$\exp(i\mathscr{H}_0 t/\hbar)M_x \exp(-i\mathscr{H}_0 t/\hbar) = M_x \cos\omega_0 t + M_y \sin\omega_0 t, \tag{5.58}$$

we have

$$\begin{aligned} G(t) = \mathrm{Tr}\{&[\exp(i\mathscr{H}'_{\mathrm{dd}}t/\hbar)M_x \exp(-i\mathscr{H}'_{\mathrm{dd}}t/\hbar)M_x]\cos\omega_0 t \\ &+ [\exp(i\mathscr{H}'_{\mathrm{dd}}t/\hbar)M_y \exp(-i\mathscr{H}'_{\mathrm{dd}}t/\hbar)M_x]\sin\omega_0 t]. \end{aligned} \tag{5.59}$$

The second term on the right-hand side of Eq. (5.59) vanishes because of symmetry. Thus

$$G(t) = F(t)\cos\omega_0 t, \tag{5.60}$$

where $F(t)$, the reduced time-correlation function, is given by

$$F(t) = \mathrm{Tr}[\exp(i\mathscr{H}'_{\mathrm{dd}}t/\hbar)M_x \exp(-i\mathscr{H}'_{\mathrm{dd}}t/\hbar)M_x]. \tag{5.61}$$

We now show that the function $F(t)$ is proportional to the free-induction decay signal in the pulse experiment (this is the nuclear induction signal after the magnetization is applied to the plane perpendicular to the z axis by a 90° rf pulse).

Before the application of a 90° pulse, the spin system that is polarized by the static field $\mathbf{H}_0$ is at equilibrium with the equilibrium-density operator given by $\rho(0) \simeq [1/(2I+1)^N](\varepsilon - \beta\mathscr{H}_0)$. At the beginning of the experiment ($t = 0$), a strong-resonance rf pulse with a field strength much greater than the local field is applied for a duration τ along the y axis of the laboratory coordinate system. The density operator describing the spin system in the presence of the static and rf fields will develop according to the Liouville equation of motion, given by

$$\partial\rho(t)/\partial t = -i[\mathscr{H}, \rho(t)], \tag{5.62}$$

where the Hamiltonian (in frequency units) in this case is given by

$$\mathscr{H} = \omega_0 I_z - 2\omega_1 I_y \cos\omega_0 t + \mathscr{H}'_{\mathrm{dd}}. \tag{5.63}$$

We can transform Eq. (5.62) into a rotating frame according to

$$\rho(t) = \exp(i\omega_0 I_z t)\sigma(t)\exp(-i\omega_0 I_z t). \tag{5.64}$$

Substituting Eq. (5.64) into (5.62) and neglecting terms that oscillate with $2\omega_0$, we obtain

$$\partial\sigma(t)/\partial t = -i[-\omega_1 I_y + \mathscr{H}'_{\mathrm{dd}}, \sigma(t)]. \tag{5.65}$$

When the rf field is much stronger than the local field, $\mathscr{H}'_{\rm dd}$ in Eq. (5.65) can be ignored. In this approximation, the solution of Eq. (5.65) during the application of the rf pulse is given by

$$\begin{aligned}\sigma(t) &= \exp(i\omega_1 I_y t)\sigma(0)\exp(-i\omega_1 I_y t)\\ &= [1/(2I+1)^N][\varepsilon + \beta\hbar\omega_0 \exp(i\omega_1 I_y t)I_z \exp(-i\omega_1 I_y t)]. \quad (5.66)\end{aligned}$$

If t is adjusted to be equal to τ such that $\omega_1\tau = \pi/2$ (the 90° pulse), then the density operator after the 90° pulse is

$$\sigma_{\pi/2} = [1/(2I+1)^N](\varepsilon + \beta\hbar\omega_0 I_x). \quad (5.67)$$

After the 90° pulse, the density operator will develop according to $\mathscr{H}'_{\rm dd}$. Thus, the density operator at time t after the 90° pulse is given by

$$\begin{aligned}\sigma(t) &= \exp(-i\mathscr{H}'_{\rm dd}t)\sigma_{\pi/2}\exp(i\mathscr{H}'_{\rm dd}t)\\ &= [1/(2I+1)^N][\varepsilon + \beta\hbar\omega_0 \exp(-i\mathscr{H}'_{\rm dd}t)I_x \exp(i\mathscr{H}'_{\rm dd}t)]. \quad (5.68)\end{aligned}$$

The observed induction signal is proportional to $\langle M_x(t)\rangle$ and equal to

$$\begin{aligned}\langle M_x(t)\rangle &= \mathrm{Tr}[\sigma(t)M_x]\\ &= \beta\hbar\omega_0/(2I+1)^N\,\mathrm{Tr}[\exp(-i\mathscr{H}'_{\rm dd}t)I_x\exp(i\mathscr{H}'_{\rm dd}t)M_x]\\ &= C\,\mathrm{Tr}(\exp(-\mathscr{H}'_{\rm dd}t)M_x\exp(i\mathscr{H}'_{\rm dd}t)M_x) = CF(t), \quad (5.69)\end{aligned}$$

where C is a constant equal to $\beta\mathscr{H}_0/(2I+1)^N$.

Since $\langle M_x(t)\rangle$ will diminish in time as the phase coherence of individual spins decay, the time-correlation function $F(t)$ is called the free-induction decay (FID). We shall now show that the FID is proportional to the Fourier transform having the line shape of the absorption envelope (Lowe and Norberg, 1957).

Define a line-shape function as

$$g(\omega) = [\chi''_{\rm MM}(\omega)/\omega]2kT(2I+1)^N. \quad (5.70)$$

Taking the inverse Fourier transform of Eq. (5.55) and using (5.60), we obtain

$$F(t)\cos\omega_0 t = \frac{1}{2\pi}\int_{-\infty}^{\infty} d\omega\, e^{i\omega t}g(\omega). \quad (5.71)$$

Recalling that $g(\omega)$ is an even function of ω, we can rewrite Eq. (5.71) as

$$F(t)\cos\omega_0 t = \frac{1}{2\pi}\int_{-\infty}^{\infty} d\mu\,\cos[(\omega_0+u)t]g(\omega_0+u),$$

which reduces to

$$F(t) = 1/2\pi\int_{-\infty}^{\infty} du\,\cos(ut)f(u) = 1/2\pi\int_{-\infty}^{\infty} e^{iut}f(u). \quad (5.72)$$

In Eq. (5.71), we write $f(u) = g(\omega_0 + u)$ and also use the fact that $g(\omega_0 + u) = g(\omega_0 - u)$.

Equation (5.71) is a basic NMR line-shape equation. It shows that the FID is the Fourier transform of the absorption line-shape. In general, for solids the FID decays in a time T_2 (a few microseconds), owing to a broad distribution in the local field. In liquids, T_2 is much longer (a few milliseconds) because of the molecular motion that reduces and narrows the distribution of the local field.

The NMR line shape or the FID can readily be measured with steady-state absorption or with a pulse experiment. However, calculating the line shape or FID is a formidable task. No general technique has yet been developed to calculate the line-shape function to compare with the experiment. Nevertheless, it is possible to compare theory with the experiment using the method of spectral moments discussed in Chapter 1. Expressions for the spectral moments for dipolar-broadened crystallites in random orientation, or for crystal lattices of special symmetry, have been constructed and are available (Abragam, 1961).

5.7 SPIN ECHOES IN DIPOLAR SOLIDS

The dipole–dipole interaction discussed in the previous section results in a homogeneous broadening of the NMR resonance line. For this reason, the dipolar interaction is often thought to cause an irreversible decay of the transverse magnetization. However, it is possible to stimulate a spin echo or a train of multiple spin echoes in a dipolar-broadened solid by 90° rf pulses (Mansfield and Ware, 1966; Ostroff and Waugh, 1966). The effect is analogous to (but quite different in character) the familiar "classical" spin echoes induced in inhomogeneously broadened spin systems first observed by Hahn (1950). The observation of the spin echo from a dipolar-broadened solid depends very much on the phase of each applied rf pulse. In a single dipolar-echo case, a first 90° pulse is applied along the y axis to bring the nuclear spins to the x axis of the rotating frame. If no further rf pulse is applied, the magnetization will simply decay to a vanishing value and we have the usual FID. However, if a second 90° pulse is applied along the x (not y) axis within T_2 (the lifetime of the FID), then a dipolar echo is observed.

Multiple spin echoes are induced by a periodic train of 90° pulses following the second 90° pulse. This greatly prolongs the decay of transverse nuclear magnetization in dipolar solids. Detailed analysis of this effect, including the dependence of the decay time T^* for the echo envelope on the pulse spacing, was given by Waugh and Wang (1967). The pulse sequence that yields the effect can be represented symbolically as $P_y, \tau, P_x, (2\tau, P_x)_n$, where P_α ($\alpha = x, y$) denotes a 90° pulse along the α axis of a reference frame rotating at the Larmor frequency ω_0 (or γH_0). The first pulse P_y serves merely to establish a suitable initial condition for the remainder of the sequence. This pulse is followed, after a time τ, by a train of n p_x pulses ($n \simeq 10^3$) spaced apart by a time 2τ ($\tau < T_2$).

The echo train persists for a time several orders of magnitude longer than that of T_2. In fact, by reducing the pulse spacing, the time constant T^* characterizing the decay of the echo envelope can be made to approach $T_{1\rho}$, the spin-lattice relaxation time in the rotating frame, which we shall discuss in a subsequent section.

It should be noted that the principal significance of the prolonged decay time in the multiple dipolar-echo envelope is the line narrowing in the frequency space (in contrast to the line broadening predicted by the Bloch equations). Motivated by the possibility this suggested for accomplishing high-resolution NMR in solids, Waugh and coworkers soon discovered that the decay of the magnetization in a wide class of multipulse experiments could be easily understood by introducing the concept of an "average Hamiltonian," which is determined by the symmetry properties of the true Hamiltonian under pulse transformations (Waugh *et al.*, 1968b). Once this concept was clearly understood, it became possible to design pulse sequences specifically to suppress the dipole–dipole interactions, enabling the structure the spectral fine structure resulting from chemical shifts, scalar coupling, etc., to be resolved. We shall discuss this type of multipulse experiment in the next section. In this section, we shall analyze the behavior of the transverse magnetization under the action of a train of rf pulses. We consider the evolution of a system described by the density operator $\rho(t)$ under the influence of a time-dependent Hamiltonian $\mathscr{H}(t)$. For simplicity, the spin–lattice-relaxation effect will be neglected. Before the application of the rf pulses, the spin system is at equilibrium. After the first 90° pulse, the density operator in the high-temperature approximation is given by Eq. (5.67). Since the unit operator does not contribute to the final result, the significant part of the density operator after this initial 90° pulse is proportional to I_x, which will be considered as an initial condition, i.e., $\rho(0) = I_x$.

The subsequent evolution of $\rho(0)$ is then governed by the internal Hamiltonian and the remainder of the rf pulses. The density operator that describes the FID at a time τ after the first pulse is

$$\rho(\tau) = D(\tau) I_x D^{-1}(\tau), \tag{5.73}$$

where

$$D(\tau) = \exp(-i\mathscr{H}'_{\text{dd}}\tau) \tag{5.74}$$

In Eq. (5.74), we adopted the convention of writing $\mathscr{H}'_{\text{dd}}$ in units of angular frequency.

But instead of observing the free-induction decay, another pulse is applied at a time $\tau = \tau_0$. The system is then left to evolve freely for a time τ_1 before another pulse P_2 is applied. We continue in this manner until a sequence of N pulses $P_1 P_2 \cdots P_N$ has been applied, the spacing between successive pulses P_i and P_{i+1} being denoted by τ_i. At a time τ_N after the pulse P_N, the density operator becomes

$$\rho_N = K I_x K^{-1}, \tag{5.75}$$

where

$$K = D(\tau_N) P_N D(\tau_{N-1}) P_{N-1} \cdots P_1 D(\tau_0). \tag{5.76}$$

It should be pointed out that in this formulation the pulses are considered to be δ functions, which can be realized experimentally if the rf pulse field is made much greater than the local dipolar field. Haeberlen and Waugh (1968) have analyzed the situation when the δ-function pulses are not realized.

There exists no simple way of calculating ρ_N for an arbitrary rf pulse sequence with random pulse spacings. However, a considerable simplification results if the pulse sequence is designed in such a way that a set of m successive pulses ($m \ll N$) recovers the original Hamiltonian, i.e.,

$$P_{j+m} \cdots P_{j+1} \mathscr{H}'_{\text{dd}} P_{j+1}^{-1} \cdots P_{j+m}^{-1} = \mathscr{H}'_{\text{dd}}; \tag{5.77}$$

and if the pulse spacings are so adjusted that the pulse sequence becomes periodic in time, each period t_c containing m pulses. Pulse sequences satisfying these conditions are said to be cyclic; the period t_c, called the cycle time, is equal to $\tau_0 + \tau_1 + \cdots + \tau_m$.

If the pulse sequence is cyclic and N is an integral multiple of m (say $N = nm$), then Eq. (5.75) reduces to

$$\rho_N(N = nm) = \mathscr{L}^n(t_c) I_x \mathscr{L}^{-n}(t_c), \tag{5.78}$$

where the one-cycle evolution operator $\mathscr{L}(t_c)$ is given by

$$\mathscr{L}(t_c) = D(\tau_m) P_m D(\tau_{m-1}) \cdots P_1 D(\tau_0). \tag{5.79}$$

The x component (in the rotating frame) of the transverse magnetization at the end of the nth pulse cycle is then proportional to

$$\langle I_x(nt_c)\rangle = \operatorname{Tr}[I_x L^n(t_c) I_x \mathscr{L}^{-n}(t_c)]. \tag{5.80}$$

The one-cycle propagator $\mathscr{L}(t_c)$ defined in Eq. (5.79) can be written in the form

$$\begin{aligned} \mathscr{L}(t_c) = {} & D(\tau_m) D_m(\tau_{m-1}) D_{m,m-1}(\tau_{m-2}) \cdots \\ & \times D_{m,m-1,\ldots,2,1}(\tau_0)(P_m P_{m-1} \cdots P_2 P_1), \end{aligned} \tag{5.81}$$

where

$$D_{i,j}\cdots(\tau) = \exp[-i\tau(P_iP_j\cdots\mathscr{H}_{dd}\cdots P_j^{-1}P_i^{-1})]. \tag{5.82}$$

Since m pulses form a cycle, the effect of $P_mP_{m-1}\cdots P_2P_1$ cancels out if Eq. (5.80) is substituted into (5.79), and the one-cycle operator $\mathscr{L}(t_c)$ can be written as

$$\mathscr{L}(t_c) = D(\tau_m)D_m(\tau_{m-1})\cdots D_{m,m-1,\ldots,1}(\tau_0). \tag{5.83}$$

It is convenient to define a t_c-dependent effective Hamiltonian $\mathscr{H}_e(t_c)$ by the equation

$$\mathscr{L}(t_c) = \exp[-it_c\mathscr{H}_e(t_0)]. \tag{5.84}$$

Since $\mathscr{L}(t_c)$ is unitary, $\mathscr{H}_e(t_c)$ must be Hermitian. It can be evaluated in series form by the Magnus expansion (Evans, 1968; Magnus, 1954; Pechukas and Light, 1966), which yields

$$\mathscr{H}_e(t_c) = \overline{\mathscr{H}} + \sum_{k=1}^{\infty}\mathscr{H}^{(k)}(t_c), \tag{5.85}$$

together with a prescription for determining the quantities $\overline{\mathscr{H}}$ and $\mathscr{H}^{(k)}(t_c)$. The quantity $\overline{\mathscr{H}}$ is called the average Hamiltonian, as indicated by the notation that it is independent of t_c. If the pulse spacings are fixed fractions of t_c, then $\mathscr{H}^{(k)}(t_c) = t_c^kF_k$, where F_k is a Hermitian operator independent of t_c. The quanities F_k can be determined either by the Magnus prescription or by simply expanding both sides of Eq. (5.83) in powers of t_c and equating coefficients.

We see from Eqs. (5.84) and (5.85) that in the limit of $t_c/T_2 \to 0, n \to \infty$, and $nt_c \to t$, $\mathscr{L}(t_c)$ approaches $\exp(-it\overline{\mathscr{H}})$ exactly. Historically, this fact was realized before it became clear how to introduce correction terms that allowed for finite t_c (Waugh *et al.*, 1968b). These correction terms are important since the mathematical limit t_c/T_2 cannot be realized in practice. In the present context, these corrections are of vital importance to the calculation of the decay of $\langle I_x(nt_c\rangle$ [see Eq. (5.80)], since, if they were omitted, there would be no decay of the magnetization.

The foregoing discussion is essentially a capsule summary of what has become known as average Hamiltonian theory. The reader desiring further details is referred to the original sources (Haeberlen and Waugh, 1968; Waugh *et al.*, 1988b).

According to Eq. (5.84), we can rewrite Eq. (5.80) as

$$\begin{aligned}\langle I_x(t)\rangle &= \mathrm{Tr}\{I_x\exp[-it\mathscr{H}_e(t_c)]I_x\exp[it\mathscr{H}_e(t_c)]\}\\ &= \mathrm{Tr}[I_x(t)I_x] = \langle I_x(t)I_x\rangle,\end{aligned} \tag{5.86}$$

where we have set $t = nt_c$ and

$$I_x(t) = \exp[it\mathscr{H}_e(t_c)]I_x \exp[-it\mathscr{H}_e(t_c)]. \tag{5.87}$$

Equation (5.86) now appears in a form resembling an ordinary time-correlation function, and we can now use the technique of statistical mechanics to evaluate it. We first differentiate $I_x(t)$, given in Eq. (5.87), with respect to t, obtaining

$$\partial I_x(t)/\partial t = i[\mathscr{H}_e(t_c), I_x(t)] \equiv i\mathscr{L}(t_c)I_x(t) \tag{5.88}$$

with the initial condition $I_x(0) = I_x$. Note that $\mathscr{H}_e(t_c)$ or $\mathscr{L}(t_c)$ depends on t_c, which is a fixed quantity in a multipulse experiment. Thus, Eq. (5.88) is a Liouville equation with a time-independent Hamiltonian, which is the effective Hamiltonian $\mathscr{H}_e(t_c)$; $\mathscr{L}(t_c)$ is a superoperator that yields the commutation $[\mathscr{H}_e(t_c), A]$ when it operates on an operator A.

We are not really interested in the complete solution to Eq. (5.88); what is needed is the component of $I_x(t)$ that is necessary to calculate $\langle I_x(t)\rangle$. Thus, the projection-operator technique introduced in Chapter 4 can be applied here. The appropriate projection operator here is defined as

$$PA = \langle AI_x\rangle\langle I_x^2\rangle^{-1}I_x, \tag{5.89}$$

where A is any quantum-mechanical operator. Using Eq. (5.89), it is easy to show that $P^2 = P$; hence, P is a projection operator.

Having obtained Eq. (5.88) and defined the projection operator, we can now follow the various steps starting from Eq. (4.34) that lead to Eq. (4.56) of Chapter 4 by identifying A with I_x and $C_A(t)$ with $\langle I_x(t)\rangle$ to obtain (noting $\Omega = 0$)

$$\partial\langle I_x(t)\rangle/\partial t = -\int_0^t dt'\, M(t')\langle I_x(t - t')\rangle, \tag{5.90}$$

where the memory function $M(t')$ is an implicit function of t_c and will be written as

$$M(t'\,|\,t_c) = \langle I_x^2\rangle^{-1}\langle[\mathscr{H}_e(t_c), I_x] \exp[t'(1 - P)i\mathscr{L}(t_c)][\mathscr{H}_e(t_c), I_x]\rangle. \tag{5.91}$$

In the $P_y\tau$, P_x, $(2\tau, P_x)_2$ pulse cycle, the one-cycle propagator is given by

$$\mathscr{L}(t_c) = \exp(-i\mathscr{H}'_{dd}\tau) \exp(-2i\tilde{\mathscr{H}}'_{dd}\tau) \exp(-i\mathscr{H}'_{dd}\tau), \tag{5.92}$$

where $t_c = 4\tau$ and

$$\tilde{\mathscr{H}}'_{dd} = P_x\mathscr{H}'_{dd}P_x^{-1} = \sum_{i>j}\gamma^2\hbar r_{ij}^{-3}P_2(\cos\theta_{ij})(\mathbf{I}_i\mathbf{I}_j - 3I_{iy}I_{jy}).$$

Note that the one-cycle propagator is symmetric [i.e., $\mathscr{L}^+(t_c) = \mathscr{L}(-t_c)$]. It follows that all the odd-order correction terms in Eq. (5.85) vanish, and Eq. (5.92) can be written as (Wang and Ramshaw, 1972)

$$\mathscr{L}(t_c) = \exp\{-it_c[\overline{\mathscr{H}} + V(t_c)]\}, \tag{5.93}$$

where

$$\overline{\mathscr{H}} = \tfrac{1}{4}i\mathscr{L}_1 = \tfrac{1}{2}(\mathscr{H}'_{dd} - \tilde{\mathscr{H}}'_{dd}), \tag{5.94}$$

$$V(t_c) = \sum_{k=1}^{\infty} t_c^{2k} F_{2k}, \tag{5.95}$$

$$F_2 = \tfrac{1}{4}i[\mathscr{L}_3 - \tfrac{1}{3}\mathscr{L}_1^3]. \tag{5.96}$$

Here $\mathscr{L}_k$ is the coefficient τ^k in a power-series expansion of $\mathscr{L}(4\tau)$ [Eq. (5.92)].

Next we note that $[\overline{\mathscr{H}}, I_x] = 0$. This implies that $P[\overline{\mathscr{H}}, A] = 0$ for any operator A. As a result, the memory function $M(t'|t_c)$ given in Eq. (5.91) reduces to

$$M(t'|t_c) = \langle I_x^2 \rangle^{-1} \langle [V(t_c), I_x] \exp[i\mathscr{L}_0 t' + i(1 - P)\mathscr{L}'t'][V(t_c), I_x] \rangle, \tag{5.97}$$

where $\mathscr{L}_0 A \equiv [\overline{\mathscr{H}}, A]$ and $\mathscr{L}'A = [V(t_c), A]$ for any operator A.

Equations (5.90) and (5.97) completely determine the evolution of magnetization at the end of the nth pulse cycle. These equations are valid for both small and large values of time. The quantities $\overline{\mathscr{H}}$ and $V(t_c)$ play the roles of secular and nonsecular Hamiltonians, respectively. If $t_c \ll T_2$, the nonsecular effect becomes negligible, and the motion of the spin system is described by the average Hamiltonian $\overline{\mathscr{H}}$. In this case, there will be no decay of the echo envelope, i.e., the magnetization will be locked along the x axis of the rotating frame.

The behavior of magnetization at both small and large values of time, in the case of the P_y, τ, P_x, $(2\tau, P_x)_k$ pulse cycle, has been discussed in papers by Waugh and Wang (1967) and Wang and Ramshaw (1972).

5.8 HIGH-RESOLUTION NMR OF SOLIDS

The effective-Hamiltonian concept introduced here provides a powerful theoretical tool to describe motions of the spin systems in which a static Hamiltonian (e.g., $\mathscr{H}'_{dd}$) is made to appear, in a sense, time dependent. The time dependence is introduced by applying a train of intense magnetic-field pulses as the characteristic repetition period t of the pulses becomes small compared with T_2; over a long period of time, the spin system again behaves as though under the exclusive influence of a time-dependent average Hamiltonian. The experimenter controls the value of this average by manipulating the intensity, timing, and phase of the pulses. Another method for controlling Hamiltonian of the spin system is by sample spinning. Both methods induce a coherent motion in the spin system and provide very useful ways to eliminate a large part of the spectral broadening that results from dipolar interaction, which in the usual case tends to obscure potentially useful information contained in the chemical shift.

As mentioned earlier, the local field responsible for the homonuclear dipolar-broadening of NMR spectra in solids originates from the term $\mathscr{H}'_{\mathrm{dd}}$, and the information sought in the chemical shift from a term of the form

$$\mathscr{H}_{\mathrm{c}} = \omega_0 \sum_i \sigma_{zzi} I_{zi}, \tag{5.98}$$

where σ_{zzi} is the zz component of the chemical-shift tensor of spin i in the laboratory coordinate frame, where the direction of the static field is along the z axis. In terms of the principal-axis system (xyz) of the chemical-shift tensor, related to the laboratory system by the Euler angles θ, ϕ, ψ, we have

$$\sigma_{zz} = \sigma_{xx} \sin^2 \theta \cos^2 \phi + \sigma_{yy} \sin^2 \theta \sin^2 \phi + \sigma_{zz} \cos^2 \theta. \tag{5.99}$$

We introduce the anisotropy $\Delta\sigma$ and asymmetry parameter η of the chemical shift by

$$\sigma_{xx} = -\Delta\sigma(1 - \tfrac{1}{2}\eta) + \bar{\sigma}, \qquad \sigma_{yy} = -\Delta\sigma(1 + \tfrac{1}{2}\eta) + \bar{\sigma}, \qquad \sigma_{zz} = 2\Delta\sigma + \bar{\sigma}, \tag{5.100}$$

where $\bar{\sigma} = \frac{1}{3}\,\mathrm{Tr}\,\boldsymbol{\sigma}$ is the average of the chemical-shift tensor, and we rewrite Eq. (5.99) as

$$\sigma_{zz} = \bar{\sigma} + \Delta\sigma\sqrt{\pi/5}\,\{4Y_{2,0}(\theta, \phi) + \eta\sqrt{2/3}\,[Y_{2,2}(\theta, \phi) + Y_{2,-2}(\theta, \phi)]\}. \tag{5.101}$$

The experiment utilizing the P_y, τ, P_x, $(2\tau, P_x)_k$ pulse sequence, as previously discussed, achieves the line-narrowing effect, but it has no application in obtaining the high-resolution spectra in solids because this particular pulse sequence also averages the chemical-shift information to zero. To maximize the averaging of the dipolar interaction, a pulse cycle $(\tau, P_y, 2\tau, P_{-y}, \tau, P_{-x}, 2\tau, P_x)_n$, known as the WAHUHA cycle, was introduced (Waugh *et al.*, 1968a). This pulse cycle causes the dipolar contribution to $\bar{\mathscr{H}}$ to vanish, yet retains the information on chemical shifts.

The original WAHUHA cycle, however, removes only part of the dipolar contribution to $\mathscr{H}^{(1)}$ in the Magnus expansion of the one-cycle propagator; this leaves a nonzero coupling term between the dipolar and chemical-shift Hamiltonians. Better resolution of chemical shift and other spectral details, however, is found to use symmetric pulse sequences (Mehring, 1976; Rhim *et al.*, 1971; Wang, 1973), such as $(\tau, P_y, \tau, P_{-x}, 2\tau, P_x, \tau, P_y, \tau)_n$. This is the symmetrized WAHUHA pulse cycle.

The one-cycle propagator for this cycle is (assuming δ-function pulses)

$$\mathscr{L}(t_c) = \exp(i\mathscr{H}_z\tau)\exp(i\mathscr{H}_x\tau)\exp(2i\mathscr{H}_y\tau)\exp(i\mathscr{H}_x\tau)\exp(i\mathscr{H}_z\tau)$$

$$= \exp\left\{it_c\left[\overline{\mathscr{H}} + \sum_{k=1} \mathscr{H}^{(k)}(t_c)\right]\right\}, \tag{5.102}$$

where $\tau = t_c/6$ and

$$\mathscr{H}_\alpha = \sum_{i<j} \gamma^2\hbar r_{ij}^{-3} P_2(\cos\theta_{ij})(\mathbf{I}_i - \mathbf{I}_j - 3I_{\alpha i}I_{\alpha j}) + \omega_0 \sum_i \sigma_{zzi} I_{\alpha i},$$

$$\alpha = x, y, z \tag{5.103}$$

Wang and Ramshaw (1972) demonstrated a theorem that states if the one-cycle propagator is symmetric (i.e., $\mathscr{L}^+(t_c) = \mathscr{L}(-t_c)$), then all the odd-order correction terms vanish, i.e., $\mathscr{H}^{(2k+1)} = 0$ for $k = 0, 1, 2, \ldots$. The proof of this theorem is elementary: If $\mathscr{L}(t_c)$ is symmetrical, then the fact that it is also unitary implies that the quantity in square brackets in Eq. (5.102) must be an even function of t_c. This means that $\mathscr{H}^{(k)}$ vanishes for odd values of k.

This theorem is useful as a guide to design pulse sequences that will yield the maximum feasible suppression of dipolar broadening. In particular, in the present case it is only necessary to make $\mathscr{H}^{(2)}$ small, since $\mathscr{H}^{(1)}$ and $\mathscr{H}^{(3)}$ are zero for symmetrical cycles.

Neglecting $\mathscr{H}^{(2)}$ and higher even-order terms, according to Eq. (5.102) we find that the average Hamiltonian for the symmetrized WAHUHA pulse cycle is given by

$$\overline{\mathscr{H}} = \tfrac{1}{3}(\mathscr{H}_x + \mathscr{H}_y + \mathscr{H}_z) = \tfrac{1}{3}\omega_0 \sum_i \sigma_{zzi}(I_{xi} + I_{yi} + I_{zi}). \tag{5.104}$$

Note that the dipolar contribution disappears, leaving only the chemical-shift contribution. Equation (5.104) is more easily interpreted if transformed to a tilted frame with its axis along the body diagonal direction. In this frame, the average Hamiltonian becomes

$$\overline{\mathscr{H}}' = \frac{1}{\sqrt{3}}\omega_0 \sum_i \sigma_{zzi} I_{zi}. \tag{5.105}$$

This corresponds to an apparent precession of the spins about the (111) direction of the rotating frame, each at a rate corresponding to its chemical shift reduced by a factor $\sqrt{3}$. Except for this reduction, the information about the entire chemical-shift tensor is retained.

Consider next the technique of sample spinning. With this technique, $\mathscr{H}'_{\mathrm{dd}}$ is eliminated from the average Hamiltonian by making θ_{12} time dependent by mechanical motion of the sample in such a way that $P_2(\cos\theta_{ij})$ in the expression of $\mathscr{H}'_{\text{dd}}$ averages to zero. The effect of sample spinning can be explained with the help of the relation

$$P_2(\cos\theta_{ij}) = \sum_m D_{0m}^{(2)}(\alpha, \beta, \gamma) Y_{2m}(\theta'_{ij}, \phi'_{ij}), \tag{5.106}$$

where $D_{mm'}^{(2)}(\alpha, \beta, \gamma)$ are elements of the rotation matrix defined in Chapter 3. Here the Euler angles define the orientation of the spinning axis with respect to the external magnetic field.

Rotation of a sample about its cylindrical axis eliminates the $m \neq 0$ terms in Eq. (5.106) and gives an average of $P_2(\cos\theta_{ij})$ as

$$\overline{P_2(\cos\theta_{ij})} = (1 - 3\cos^2\beta) P_2(\cos\theta'_{ij}). \tag{5.107}$$

If the angle β between the rotation axis and the field is chosen to be the "magic angle" of 54°42′ (i.e., $\beta = \cos^{-1}\sqrt{1/3}$), then $\overline{P_2(\cos\theta_{ij})} = 0$.

However, in a similar manner the $Y_{2,2}$ and $Y_{2,-2}$ terms in Eq. (5.101) also average to zero, and the average Hamiltonian reduces to

$$\overline{\mathscr{H}} = (1/\sqrt{3})\omega_0 \bar{\sigma} I_z; \tag{5.108}$$

only the isotropic-average part similar to that observed in solutions present in the solid spectrum enters the expression. Thus, the sample-spinning technique yields only a single line for each type of nuclear environment. In gneral, the chemical shift owing to coupling of the nuclear moments with the magnetic field, induced by the Larmor procession of electrons in the presence of the externally applied field, has a complex angular dependence. The multiple pulse NMR technique has opened the door for studying the complete tensorial property of the chemical-shift interaction.

Finally, consider the high-resolution spectra of solids consisting of unlike spins. For dipolar broadening owing to unlike spins, the pertinent dipolar Hamiltonian is simpler. Instead of Eq. (5.48), the interaction Hamiltonian between two unlike nuclear species is given by

$$\mathscr{H}'_{\text{dd}} = 2\gamma_I \gamma_S \hbar \sum_{i,m} r_{im}^{-3} P_2(\cos\theta_{im}) I_{iz} S_{mz}, \tag{5.109}$$

where the S spin is considered a rare species and I spin the abundant species.

In this case, saturation of I_z will make the average of $I_z = 0$. This is the situation in ^{13}C experiments where ^{13}C is dilute and the local dipolar field is owing to the abundant protons. The suppression of dipolar coupling to observe the dilute spin by saturation of the abundant spins is known as spin decoupling.

5.9 SPIN TEMPERATURE AND DOUBLE RESONANCE IN THE ROTATING FRAME

The NMR signal can be observed in a solid because of its weak coupling to the nonspin degrees of freedom (the lattice). Since the coupling between the spin system and the lattice is weak and the spins are further coupled by the dipole–dipole interaction, it is possible to view the spin system as a thermal entity, describable by a temperature that may or may not be the same as the thermodynamic temperature describing the lattice.

The introduction of the concept of spin temperature to characterize the spin system provided the basis of many subsequent developments in NMR spectroscopy of solids (Abragam, 1961). It should be mentioned, however, that the spin-temperature concept cannot be used to treat the coherence phenomena induced by a burst of coherent resonant radiation (which was earlier discussed in relation to the multipulse experiment). However, since phase coherence in the precession of nuclear spins vanishes with a time constant approximately equal to the spin–spin relaxation time T_2, the spin can then be characterized by a spin temperature for values of time subsequent to excitation that are large compared with T_2.

In this section we shall define spin temperature and then use the concept in discussion of nuclear double resonance in the rotating frame, a technique first introduced by Hartmann and Hahn (1962). This has proven to be a powerful technique for enhancement of the sensitivity of detecting rare nuclear species coupled to abundant species through the dipolar Hamiltonian given by Eq. (5.109).

In the double-resonance experiment, resonant radiation is applied to each spin simultaneously. Hartmann and Hahn (1962) showed that if the amplitude of the rf field applied to the abundant spins is $H_{1\mathrm{I}}$ and that applied to the rare spin is $H_{1\mathrm{S}}$, then the flow of energy from rare spins to abundant spins will be maximum under the condition

$$\gamma_I H_{1I} = \gamma_S H_{1S}. \tag{5.110}$$

This results in lowering the spin temperature of the rare spins and hence, increasing the magnetization of the rare spins or the signal-to-noise (S/N) ratio.

Formulating the double-resonance problem proceeds by writing the Hamiltonian in the laboratory frame as

$$\mathscr{H} = \mathscr{H}_I + \mathscr{H}_S + \mathscr{H}_{IS} + \mathscr{H}_{\mathrm{rf}}(t), \tag{5.111}$$

with the Hamiltonian $\mathscr{H}_I$ defined as

$$\mathscr{H}_I = \mathscr{H}_{ZI} + \mathscr{H}_{CI} + \mathscr{H}_{II}, \tag{5.112}$$

where $\mathscr{H}_{ZI} = -\omega_{0I} I_z$, $\omega_{0I} = \gamma_I H_0$ represents the Zeeman Hamiltonian; $\mathscr{H}_{CI}$ is the chemical-shift Hamiltonian, and $\mathscr{H}_{II}$ is equal to $\mathscr{H}'_{\text{dd}}$ given by Eq. (5.48). In practice, $\mathscr{H}_{CI}$ is very small compared with $\mathscr{H}_{ZI}$ or $\mathscr{H}_{II}$ and can be neglected. The Hamiltonian $\mathscr{H}_S$ that characterizes the dilute spin system is given by

$$\mathscr{H}_S = \mathscr{H}_{ZS} + \mathscr{H}_{CS} + \mathscr{H}_{SS}, \tag{5.113}$$

where the various terms in Eq. (5.113) have the same significance as in Eq. (5.112). However, the term $\mathscr{H}_{SS}$ can be neglected, because the rare spins are separated at greater distances. For convenience, the chemical-shift term $\mathscr{H}_{CS}$ can also be combined with the Zeeman term, because the dilute spin species can be considered to consist of a number of resolved lines in the high-resolution spectrum.

The Hamiltonian $\mathscr{H}_{IS}$ plays an important role. It provides the mechanism for interactions between I and S spins. We shall equate $\mathscr{H}_{IS}$ to $\mathscr{H}'_{\text{dd}}$ given in Eq. (5.110). The Hamiltonian concerned with the rf part is $\mathscr{H}_{\text{rf}}(t)$. It is given by

$$\mathscr{H}_{\text{rf}}(t) = -2\omega_{IZ} I_x \cos \omega_\text{I} t - 2\omega_{IS} S_X \cos \omega_S t, \tag{5.114}$$

where $\omega_{1I} = \gamma_I H_{1I}$ and $\omega_{1S} = \gamma_S H_{1S}$.

To describe the dynamic behavior of the combined spin system, we again start with the Liouville–von Neumann equation,

$$\partial\rho(t)/\partial t = -i[\mathscr{H}, \rho(t)]. \tag{5.115}$$

To obtain the solution of Eq. (5.115), it is convenient to first carry out two successive canonical transformations. We first transform $\rho(t)$ to the respective rotating frames of I and S spins by

$$\sigma(t) = R\rho(t)R^{-1}, \tag{5.116}$$

where

$$R = e^{-i\omega_I I_Z t} e^{-i\omega_S S_Z t}. \tag{5.117}$$

This transformation removes the time dependence in the rf field, and Eq. (5.115) becomes

$$\partial\sigma(t)/\partial t = -i[-\Delta_I I_Z - \omega_{1I} I_X - \Delta_S S_Z - \omega_{1S} S_X + \mathscr{H}_{II} + \mathscr{H}_{IS}, \sigma(t)], \tag{5.118}$$

where $\Delta_I = \omega_{0I} - \omega_I$ and $\Delta_S = \omega_{0S} - \omega_S$. We next transform Eq. (5.118) to the tilted frames of the I and S spins by

$$\eta(t) = T\sigma(t)T^{-1}, \tag{5.119}$$

where

$$T = e^{-i\theta_I I_y} e^{-i\theta_S S_y}. \tag{5.120}$$

This transformation defines a new z axis through the angles $\theta_I = \tan^{-1}(\omega_{1I}/\Delta_I)$ and $\theta_S = \tan^{-1}(\omega_{1S}/\Delta_S)$. The resulting equation for $\eta(t)$ is

$$\partial\eta(t)/\partial t = -i[\mathscr{H}_{\mathrm{TR}}, \eta(t)] \tag{5.121}$$

with the transformed Hamiltonian $\mathscr{H}_{\mathrm{TR}}$ given by

$$\mathscr{H}_{\mathrm{TR}} = -\omega_{eI} I_z - \omega_{eS} S_z + P_2(\cos\theta_I)\mathscr{H}^0_{II} + \mathscr{H}^{(ns)}_{II} + \mathscr{H}_p, \tag{5.122}$$

where $\omega_{0I} = (\omega_{1I}^2 + \Delta_I^2)^{1/2}$, $\omega_{eS} = (\omega_{1S}^2 + \Delta_S^2)^{1/2}$, and

$$\mathscr{H}^0_{II} = \sum_{i<j} b_{ij}(\mathbf{I}_i \mathbf{I}_j - 3I_{iz}I_{jz}), \tag{5.123}$$

with the interaction factor b_{ij} given by

$$b_{ij} = \gamma_I^2 \hbar r_{ij}^{-3} P_2(\cos\theta_{ij}). \tag{5.124}$$

The nonsecular part of the dipolar Hamiltonian $\mathscr{H}^{(ns)}_{II}$ is given by

$$\mathscr{H}^{(ns)}_{II} = \tfrac{3}{2}\sin^2\theta_I \sum_{i<j} b_{ij}[(I_{ix}I_{jz} + I_{iz}I_{jx}) + (I_{ix}I_{jx} - I_{iy}I_{jx})], \tag{5.125}$$

which can be neglected because it does not commute with I_z. However, this term is responsible for the coupling of the Zeeman (the Zeeman reservoir) and the secular dipolar term (the dipolar reservoir). Finally, the Hamiltonian $\mathscr{H}_p$ originates from the dipolar coupling of the I and S spins and has the general form

$$\begin{aligned}\mathscr{H}_p = {} & \cos\theta_I \cos\theta_S \sum_{i,m} b_{im} I_{iz} S_{mz} + \sin\theta_I \sin\theta_S \sum_{i,m} b_{im} I_{ix} S_{mx} \\ & - \sin\theta_I \cos\theta_S \sum_{i,m} b_{im} I_{ix} S_{mz} - \cos\theta_I \sin\theta_S \sum_{i,m} b_{im} I_{iz} S_{mx},\end{aligned} \tag{5.126}$$

where $b_{im} = -2\gamma_I \gamma_S \hbar r_{im}^{-3} P_2(\cos\theta_{im})$.

The spin dynamics in the double-resonance experiment can now be analyzed by considering two subsystems characterized by $\mathscr{H}_1$ and $\mathscr{H}_2$, defined by

$$\mathscr{H}_1 = \omega_{eI} I_z + P_2(\cos\theta_I)\mathscr{H}^0_{II}, \qquad \mathscr{H}_2 = -\omega_{eS} S_z, \tag{5.127}$$

coupled by the Hamiltonian $\mathscr{H}_P$ given in Eq. (5.126). By expressing $\mathscr{H}_1$ in the form of Eq. (5.127), we assume that the Zeeman and dipolar reservoirs have reached equilibrium.

In the absence of $\mathscr{H}_P$, $\mathscr{H}_1$ and $\mathscr{H}_2$ are associated with constants of motion. Thus, it is useful to define the thermodynamic functions by

$$\beta_1(t) = \langle \mathscr{H}_1 \rangle_t / \mathrm{Tr}\{\mathscr{H}_1^2\},\ \beta_2(t) = \langle \mathscr{H}_2 \rangle_t / \mathrm{Tr}\{\mathscr{H}_2^2\}, \tag{5.128}$$

where

$$\langle \mathscr{H}_i \rangle_t = \mathrm{Tr}\{\mathscr{H}_i \eta(t)\}, \qquad i = 1, 2 \tag{5.129}$$

In the high-temperature approximation, $\beta_1(t)$ and $\beta_2(t)$ have the dimensions of an inverse temperature ($k_B = 1$). In the indirect procedure of double resonance, the signal detected in the rotating reference frame of the abundant I-spin system is proportional to $\beta_1(t)$ (Bleich and Redfield, 1971; Grannell *et al.*, 1973; Hartman and Hahn, 1962; Lurie and Slichter, 1964; McArthur *et al.*, 1969), whereas in the direct-detection version of double resonance in the rotating frame of dilute spins (Pines *et al.*, 1973), the signal is directly related to $\beta_2(t)$.

The projection-operator technique introduced in Chapter 4 is useful to derive the dynamic equation for $\beta_1(t)$ and $\beta_2(t)$. To accomplish this, we recognize that the formal solution of Eq. (5.121) is

$$\eta(t) = e^{-i\mathscr{H}_{\mathrm{TR}}t} \eta(0) e^{i\mathscr{H}_{\mathrm{TR}}t}, \tag{5.130}$$

where $\eta(0)$ corresponds to the initial condition of the ensemble. Substituting the expression for $\eta(t)$ into Eq. (5.129) and rearranging the order of the operator, we obtain

$$\langle \mathscr{H}_i \rangle_t = \mathrm{Tr}[\mathscr{H}_i(t) \eta(0)], \qquad i = 1, 2 \tag{5.131}$$

where $\mathscr{H}_i(t)$ is given by

$$\mathscr{H}_i(t) = e^{i\mathscr{H}_{\mathrm{TR}}t} \mathscr{H}_i e^{-i\mathscr{H}_{\mathrm{TR}}t} \equiv e^{i\mathscr{H}_{\mathrm{TR}}^{\times}t} \mathscr{H}_i. \tag{5.132}$$

The time derivative of $\mathscr{H}_i(t)$ is

$$\partial \mathscr{H}_i(t)/\partial t = e^{i\mathscr{L}t} i\mathscr{L}\mathscr{H}_i. \tag{5.133}$$

In Eq. (5.132), we introduce the superoperator notation. The superoperator $\mathscr{H}_{\mathrm{TR}}^{\times}$ is set equal to the Liouville operator $\mathscr{L}$ in Eq. (5.133).

Equation (5.133) is in the same form as Eq. (4.35). Given the equation of motion for $\mathscr{H}_i(t)$ by Eq. (5.133), it follows that the generalized Langevin equation for $\mathscr{H}_i(t)$ is [see Eq. (4.60)]

$$\partial \mathscr{H}_j(t)/\partial t = i \sum_{k=1}^{2} \Omega_{ik} \mathscr{H}_k(t) - \int_0^t d\tau \sum_{k=1}^{2} K_{jk}(\tau) \mathscr{H}_k(t - \tau) + F_j(t), \qquad j = 1, 2. \tag{5.134}$$

Here Ω_{jk}, the jk component of the frequency matrix, vanishes owing to symmetry. The random force $F_j(t)$ is given by $e^{i(1-P)\mathscr{L}}(1 - P)\, i\mathscr{L}\mathscr{H}_j$. The term P is the projection operator defined in accordance with Eq. (4.59).

Multiplying both sides of Eq. (5.134) by $\eta(0)$ and then taking the ensemble average, we obtain the dynamic equation for $\beta_j(t)$ as

$$\frac{\partial \beta_j(t)}{\partial t} = -\sum_{k-1}^{2} \int_0^t d\tau \, \langle F_j(\tau) F_k^+ \rangle \beta_k(t-\tau), \qquad j = 1, 2. \tag{5.135}$$

Equation (5.135) is exact. It shows the memory effect in that the values of thermodynamic variables $\beta_j(t)$ at the moment t are affected by all earlier values. However, in the limit of $t \gg \tau_c$, the correlation time of the correlation function $\langle F_j(\tau) F_k^+ \rangle$, we can neglect the memory effect by replacing $t - \tau$ by t and also setting the upper limit of the integral to infinity. Under this approximation, we find specifically that the equation for the dilute-spin species is given by

$$\partial \beta_2(t)/\partial t = -\beta_2(t)/T_{22} - \beta_1(t)/T_{21}, \tag{5.136}$$

where

$$T_{jk}^{-1} = \int_0^{\infty} d\tau \langle F_j(\tau) F_k^+ \rangle. \tag{5.137}$$

The random force F_j given previously can be simplified as

$$\begin{bmatrix} F_1 \\ F_2 \end{bmatrix} = e^{iQ\mathscr{L}t}(1 - P)i\mathscr{L}\begin{bmatrix} \mathscr{H}_1 \\ \mathscr{H}_2 \end{bmatrix} = ie^{iQ\mathscr{L}t}\begin{bmatrix} [\mathscr{H}_P, \mathscr{H}_1] \\ [\mathscr{H}_P, \mathscr{H}_2] \end{bmatrix}. \tag{5.138}$$

The second identity follows because $Pi\mathscr{L}\mathscr{H}_i = 0$. The second (lowest) order in $\mathscr{H}_p$, the correlation function of the random force, will have the form

$$\begin{aligned} \langle F_j(t) F_k^+ \rangle &= \langle \{\exp[i(\mathscr{H}_1^x + \mathscr{H}_2^x)t][\mathscr{H}_P, \mathscr{H}_j]\}[\mathscr{H}_P, \mathscr{H}_k]^+ \rangle \\ &= \langle [\mathscr{H}_P(t), \mathscr{H}_j][\mathscr{H}_P, \mathscr{H}_k]^+ \rangle, \end{aligned} \tag{5.139}$$

where $\mathscr{H}_P(t) = \exp[i(\mathscr{H}_1 + \mathscr{H}_2)t]\mathscr{H}_P \exp[-i(\mathscr{H}_1 + \mathscr{H}_2)t]$. From the foregoing, we find that the following relation is valid:

$$\int_0^{\infty} d\tau \, \langle [\mathscr{H}_P(t), \mathscr{H}_j][\mathscr{H}_P, \mathscr{H}_1 + \mathscr{H}_2]^+ \rangle = 0. \tag{5.140}$$

As a result, T_{22} and T_{21} in Eq. (5.136) can be combined to yield,

$$\partial \beta_2(t)/\partial t = -[\beta_2(t) - \beta_1(t)]/T_{IS}, \tag{5.141}$$

where the cross-relaxation time T_{IS} is given by

$$T_{IS}^{-1} = [\mathrm{Tr}(\mathscr{H}_2^2)]^{-1} \int_0^{\infty} d\tau \, \langle [\mathscr{H}_P(\tau), \mathscr{H}_2][\mathscr{H}_P, \mathscr{H}_2]^+ \rangle. \tag{5.142}$$

Further evaluation of T_{IS}^{-1} will require the specific forms of $\mathscr{H}_P$, $\mathscr{H}_1$, $\mathscr{H}_2$, which are given in Eqs. (5.126) and (5.127). Evaluating the commutator, we obtain, after some algebra

$$T_{IS}^{-1} = \sin^2\theta_S \cos^2\theta_I M_{2,SI} J_z(\omega_{eS}) + \tfrac{1}{2}\sin^2\theta_S \sin^2\theta_I M_{2,SI}\{J_x(\omega_{eS} - \omega_{eI}) + J_x(\omega_{eS} + \omega_{eI})\}, \quad (5.143)$$

where $M_{2,SI}$ is the van Vleck second moment of the magnetic resonance (Abragam, 1961) calculated in accordance with Eq. (5.108). The spectral power density functions $J_z(\omega)$ and $J_x(\omega)$ are given by

$$J_x(\omega) = \int_0^\infty d\tau\,(\cos\omega\tau) C_x(\tau), \qquad J_z(\omega) = \int_0^\infty d\tau\,(\cos\omega\tau) C_z(\tau), \quad (5.144)$$

where

$$C_x = \left\langle \left(\sum_i b_i I_{ix}\right) \exp[-iP_2(\cos\theta_I)^x \mathscr{H}_{II,0}\tau] \left(\sum_i b_i I_{ix}\right)\right\rangle \Big/ \left\langle \left(\sum_j b_j I_{jx}^2\right)\right\rangle, \quad (5.145)$$

$$C_z = \left\langle \left(\sum_i b_i I_{iz}\right) \exp[-iP_2(\cos\theta_I)\mathscr{H}_{II,0}^x\tau] \left(\sum_i b_i I_{iz}\right)\right\rangle \Big/ \left\langle \left(\sum_j b_i I_{jx}\right)^2\right\rangle, \quad (5.146)$$

where the cross-coupling interaction factors b_j are the shorthand notations for b_{jm}, because we assume that the dilute spins are magnetically equivalent and b_{jm} is independent of m.

Equation (5.143) simplifies in the case of adiabatic demagnetization in the rotating frame (*ADRF*) (Slichter and Holton, 1961). In *ADRF*, $\mathscr{H}_{1I}$ is changed adiabatically from the value $H_{1I} \gg H_{LI}$ to the value $H_{1I} \ll H_{LI}$, where H_{LI} is the local field at the site of I spins in the rotating frame. In such case, the tilted angle $\theta_I \simeq 0$, and T_{IS}^{-1} becomes

$$T_{IS}^{-1} = (\sin^2\theta_S) M_{2,SI} J_z(\omega_{eS}). \quad (5.147)$$

Another situation that will simplify Eq. (5.143) is the high-field spin-lock case (Solomon, 1959) in which H_{1I} is made to be much larger than H_{LI}. In this case, $\omega_{eI} > \Delta_I \gg T_2^{-1} \simeq \tau_c^{-1}$, we have $\theta_I \simeq 90°$ and Eq. (5.143) reduces to

$$T_{IS}^{-1} = \tfrac{1}{2}\sin^2\theta_S\, M_{2,SI} J_x(\Delta\omega_e), \quad (5.148)$$

where $\Delta\omega_e = \omega_{eI} - \omega_{eS}$.

The term $J_x(\omega_{eS} + \omega_{eI})$ drops out because the correlation function $C_x(\tau)$ decays monotonically with time, and this makes $J_x(\omega)$ approach zero for $\omega_{eI}\tau_c \gg 1$.

The spectral power density $J_x(\Delta\omega_e)$ is maximum when $\Delta\omega_e = 0$. When the rf fields applied to the I and S spins are at resonance with the respective Larmor frequencies, we have $\Delta_I = \Delta_S = 0$, and this means that T_{IS}^{-1} is maximum when $\Delta\omega_e = 0$ is equivalent to the Hartmann–Hahn condition given in Eq. (5.110).

McArthur *et al.* (1969) have studied the dependence of T_{IS}^{-1} on ω_{1S}. They referred such frequency dependence to the dipolar-fluctuation spectrum. A detailed experimental and theoretical study of the frequency dependence of T_{IS}^{-1} for CaF_2 was also provided by Demco *et al.* (1975).

Equation (5.141) essentially states that the rare spins (e.g., C^{13}) are cooled by cross-relaxation to the abundant spins (e.g., H^1). Signal enhancement using the cross-relaxation technique is substantial.

A substantial part of the dipolar linewidth is removed by irradiating the abundant spins (spin decoupling); further reduction of the linewidth is accomplished by magic-angle spinning (Schaefer, *et al*, 1974 and 1977). As pointed out earlier, the magic-angle spinning technique also removes the anisotropy of the chemical shift. Thus, the combination of heteronuclear decoupling and sample spinning produces well-resolved high-resolution spectra in solids.

The materials discussed so far assume no spin–lattice relaxation. In the next section, we shall consider the influence of spin–lattice relaxation on the spin temperature.

5.10 SPIN–LATTICE RELAXATION IN SOLIDS AND ITS EFFECT ON SPIN TEMPERATURE

As previously mentioned, success in observing the NMR signal requires effective relaxation of the Zeeman levels. The relaxation mechanisms are closely associated with spectral line shape and relaxation times (Abragam, 1961). If we characterize the molecular motion in a solid by a correlation time τ_c, we generally observe the narrowing of the spectral line when τ_c becomes less than T_2 (the spin–spin relaxation time) and a minimum in T_1 (the spin–lattice relaxation time in the laboratory frame) when τ_c is near the Larmor precession period (ω_0^{-1}). In a solid T_2 is much smaller than T_1.

The BBP analysis of T_1 and T_2 owing to dipolar interactions is a convenient starting point for many of the theories in the literature (Bloembergen *et al.*, 1948). The Kubo–Tomita approach provides an attractive alternative (Kubo and Tomita, 1954). In this section, we shall consider only the spin–lattice relaxation in solids; relaxation in liquids is well treated in Chapter VIII of Abragam (1961).

Consider first the spin–lattice relaxation in the laboratory frame. We assume that the heat capacity of the lattice is much larger than that of the spin system. The spin–lattice coupling provides a pathway to warm up the spin system. However, since it is assumed that the heat capacity of the lattice is large, the lattice always remains at equilibrium at constant temperature.

To begin, we write the Hamiltonian of the system as

$$\mathscr{H} = \mathscr{H}_0 + \mathscr{H}_i(t), \tag{5.149}$$

where $\mathscr{H}_0$ is the time-independent spin Hamiltonian, corresponding to Zeeman and the time-independent part of the spin–spin interactions, and $\mathscr{H}_i(t)$ the spin–lattice coupling term, responsible for the relaxation toward thermal equilibrium with lattice.

The spin–lattice coupling is assumed to be a random function of time. This assumption is equivalent to treating the lattice motion classically, in which case magnetization will approach the zero value corresponding to an infinite spin temperature. This difficulty can be remedied by using a quantum-mechanical description of the lattice, but the general treatment gives the same result for the spin–lattice relaxation time as does the classical treatment (Abragam, 1961).

There are several methods that can be used to calculate relaxation times. We start by using the master equation governing the evolution of the density matrix given by [Eq. (42‴), Chapter VIII, of Abragam, (1961)]

$$\partial\rho(t)/\partial t = i[\mathscr{H}_0, \rho(t)] - \mathrm{Re}\int_0^\infty d\tau\, \overline{[\mathscr{H}_1(t), [\mathscr{H}_1^*(t-\tau), \rho(t) - \rho_{\mathrm{L}}]]}, \tag{5.150}$$

where the overbar denotes the average over the lattice variable and Re denotes the real part. The quantity ρ_{L} is the density operator of the spin system at equilibrium, at the lattice temperature. The operator $\mathscr{H}_1^*(t-\tau)$ is given by

$$\mathscr{H}_1^*(t-\tau) = e^{-i\mathscr{H}_0\tau}\mathscr{H}_1(t-\tau)e^{i\mathscr{H}_0\tau}. \tag{5.151}$$

It should be emphasized that Eq. (5.150) is valid only when the correlation time τ_c for the lattice motion is much smaller than T_2 ($\tau_c \ll T_2$) and the heat capacity of the spin system is much smaller than that of the lattice.

For practically all relaxation mechanisms in NMR, the time-dependent operator $\mathscr{H}_1(t)$ can be written as

$$\mathscr{H}_1(t) = \sum_\mu A_\mu F_\mu(t), \tag{5.152}$$

where A_μ are spin operators and $F_\mu(t)$ are classical functions of the lattice variable. The spin operators are Hermitian; i.e., $A_\mu^+ = (-1)^\mu A_{-\mu}$.

We assume that the classical functions $F_\mu(t)$ are statistically independent, stationary, and symmetric upon time reversal. That is,

$$\overline{F_\mu(t)F_\nu^*(t')} = \delta_{\mu\nu} G_\mu(|t - t'|), \tag{5.153}$$

where the asterisk denotes the complex conjugate.

Substituting Eq. (5.152) into (5.150) and using the property of Eq. (5.153), we obtain

$$\partial\rho(t)/\partial t = [\mathscr{H}_0, \rho(t)] - \mathrm{Re} \sum_\mu \int_0^\infty d\tau [A_\mu, [A_\mu^+(\tau), \rho(t) - \rho_\mathrm{L}] G_\mu(\tau), \tag{5.154}$$

where

$$A_\mu^+(\tau) = e^{-i\mathscr{H}_0\tau} A_\mu^+ e^{i\mathscr{H}_0\tau}. \tag{5.155}$$

We now use Eq. (5.154) to derive an expression for the spin–lattice relaxation times in low and high external magnetic fields.

Consider first the low-field case. Since $\mathscr{H}_0$ consists of Zeeman ($Z = -\omega_0 I_z$) and time-independent spin–spin interaction ($\mathscr{H}_\mathrm{D}$) parts, we can write

$$\mathscr{H}_0 = Z + \mathscr{H}_\mathrm{D}. \tag{5.156}$$

For the field low enough to be comparable with the local field, the time dependence in $A_\mu^+(\tau)$ of Eq. (5.155) can be neglected. Under this approximation, Eq. (5.154) reduces to

$$\partial\rho(t)/\partial t = -i[\mathscr{H}_0, \rho(t)] - \sum_\mu \tau_{c\mu} G_\mu(0) \{A_\mu, [A_\mu^+, \rho(t) - \rho_\mathrm{L}]\}, \tag{5.157}$$

where

$$\tau_{c\mu} = \int_0^\infty d\tau\, G_\mu(\tau)/G_\mu(0). \tag{5.158}$$

Multiplying Z and taking the traces on both sides of Eq. (5.157), we obtain

$$\partial\langle Z\rangle_t/\partial t = -i\langle[Z, \mathscr{H}_0]\rangle_t - \sum_\mu \tau_{c\mu} G_\mu(0)\, \mathrm{Tr}\{[Z, A_\mu][A_\mu^+, \rho(t) - \rho_\mathrm{L}]\}, \tag{5.159}$$

where

$$\langle O\rangle_t = \mathrm{Tr}[O\rho(t)] \tag{5.160}$$

Similarly, multiplying both sides of Eq. (5.157) by $\mathscr{H}_D$ and then taking the trace, we obtain

$$\partial\langle\mathscr{H}_D\rangle_t/\partial t = i\langle[Z, \mathscr{H}_D]\rangle_t - \sum_\mu \tau_{c\mu} G_\mu(c)\,\mathrm{Tr}\{[\mathscr{H}_D, A_\mu][A_\mu^+, \rho(t) - \rho_L]\}, \tag{5.161}$$

In solids, T_2 is very much smaller than the spin–lattice relaxation time. Thus, for $t > T_2$ we can assume that the spin system has reached a state very close to interval equilibrium, and we can approximate $\rho(t) \simeq 1 - \beta(t)\mathscr{H}_0$. Similarly, the equilibrium-density operator will be approximated as $\rho_L \simeq 1 - \beta_L \mathscr{H}_0$.

As a result, Eqs. (5.159) and (5.161) can be written as

$$\partial\langle Z\rangle_t/\partial t = -i\langle[Z, \mathscr{H}_0]\rangle_t - [\beta(t) - \beta_L] \sum_\mu \tau_{c\mu} G_\mu(0)\,\mathrm{Tr}\{[Z, A_\mu][Z, A_\mu^+]\} \tag{5.162}$$

and

$$\partial\langle\mathscr{H}_D\rangle_t/\partial t = i\langle[Z, \mathscr{H}_D]\rangle_t - [\beta(t) - \beta_L] \sum_\mu \tau_{c\mu} G_\mu(0)\,\mathrm{Tr}\{[\mathscr{H}_D, A_\mu][\mathscr{H}_D, A_\mu^+]\}. \tag{5.163}$$

Consistent with the approximations are

$$\langle Z\rangle_t = \mathrm{Tr}[Z\rho(t)] = -\beta(t)\,\mathrm{Tr}(Z^2), \tag{5.164}$$

$$\langle Z\rangle_L = \mathrm{Tr}(Z\rho_L) = -\beta_L\,\mathrm{Tr}(Z^2), \tag{5.165}$$

$$\langle\mathscr{H}_D\rangle_t = \mathrm{Tr}[\mathscr{H}_D\rho(t)] = -\beta(t)\,\mathrm{Tr}(\mathscr{H}_D^2), \tag{5.166}$$

$$\langle\mathscr{H}_D\rangle_L = \mathrm{Tr}(\mathscr{H}_D\rho_L) = -\beta_L\,\mathrm{Tr}(\mathscr{H}_D^2). \tag{5.167}$$

Utilizing the results of Eqs. (5.164)–(5.167), we can write Eq. (5.162) as

$$\partial\langle Z\rangle_t/\partial t = -i\langle(Z, \mathscr{H}_D)\rangle_t - T_Z^{-1}\{\langle Z\rangle_t - \langle Z\rangle_L\}, \tag{5.168}$$

and

$$\partial\langle\mathscr{H}_D\rangle_t/\partial t = i\langle(Z, \mathscr{H}_D)\rangle_t - T_D^{-1}(\langle\mathscr{H}_D\rangle_t - \langle\mathscr{H}_D\rangle_L); \tag{5.169}$$

where

$$T_Z^{-1} = \sum_\mu \tau_{c\mu} G_\mu(0)\,\mathrm{Tr}\{[Z, A_\mu][A_\mu^+, Z]\}/\mathrm{Tr}(Z^2), \tag{5.170}$$

and

$$T_D^{-1} = \sum_\mu \tau_{c\mu} G_\mu(0)\,\mathrm{Tr}\{[\mathscr{H}_D, A_\mu][A_\mu^+, \mathscr{H}_D]\}/\mathrm{Tr}(\mathscr{H}_D^2). \tag{5.171}$$

Adding Eq. (5.168) to (5.169) and also using Eqs. (5.164)–(5.167), we obtain the relaxation equation for the inverse-spin temperature $\beta(t)$ as

$$\partial\beta(t)/\partial t = -T_1^{-1}[\beta(t) - \beta_L], \tag{5.172}$$

where

$$T_1^{-1} = \frac{[\text{Tr}(Z^2)/T_Z + \text{Tr}(\mathscr{H}_D^2)/T_D]}{[\text{Tr}(Z^2) + \text{Tr}(\mathscr{H}_D^2)]}. \tag{5.173}$$

We define the local field as

$$H_L^2 = \text{Tr}(\mathscr{H}_D^2)/\text{Tr}(M_Z^2) \tag{5.174}$$

and rewrite Eq. (5.173) in terms of the weighted contribution of the local field and the external fields as

$$T_1^{-1} = \frac{H^2 + (T_Z/T_D)H_L^2}{H^2 + H_L^2}. \tag{5.175}$$

Thus, by measuring T_1 as a function of the external field, it is possible to determine the T_Z/T_D ratio. Experimental evidence (Goldman, 1970) indicate that the T_Z/T_D ratio is between 2 and 3. In the low field, T_Z or T_D has the same order of magnitude as T_1 (Goldman, 1970).

In the high field, $\mathscr{H}_0$ in Eq. (5.155) cannot be neglected. But if the field is so large that the Zeeman is much greater than the spin–spin part, we can neglect $\mathscr{H}_D$ and approximate Eq. (5.155) as

$$A_\mu^+(\tau) \simeq e^{-iZt}A_\mu^+ e^{iZt} = e^{i\mu\omega_0 t}A_\mu^+. \tag{5.176}$$

The last identity is the result of the property of A_μ, which obeys the identity

$$[I_z, A_\mu] = \mu A_\mu. \tag{5.177}$$

We next assume that the time-correlation function $G_\mu(t)$ decays exponentially with a correlation time $\tau_{c\mu}$; i.e., $G_\mu(t) = G_\mu(0)e^{-t/\tau_{c\mu}}$. As a result, Eq. (5.154) becomes

$$\frac{\partial\rho(t)}{\partial t} = -i[\mathscr{H}_0, \rho(t)] - \sum_\mu \frac{G_\mu(0)\tau_{c\mu}}{1 + (\mu\omega_0\tau_{c\mu})^2}[A_\mu, [A_\mu^+, \rho(t) - \rho_L]\}. \tag{5.178}$$

Equation (5.178) has the same form as Eq. (5.157) for the low-field case if $\tau_{c\mu}$ in Eq. (5.157) is replaced by $\tau_{c\mu}/[1 + (\mu\omega_0\tau_{c\mu})^2]$. The two equations become identical if the factor $\mu\omega_0\tau_{c\mu}$ in the denominator is neglected. By multiplying Z (or $\mathscr{H}_D$) on both sides of Eq. (5.178) and then taking the trace, we obtain an equation similar to Eq. (5.159) [or Eq. (5.161)]. However, in the high field, because of nonsecular spin–spin interaction, the rate of thermal mixing between Z and $\mathscr{H}_D$ may become slower than the spin–lattice relaxation rate (T_1^{-1}). Under such circumstances, $\langle Z \rangle_t$ and $\langle \mathscr{H}_D \rangle_t$ are expected to relax more or less independently toward the lattice equilibrium temperature.

Hence, we approximate $\rho(t) = 1 - \beta_1 Z - \beta_2 \mathscr{H}_D$ and substitute it into Eq. (5.178) to obtain

$$\partial \beta_1 / \partial t = -T_Z^{-1}(\beta_1 - \beta_L), \tag{5.179}$$

and

$$\partial \beta_2 / \partial t = -T_D^{-1}(\beta_2 - \beta_L), \tag{5.180}$$

where

$$T_Z^{-1} = \frac{1}{\mathrm{Tr}(Z^2)} \sum_\mu \frac{G_\mu(0)\tau_{c\mu}}{1 + (\mu\omega_0\tau_{c\mu})^2} \mathrm{Tr}\{[Z, A_\mu][A_\mu^+, Z]\} \tag{5.181}$$

$$= \frac{1}{\mathrm{Tr}(I_z^2)_\mu} \sum_\mu \frac{G_\mu(0)\tau_{c\mu}\mu^2}{1 + (\mu\omega_0\tau_{c\mu})^2} \mathrm{Tr}(A_\mu^2), \tag{5.182}$$

and

$$T_D^{-1} = \frac{1}{\mathrm{Tr}(\mathscr{H}_D^2)} \sum_\mu \frac{G_\mu(0)\tau_{c\mu}}{1 + (\mu\omega_0\tau_{c\mu})^2} \mathrm{Tr}\{[\mathscr{H}_D, A_\mu][A_\mu^+, \mathscr{H}_D]\}. \tag{5.183}$$

Note that the term with $\mu = 0$ does not contribute to the Zeeman relaxation time T_Z but does contribute to T_D. Clearly, we have $T_Z > T_D$. For $\omega_0\tau_c \gg 1$, T_Z may become much greater than T_D (Solomon and Ezratly, 1962; Goldman, 1970).

5.11 SPIN–LATTICE RELAXATION IN THE ROTATING FRAME

The Hamiltonian for the spin system irradiated with an rf field of frequency ω with Zeeman, spin–spin, and spin–lattice interactions is given by

$$\mathscr{H} = -\omega_0 I_z + \mathscr{H}_D - 2\omega_1 I_x \cos \omega t + \sum_\mu A_\mu F_\mu(t). \tag{5.184}$$

In the frame rotating at frequency ω, the density matrix σ is related to that of the laboratory frame by

$$\sigma(t) = e^{-i\omega I_z t} \rho(t) e^{i\omega I_z t}. \tag{5.185}$$

The transformation removes the time dependence in the rf part but gives the time dependence in the spin–lattice relaxation term:

$$\mathscr{H}^* = \Delta I_z + \omega_1 I_x + \mathscr{H}_D + \sum_{\mu} F_\mu(t) A_\mu e^{i\mu\omega t} = \mathscr{H}_0 + \sum_{\mu} F_\mu(t) A_\mu e^{i\mu\omega t}, \tag{5.186}$$

where $\Delta = \omega_0 - \omega$.

With the Hamiltonian given by Eq. (5.186) we can derive a master equation for σ. The master equation is identical to Eq. (5.176), with ρ replaced by σ, assuming an exponential time-correlation function for the random classical function $F_\mu(t)$. That is,

$$\frac{\partial \sigma}{\partial t} = i[\mathscr{H}_0, \sigma] - \sum_{\mu} \frac{G_\mu(0)\tau_{c\mu}}{1 + (\mu\omega\tau_{c\mu})^2} \{A_\mu, [A_\mu^+, \sigma - \sigma_2]\}. \tag{5.187}$$

Redfield (1955) showed that if the rf amplitude is strong enough to saturate the absorption (i.e., $\omega_1^2 T_1 T_2 \gg 1$), then the spin system in the rotating frame will quickly establish an inverse-spin temperature (β) in a time T_2; thus, σ can be approximated as

$$\sigma \simeq 1 - \beta \mathscr{H}_0, \tag{5.188}$$

and it is a good approximation to write $\sigma_L \simeq 1 + \beta_L \omega_0 I_Z$. Using the technique previously described, we can immediately write the expectation values of Z, $\mathscr{H}_D$, and X as follows:

$$\partial\langle Z\rangle_t/\partial t = i\langle Z, \mathscr{H}_0)\rangle - T_Z^{-1}(\langle Z\rangle_t - \langle Z\rangle_L), \tag{5.189}$$

$$\partial\langle \mathscr{H}_D\rangle_t/\partial t = -i\langle(\mathscr{H}_D, \mathscr{H}_0)\rangle - T_D^{-1}\langle \mathscr{H}_D\rangle_t, \tag{5.190}$$

$$\partial\langle X\rangle_t/\partial t = i\langle(X, \mathscr{H}_0)\rangle - T^{-1}\langle X\rangle_t, \tag{5.191}$$

where $Z = -\Delta I_Z$, $X = -\omega_1 I_X$, and the relaxation times T_Z and T_D are given by Eqs. (5.170) and (5.171), with the exception that ω_0 is replaced by ω. Since $\Delta \ll \omega_0$, there is practically no difference between these equations.

The expression for the relaxation time T_X is given by

$$T_X^{-1} = \frac{1}{\mathrm{Tr}(I_X^2)} \sum_{\mu} \frac{G_\mu(0)\tau_{c\mu}}{1 + (\mu\omega\tau_{c\mu})^2} \mathrm{Tr}\{[I_X, A_\mu][A_\mu^+, I_X]\}. \tag{5.192}$$

If we multiply Eq. (5.189) by $-\Delta$ and (5.191) by ω_1 and then sum up these equations and Eq. (5.190), we obtain

$$\begin{aligned}(\partial/\partial t)\{\langle Z\rangle_t + \langle \mathscr{H}_D\rangle_t + \langle X\rangle_t\} \\ = -[T_Z^{-1}(\langle Z\rangle_t - \langle Z\rangle_L) + T_D^{-1}\langle \mathscr{H}_D\rangle + T_X^{-1}\langle X\rangle]. \end{aligned} \tag{5.193}$$

From this we obtain the basic equation for the spin–lattice relaxation in the rotating frame:

$$\partial\beta/\partial t = -T_{1\rho}^{-1}(\beta - \beta'_{\mathrm{L}}) \tag{5.194}$$

where the spin–lattice relaxation time in the rotating frame ($T_{1\rho}$) is given by

$$\begin{aligned} T_{1\rho}^{-1} &= \frac{\langle Z^2\rangle/T_{\mathrm{Z}} + \langle \mathscr{H}_{\mathrm{D}}^2\rangle/T_{\mathrm{D}} + \langle X^2\rangle/T_{\mathrm{X}}}{\langle Z^2\rangle + \langle \mathscr{H}_{\mathrm{D}}^2\rangle + \langle X^2\rangle} \\ &= \frac{h^2/T_{\mathrm{Z}} + H_1^2/T_{\mathrm{X}} + H_{\mathrm{L}}^2/T_{\mathrm{D}}}{h^2 + H_1^2 + H_{\mathrm{L}}^2} \end{aligned} \tag{5.195}$$

and β'_{L} is the effective inverse-lattice temperature related to the true inverse-lattice temperature β_{L} by

$$\beta_{\mathrm{L}}^1 = \frac{\beta_{\mathrm{L}} h H_0}{h^2 + \lambda H_1^2 + \delta H_{\mathrm{L}}^2}, \tag{5.196}$$

where $\lambda = T_{\mathrm{Z}}/T_{\mathrm{X}}$ and $\delta = T_{\mathrm{Z}}/T_{\mathrm{D}}$.

Note that the effective inverse equilibrium spin temperature is a function of h. It becomes zero (infinite spin temperature) at exact resonance ($h = 0$); it is positive when $H > 0\,(|\omega| < |\omega_0|)$ and negative when $h < 0\,(|\omega| > |\omega_0|)$, β'_{L} is maximum when $h^2 = \lambda H_1^2 + \delta H_{\mathrm{L}}^2$, given by

$$(\beta'_{\mathrm{L}})_{\max} = \tfrac{1}{2}\beta_{\mathrm{L}} H_0/(\lambda H_1^2 + \delta H_{\mathrm{L}}^2). \tag{5.197}$$

Since $H_0 \gg H_1 \geq H_{\mathrm{L}}$, clearly β'_{L} is much larger than β_{L}. In other words, in the rotating frame the effective lattice temperature is much lower than the true lattice temperature. This is because the effective magnetic field along the z axis in the rotating frame is not H_0 but $(H_0 - \omega/\gamma) = h$. Thus, under the effect of spin–lattice relaxation, the density matrix relaxes to $\sigma_{\mathrm{L}} = 1 + \beta_{\mathrm{L}}\omega_0 I_{\mathrm{Z}} = 1 - \beta'_{\mathrm{L}}\Delta I_{\mathrm{Z}}$. Since $\beta'_{\mathrm{L}}\Delta = \beta_{\mathrm{L}}\omega_0$, we have $\beta'_{\mathrm{L}} = \beta_{\mathrm{L}}(H_0/h) \gg \beta_{\mathrm{L}}$. In other words, the spin–spin relaxation establishes equilibrium for the Zeeman reservoir in the rotating frame, which is cooled down to a temperature much lower than the true lattice temperature.

Slichter and Ailion (1964, 1965) devised an important $T_{1\rho}$ experiment and provided analysis that describes situations that do not fall within the criteria assumed for Eq. (5.195). They derived a "strong-collision" theory that applies when $T_1 > \tau_{\mathrm{c}} > T_2$ and $H_1 \simeq H_{\mathrm{L}}$. In this case, almost every step of molecular motion results in relaxation, and the appropriate expression is (Redfield, 1957)

$$T_{1\rho}^{-1} = [2(1 - p)/\tau_c][H_L^2/(H_1^2 + H_L^2)]. \tag{5.198}$$

The parameter p depends on the details of motion. To a reasonable approximation, $1 - p$ is the fraction of the second moment made time-independent by the molecular motion under consideration. This fraction is often on the order of $\frac{1}{2}$, in which case $T_{1\rho} \simeq \tau_c$.

5.12 A GENERAL RELAXATION THEORY FOR RANDOM PERTURBATIONS

As shown previously, relaxation has an important effect on the response of the spin system to perturbations of the external magnetic field. In the presence of molecular motion, the spin system will be subject to perturbations of a stochastic nature. Over time that is long compared with the correlation time, these perturbations are responsible for the damping (relaxation) terms that were introduced in a phenomenological fashion in the Bloch equations.

We shall now provide a general theory for describing relaxation when the stochastic perturbation is weak (a weak-collision theory). We can treat the random (stochastic) perturbation semiclassically. Consider that the spin system is subject to the Hamiltonian $\mathscr{H} = \mathscr{H}_0 + \mathscr{H}_1(t)$, where $\mathscr{H}_1(t)$ is the operator representing random perturbation that acts on the spin system; $\mathscr{H}_0$ represents the Zeeman Hamiltonian. The random Hamiltonian vanishes when it is averaged over the equilibrium distribution function, i.e., $\overline{\mathscr{H}_1(t)} = 0$. Thus, all matrix elements of $\mathscr{H}_1(t)$ such as $\overline{\langle m|\mathscr{H}'_1(t)|n\rangle} = 0$ vanish on the average.

As before, the system is described by the density operator $\rho(t)$, which obeys the Liouville equation of motion

$$d\rho(t)/dt = -i[\mathscr{H}_0 + \mathscr{H}_1(t), \rho(t)]. \tag{5.199}$$

In the weak-collision theory, $\mathscr{H}_1(t)$ is considered a small perturbation. The solution to Eq. (5.199) can be obtained in terms of successive powers of $\mathscr{H}_1(t)$.

Transforming to the interaction representation

$$\sigma(t) = e^{i\mathscr{H}_0 t}\rho(t)e^{-i\mathscr{H}_0 t}, \tag{5.200}$$

we obtain from Eq. (5.200) the equation for $\sigma(t)$ as

$$d\sigma(t)/dt = -i[\mathscr{H}_1^*(t), \sigma(t)], \tag{5.201}$$

where

$$\mathscr{H}_1^*(t) = e^{i\mathscr{H}_0 t}\mathscr{H}_1(t)e^{-i\mathscr{H}_0 t}. \tag{5.202}$$

We convert Eq. (5.201) into an integral equation

$$\sigma(t) = \sigma(0) - i\int_0^t dt'\,[\mathscr{H}_1^*(t'), \sigma(t')] \tag{5.203}$$

and average Eq. (5.203) over all damping histories to obtain the solution by successive approximations as

$$\sigma(t) = \sigma(0) - i\int_0^t dt'\,\overline{[\mathscr{H}_1^*(t'), \sigma(0)]} - \int_0^t dt' \int_0^t dt''\,\overline{\{\mathscr{H}_1^*(t'), [\mathscr{H}_1^*(t''), \sigma(0)]\}} + \cdots, \tag{5.204}$$

where overbars denote the average.

It is useful to consider the solution to this equation for times larger than a characteristic time τ_c, known as the correlation time. For $t > \tau_c$, the time-auto correlation functions of $\mathscr{H}_1^*(t')$ become vanishingly small. We shall also neglect the correlation between $\sigma(0)$ and $\mathscr{H}_1^*(t')$. In principle, the initial value $\sigma(0)$ is determined by the values of $\mathscr{H}_1^*(t')$ immediately preceding it; thus, the correlation between $\sigma(0)$ and $\mathscr{H}_1^*(t')$ is expected at small values of time. However, for $t > \tau_c$, the correlation between $\sigma(0)$ and $\mathscr{H}_1^*(t')$ vanishes. Hence, under this approximation the second term on the right-hand side of Eq. (5.204) vanishes. Further, we assume that the stochastic process is stationary, so that the correlation functions such as

$$\overline{\langle k|\mathscr{H}_1^*(t)|l\rangle\langle m|\mathscr{H}_1^*(t-\tau)|n\rangle} = 0$$

are independent of t. It is further assumed that the values of correlation functions become very small for t longer than τ_c.

Thus, if the higher-order terms in Eq. (5.204) can be neglected, then for $t > \tau_c$ Eq. (5.204) may be approximated as

$$\sigma(t) = \sigma(0) - \int_0^t dt' \int_0^{t'} dt''\,\overline{[\mathscr{H}_1^*(t'), [\mathscr{H}_1^*(t''), \sigma(0)]]}. \tag{5.205}$$

Converting Eq. (5.205) into a differential equation, we obtain

$$d\sigma(t)/dt = -\int_0^t dt''\,\overline{[\mathscr{H}_1^*(t), [\mathscr{H}_1^*(t''), \sigma(0)]]}. \tag{5.206}$$

Equation (5.206) can be further simplified if the mean-square perturbation is sufficiently weak and rapidly varying such that the relative change in $\sigma(t)$ is small, i.e.,

$$[\sigma(t) - \sigma(0)]/\sigma(0) \simeq [t/\sigma(0)][d\sigma(t)/dt] \ll 1. \tag{5.207}$$

Thus, for $t > \tau_c$ and when the criterion given by Eq. (5.207) is valid, we can approximate Eq. (5.206) as

$$d\sigma(t)/dt = -\int_0^\infty d\tau\,[\mathscr{H}_1^*(t), [\mathscr{H}_1^*(t-\tau), \sigma(t)]]. \tag{5.208}$$

Redfield (1957) showed that if this operator equation is written out explicitly in the representation for which $\mathscr{H}_0$ is diagonal, then the rate of change of a typical matrix element of the density operator is given by

$$d\sigma_{kl}(t)/dt = \sum_{m,n} e^{i(\omega_{km}-\omega_{ln})t} R_{kl,mn}\sigma_{mn}(t), \tag{5.209}$$

where the coefficients R are independent of time because of the stationarity approximation. Because the time-dependent exponential in Eq. (5.209) has the property of making any term unimportant unless $\omega_{km} = \omega_{ln}$, we can neglect the small contribution associated with the $\omega_{km} \neq \omega_{ln}$ terms, and obtain

$$d\sigma_{kl}(t)/dt = \sum_{mn}' R_{kl,mn}\sigma_{mn}(t), \tag{5.210}$$

where the prime on the summation indicates that we keep only those terms for which $\omega_{km} = \omega_{ln}$, where $R_{kl,mn}$ is given by

$$R_{kl,mn} = \tfrac{1}{2}[J_{km,ln}(\omega_{ln}) + J_{km,ln}(\omega_{km}) - \delta_{ln}\sum_p J_{pm,pk}(\omega_{pm}) - \delta_{km}\sum_p J_{pl,pn}(\omega_{pn})]. \tag{5.211}$$

Here $J_{km,ln}(\omega)$ are the spectral densities given by

$$J_{km,ln}(\omega) = \int_{-\infty}^{\infty} d\tau\,\langle k|\mathscr{H}_1(t)|m\rangle\langle l|\mathscr{H}_1(t-\tau)|n\rangle e^{i\omega\tau}. \tag{5.212}$$

To obtain ρ_{kl}, we use Eq. (5.200), which relates $\rho(t)$ to $\sigma(t)$; i.e.,

$$\sigma_{kl} = e^{i\omega_{kl}t}\rho_{kl} \tag{5.213}$$

or

$$d\sigma_{kl}/dt = i\omega_{kl}\sigma_{kl} + e^{i\omega klt}(d\rho_{kl}/dt). \tag{5.214}$$

Using Eqs. (5.209) and (5.213), we obtain the equation of motion for ρ_{kl} as

$$
\begin{aligned}
d\rho_{kl}/dt &= i\omega_{kl}\rho_{kl} + \sum_{mn} R_{kl,\,mn}\,\rho_{mn} \\
&= i[\mathscr{H}_0, \rho]_{kl} + \sum_{mn} R_{kl,\,mn}\,\rho_{mn},
\end{aligned}
\tag{5.215}
$$

which is the Redfield equation for the density matrix in the presence of damping effect owing to random perturbations.

Note that the diagonal matrix elements follows the equation

$$
\begin{aligned}
d\rho_{kk}/dt + \sum_{mn} R_{kk,\,mn}\,\rho_{mn} &= \sum_{m} R_{kk,\,mn}\,\rho_{mm} \\
&= \sum_{m} W_{km}\,\rho_{mm} - \left(\sum_{m} W_{mk}\right)\rho_{kk}
\end{aligned}
\tag{5.216}
$$

This corresponds to a rate equation for populations in the different energy levels. The coefficients W_{km} can be identified with the transition probability between levels k and m. From Eqs. (5.211) and (5.212), we find

$$
W_{km} = J_{km,\,km}(\omega_{km}) = \int_{-\infty}^{\infty} d\tau\, \langle k|\mathscr{H}_1(t)|m\rangle\langle k|\mathscr{H}_1(t-\tau)|m\rangle e^{i\omega_{km}\tau}. \tag{5.217}
$$

The transition probability is proportional to the spectral density of the connecting matrix element at the transition frequency ω_{km}.

Clearly, all diagonal terms or populations relax in a coupled fashion, resulting in $N - 1$ characteristic relaxation times for a system with N energy levels, because one has $\operatorname{Tr} \rho = \sum_k \rho_{kk} = 1$.

Because $W_{km} = W_{mk}$, the transition probability from k to m is equal to that from m to k. The solution to the Redfield equation is an equal population among all states. This situation corresponds to an infinite temperature. Clearly, Eq. (5.215) does not describe the approach of equilibrium for a system at finite temperature. This shortcoming results from the semiclassical treatment, in which the effect of the thermal bath is neglected. A proper treatment involving quantization of the bath coordinates is needed to find the spins and to know the temperature of the bath. Refinement of the Redfield equation is given in detail by Abragam (1961). We simply assert that for an interaction in which the lattice couples to the spin via the interaction Hamiltonian $\mathscr{H}_1$, the role of the lattice results in modifying the Redfield equation (5.215) to

$$
d\rho_{kl}/dt = i[\mathscr{H}_0, \rho]_{kl} + \sum_{mn} R_{kl,\,mn}(\rho_{mn} - \rho_{mn}^{(0)}), \tag{5.218}
$$

where, at temperature $T(=1/\beta k)$, $\rho_{mn}^{(0)}$ is given by

$$\rho_{mn}^{(0)} = \delta_{mn}\left(e^{-\beta E_m} \Big/ \sum_m e^{-\beta E_m}\right). \tag{5.219}$$

On the basis of the principle of detailed balance,

$$W_{km} = W_{mk} \exp(-\hbar\omega_{km}/kT). \tag{5.220}$$

This assures a Boltzmann distribution for the equilibrium population:

$$\rho_{kk}^{(0)} = \rho_{mm}^{(0)} \exp(-\hbar\omega_{km}/kT). \tag{5.221}$$

We can use the Redfield equation to compute the time dependence of the physical qualities of interest (such as the three components of magnetization) by the fundamental equation

$$\langle M_\alpha \rangle = \mathrm{Tr}(M_\alpha \rho) = \sum_{k,l} \langle l|M_\alpha|k\rangle \rho_{kl}, \tag{5.222}$$

where $\alpha = x, y, z$. Taking the time derivative of Eq. (5.222) and then using Eq. (5.215), we find

$$d\langle M_\alpha \rangle/dt = i\sum_{kl} \langle l|M_\alpha|k\rangle\langle k|(\mathscr{H}_0, \rho)l\rangle + \sum_{klmn} \langle l|M_\alpha|k\rangle R_{klmn}\rho_{mn}. \tag{5.223}$$

Equation (5.218) yields the Block equations, shown in detail by Slichter (1963), with T_1 and T_2 given by

$$\begin{aligned} R_{kk,mm} &= R_{mm,kk} = -T_1^{-1}, \\ R_{kl,kl} &= R_{lk,lk} = -T_2^{-1}. \end{aligned} \tag{5.224}$$

6
Linear and Nonlinear Raman Spectroscopy of the Condensed Phase

6.1 INTRODUCTION

Since the mid-1970s, studies of vibrational relaxation and vibrational energy-transfer processes have intensified. Detailed studies (Weitz and Flynn, 1974) of energy-decay routes and energy-exchange rates over various vibrational-energy levels of polyatomic molecules have yielded a wide variety of experimental techniques wherein the relevant time constants are, in general, greater than 10^{-9} sec. This permits application of standard electronic techniques.

In liquids and solids at ambient temperatures, our knowledge of vibrational relaxation processes is rather limited. Ultrasonic and Brillouin scattering techniques provide vibrational-relaxation information about low-lying states, and linewidth measurements of infrared and Raman bands are the main source of vibrational-relaxation times (Lascombe and Huong, 1982). Studies show that at room temperature vibrational processes are very rapid, with time constants on the order of 10^{-12} sec.

In general, several physical processes contribute to the observed line shape. Phase relaxation, energy relaxation, reorientation, isotopic splitting, and inhomogeneous broadening resulting from a distribution of vibrational frequencies are factors that affect the spectral linewidth and line shape. In this chapter, we shall discuss methods of Raman scattering (linear and nonlinear) for the study of relaxation processes of the vibrational modes in ground electronic states. We shall consider only nonresonant Raman scattering; the dynamics associated with the excited electronic states will not be considered.

6.2 RAMAN POLARIZABILITY TENSOR

Spontaneous Raman scattering is a powerful technique for investigating the dynamics of molecular motion in the condensed phase and the effect of molecular motion on the line shape of the Raman-scattering spectrum. To help understand the fundamental process, we shall derive an expression for the Raman polarizability tensor.

Spontaneous Raman scattering is a nonresonant process. The frequency of the excitation light can vary from near infrared to ultraviolet. This makes it possible to study a variety of vibrational modes of different molecules using a single light source, which is invariably a laser. This is in contrast with the stimulated Raman-scattering process, which is selective and strongly excites only one or a few normal modes of vibration having the largest scattering cross section.

To provide a simple derivation for the spontaneous Raman-scattering process, we use the correspondence principle by considering the electric field of the excitation radiation to be a classical quantity described by

$$\mathbf{E}(t) = \mathbf{E}^{+} e^{i\omega_0 t} + \mathbf{E}^{-} e^{-i\omega_0 t}, \tag{6.1}$$

where ω_0 is the frequency of the incident laser light and $\mathbf{E}^{+}$ and $\mathbf{E}^{-}$ are constant vectors, which may be complex. To insure that $\mathbf{E}(t)$ is real, we must have $(\mathbf{E}^{+})^{*} = \mathbf{E}^{-}$. The electric field described by Eq. (6.1) represents an elliptical polarization. In Eq. (6.1), we neglect the spatial variation of the electric field, assuming that the dimension of the molecule yielding Raman scattering is small compared with the wavelength of the excitation light.

If the interaction of the electric field of the laser light with matter results in the scattering of light without a distinctive change in the frequency, then we have Rayleigh scattering as described in Chapter 2. Another type of the scattering process with a distinctive change in the frequency also occurs. The amount of shift in the frequency of the scattered light corresponds to one of the various transitions between the energy levels of the molecular system. This is known as Raman scattering.

Raman scattering occurs because the electric dipole moment induced in the medium by the electric field given in Eq. (6.1) is modulated by molecular vibration. The induced dipole moment can be calculated by considering that before the introduction of the electric field of the excitation laser the molecular system exists in one of the stationary states described by the unperturbed Hamiltonian $\mathscr{H}_0$. That is,

$$\mathscr{H}_0 | l \rangle = \varepsilon_l | l \rangle, \qquad l = 1, 2, \ldots, \tag{6.2}$$

where $| l \rangle$ are the stationary states.

In the presence of the external electric field $\mathbf{E}(t)$, the molecular system originally existing in the stationary state $| l \rangle$ is perturbed and changes to a time-dependent nonstationary state $| \psi_l(t) \rangle$. Clearly, the nonstationary state $| \psi_l(t) \rangle$ will satisfy the time-dependent Schrödinger equation, given by

$$[\mathscr{H}_0 - \mathbf{P} \cdot \mathbf{E}(t)] | \psi_l(t) \rangle = i\hbar \, (\partial/\partial t) | \psi_l(t) \rangle, \tag{6.3}$$

where $\mathbf{P}$ is the operator representing the electric dipole moment of the system.

We assume that $| \psi_l(t) \rangle$ can be written as

$$| \psi_l(t) \rangle = (| l \rangle + | \phi_l^+ \rangle \, e^{i\omega_0 t} + | \phi_l^- \rangle \, e^{-i\omega_0 t}) e^{-i\varepsilon_l t/\hbar}, \tag{6.4}$$

where the first term on the right-hand side of Eq. (6.4) obviously satisfies the equation

$$\mathscr{H}_0 | l \rangle \, e^{-i\varepsilon_l t/\hbar} = i\hbar \, (\partial/\partial t) \, (| l \rangle \, e^{-i\varepsilon_l t/\hbar}). \tag{6.5}$$

The terms associated with $| \phi_l^+ \rangle$ and $| \phi_l^- \rangle$ are due to the effect of the electric field. We shall assume that $| \phi_l^\pm \rangle$ are time independent and can be represented by a power series of the amplitude of the electric field, the lowest order being linear in the field. Substituting Eq. (6.4) into (6.3), using Eq. (6.5), and neglecting terms higher than the first order in the electric field, we obtain

$$\begin{aligned} &\mathscr{H}_0 (| \phi_l^+ \rangle \, e^{i\omega_0 t} + | \phi_l^- \rangle \, e^{-i\omega_0 t}) - (\mathbf{P} \cdot \mathbf{E}^+ \, e^{i\omega_0 t} + \mathbf{P} \cdot \mathbf{E}^- \, e^{i\omega_0 t}) | l \rangle \\ &\quad = (\varepsilon_l - \hbar\omega_0) | \phi_l^+ \rangle \, e^{i\omega_0 t} + (\varepsilon_l + \hbar\omega_0) | \phi_l^- \rangle \, e^{-i\omega_0 t}. \end{aligned} \tag{6.6}$$

Since $e^{i\omega_0 t}$ and $e^{-i\omega_0 t}$ are independent functions, Eq. (6.6) is equivalent to the following pair of equations:

$$\mathscr{H}_0 - \varepsilon_l \pm \hbar\omega_0) | \phi_l^\pm \rangle = (\mathbf{P} \cdot \mathbf{E}^\pm) | l \rangle, \tag{6.7a}$$

or

$$| \phi_l^\pm \rangle = (\mathscr{H}_0 - \varepsilon_l \pm \hbar\omega_0)^{-1} \, (\mathbf{P} \cdot \mathbf{E}^\pm) | l \rangle. \tag{6.7b}$$

Since the stationary states $|l\rangle$ form a complete set of states, we have the closure relation

$$\sum_l |l\rangle\langle l| = \sum_n |n\rangle\langle n| = I, \tag{6.8}$$

where I is a unit operator.

Substituting Eq. (6.8) into (6.7) gives

$$\begin{aligned} |\phi_l^{\pm}\rangle &= \sum_n (\mathscr{H}_0 - \varepsilon_l \pm \hbar\omega_0)^{-1} |n\rangle\langle n|(\mathbf{P}\cdot\mathbf{E}^+)|l\rangle \\ &= \sum_n |n\rangle\langle n|\mathbf{P}\cdot\mathbf{E}^+|l\rangle \Big/ \hbar(\omega_{nl} \pm \omega_0), \end{aligned} \tag{6.9}$$

where $\omega_{nl} = (\varepsilon_n - \varepsilon_l)/\hbar$, the transition frequency. Equation (6.9) is valid to the first order in the electric field, provided that the laser frequency ω_0 is not too close to the transition frequency ω_{ml}.

As previously mentioned, Raman scattering involves a frequency shift. In other words, transitions between two quantum stationary states occur in Raman scattering. The induced-transition dipole moment involved in Raman scattering causing a transition from state $|l\rangle$ to state $|m\rangle$ is given by

$$\mathbf{P}^{ml}(t) = \langle\psi_m(t)|\mathbf{P}|\psi_l(t)\rangle + \text{c.c.} \tag{6.10}$$

Substituting Eq. (6.4) into (6.10) and ignoring the part of the dipole moment responsible for the spontaneous emission, as well as neglecting terms higher than the first order in the electric field, we obtain, with the help of Eq. (6.9), the j component of the induced dipole moment describing the Raman scattering associated with a transition $m \to l$ as

$$\begin{aligned} P_j^{ml}(t) = \sum_k \{&\alpha_{jk}^{ml}(\omega_0)E_k^+\, e^{i(\omega_0+\omega_{ml})t} \\ &+ [\alpha_{jk}^{ml}(\omega_0)]^* E_k^-\, e^{-i(\omega_0+\omega_{ml})t}\}, \end{aligned} \tag{6.11}$$

where the indices j and k cover the three Cartesian components of the electric-field polarization and

$$\alpha_{jk}^{ml}(\omega) = \sum_n \left(\frac{\langle m|P_j|n\rangle\langle n|P_k|l\rangle}{\hbar(\omega_{nl} + \omega)} + \frac{\langle m|P_k|n\rangle\langle n|P_j|l\rangle}{\hbar(\omega_{nm} - \omega)} \right). \tag{6.12}$$

The quantity $\alpha_{jk}^{ml}(\omega)$ defined in Eq. (6.12) is the jk component of the Raman polarizability tensor. Note that the Raman (polarizability) tensor is frequency dependent and, in general, complex and non-hermitian. However, if the final state $|m\rangle$ is the same as the initial state $|l\rangle$, the quantity defined in Eq. (6.12) becomes Hermitian; i.e., $\alpha_{jk}^{ll}(\omega) = [\alpha_{kj}^{ll}(\omega)]^*$. Note further that for $|l\rangle = |m\rangle$, $\alpha_{jk}^{ll}(\omega) = [\alpha_{jk}^{ll}(-\omega)]^*$; this insures that the induced dipole moment for Rayleigh scattering is real. The polarizability expression for Rayleigh scattering is related to the theory of dispersion and to the double refraction of a propagating radiation field. However, we shall not further consider that subject here.

Returning to Raman scattering, we note that unlike Rayleigh scattering, Raman scattering is incoherent, because Raman polarizability contains a random-phase factor associated with the arbitrary phases of the wave functions used in forming the matrix element. Thus, Raman scattering from two independent molecular systems is not subject to the effect of interference.

In Eq. (6.11), if $\omega_{ml} < 0$, then the scattered light is shifted to lower frequency than the incident light. This corresponds to Stokes scattering. On the other hand, if $\omega_{ml} > 0$, then the scattered light is shifted to a higher frequency, and this corresponds to anti-Stokes scattering.

The intensity of the Raman-scattering light can now be obtained. This is done by substituting the induced transition dipole moment into the Maxwell equations and then solving the inhomogeneous wave equation, similar to the procedure in Chapter 2. We skip the algebra and simply write the intensity expression for the case when the incident laser light is polarized along the k direction and the intensity of the scattered light is observed in the j direction of the polarization as

$$I_{jk}^{ml}(\omega) = (I_0\omega^4/4c^4)|\alpha_{jk}^{ml}(\omega_0)|^2\,\delta(\omega - \omega_0 - \omega_{ml}), \tag{6.13}$$

where I_0 is the incident intensity, given by

$$I_0 = (c/2\pi)E_j^+E_j^-. \tag{6.14}$$

The general Raman-tensor expression given in Eq. (6.12) is not convenient to describe Raman scattering from molecules. In molecules, nuclear motion will transfer energy from the incident radiation to the nuclei and vice versa via coupling of the electronic to the nuclear motion (vibronic coupling). When the transition frequencies for electronic motion are large compared with the vibrational or rotational energy, several useful approximations introduced by Placzek (1939) can simplify Eq. (6.12) to a form suitable for the study of vibrational, rotational, and translational motions for a great number of molecular systems.

In the first approximation, the wave function of state $|l\rangle$ is represented by $\psi_l(\mathbf{x}, \zeta^a)$, which is given as

$$\psi_l(\mathbf{x}, \zeta^a) = \phi_l(\mathbf{x}|\zeta)\chi_{la}(\zeta), \tag{6.15}$$

where $\mathbf{x}$ and ζ refer to the electron and nuclear coordinates, respectively; $\phi_l(\mathbf{x}|\zeta)$ is the electronic wave function with nuclei fixed in a configuration specified by ζ; and $\chi_{la}(\zeta)$ the nuclear wave function for the nuclei situated in the a state wherein the electronic state of the molecule is l. The symbol a refers collectively to both the vibrational and the rotational states.

The approximation given by Eq. (6.15) is known as the adiabatic approximation originally introduced by Born and Oppenheimer (Born and Huang,

1962. In the adiabatic approximation, $\phi_l(\mathbf{x}|\zeta)$ is calculated by solving the Schrödinger equation for electrons in a framework of rigid nuclei; i.e., in the calculation of $\phi_l(\mathbf{x}|\zeta)$, the nuclear coordinates ζ are located as fixed parameters.

The energy associated with state $|l\rangle$ is thus equal to

$$\varepsilon_l = \varepsilon_l^{(0)} + W_{la} = \varepsilon_l(\xi), \tag{6.16}$$

where $\varepsilon_l^{(0)}$ is the eigenenergy of the electrons with nuclei fixed in the configuration ζ. W_{la} is the eigenenergy associated with the wave function $\chi_{la}(\zeta)$.

If the molecule is in the ground electronic state before and after interaction with the incident radiation, we can set the initial state $|l\rangle$ to be $|ga\rangle$ and the final state $|m\rangle$ to be $|gb\rangle$, where a and b are sets of quantum numbers describing the state of nuclear motion, and rewrite Eq. (6.12) as

$$\alpha_{jk}^{ba}(\omega) = \sum_{fc} \left\{ \frac{\langle gb|P_j|fc\rangle\langle fc|P_k|ga\rangle}{\hbar(\omega_{fg} + \omega_{ca} + \omega)} + \frac{\langle gb|P_k|fc\rangle\langle fc|P_j|ga\rangle}{\hbar(\omega_{fg} + \omega_{cb} - \omega)} \right\} \tag{6.17}$$

where $\hbar\omega_{fg}$ is the energy difference between the electronic states $|f\rangle$ and the ground electronic state $|g\rangle$. The nuclear states associated with the electronic states $|f\rangle$ are denoted by $|c\rangle$. Since the state $|g\rangle$ can be a part of $|f\rangle$, it is useful to separate Eq. (6.17) into electronic and ionic parts. The ionic part corresponds to the contribution when $|f\rangle = |g\rangle$ and is nonvanishing only when $|g\rangle$ is degenerate, because $\langle g|\mathbf{p}|g\rangle$ vanishes identically by symmetry if $|g\rangle$ is a nondegenerate electronic state.

In addition, if the incident frequency ω is far from any electronic absorption band such that $\omega_{fg} - \omega$ is large compared with the splitting or shift of the electronic level resulting from nuclear motion, then ω_{ca} and ω_{cb} in the denominator of Eq. (6.17) can be neglected. As a result, Eq. (6.17) can be simplified as

$$\alpha_{jk}^{ba} = \langle b|\alpha_{jk}(\zeta)|a\rangle, \tag{6.18}$$

where $\alpha_{jk}(\zeta)$ is considered to be an operator that connects nuclear states $|a\rangle$ and $|b\rangle$ and is given by

$$\alpha_{jk}(\zeta) = \sum_f \left[\frac{\langle g|P_j|f\rangle\langle f|P_k|g\rangle}{\hbar(\omega_{fg} + \omega)} + \frac{\langle g|P_k|f\rangle\langle f|P_j|g\rangle}{\hbar(\omega_{fg} - \omega)} \right]. \tag{6.19}$$

Note that under the Placzek approximations, the Raman tensor operator as given by Eq. (6.19) is Hermitian; it depends on the nuclear coordinates as well as the excitation frequency, despite the fact that the frequency dependence is not written out explicitly in Eq. (6.19). The electronic wave functions involved are implicitly absorbed in the $\alpha_{jk}(\zeta)$ operator. When $\omega < \omega_{fg}$, which is the case for nonresonant Raman-scattering experiments, frequency dependence can be neglected. In this case, the Raman tensor becomes symmetric. In short, for a molecular system with a nondegenerate electronic ground state and far from resonance, the Raman tensor is real and symmetric. The various approximations that lead to Eqs. (6.18) and (6.19) are the essential points in the Placzek theory. Because of the complexity of calculating the wave functions of the excited electronic states, the Placzek theory is of great importance, since the theory removes the necessity of knowing the electronic wave functions. Equation (6.18) provides the starting point for interpreting spontaneous Raman-scattering spectra of molecular systems.

6.3 HARMONIC APPROXIMATION

To describe vibrational motion, we consider a molecule consisting of n atoms oscillating about their equilibrium positions; the atoms are subject to interatomic interactions. For such a collection of atoms, the Hamiltonian $\mathscr{H}'_{\text{osc}}$ can be written as

$$\mathscr{H}'_{\text{osc}} = \sum_i \frac{p_i^2}{2m_i} + \phi, \tag{6.20}$$

where $p_i^2/2m_i$ represents the kinetic energy of the atom i and ϕ is the interatomic potential that is a function of instantaneous positions of the atom.

If the oscillation amplitudes of the atoms are small, it is a good approximation to expand ϕ in a power series of the coordinates describing the deviation from the equilibrium positions as

$$\phi = \phi^0 + \tfrac{1}{2}\sum_{i,j} f_{\alpha\beta}(i, j)\zeta_\alpha(i)\zeta_\beta(j) + \cdots, \tag{6.21}$$

where the repeated indices over α and β indicate summation over those indices, ϕ^0 is the potential energy of the atoms at the equilibrium positions, and $f_{\alpha\beta}(i, j)$ the generalized force constant given by

$$f_{\alpha\beta}(i, j) = (\partial^2\phi/\partial x_{i\alpha}\,\partial x_{j\beta})_0, \qquad \alpha, \beta = x, y, z, \tag{6.22}$$

which forms a $3n \times 3n$ matrix. The quantity $\xi_\alpha(i)$ represents the α component of the coordinate representing the deviation of the i atom from the equilibrium position and is known as the vibrational coordinate of the i atom. The linear terms in the vibrational coordinates vanish because atoms are in equilibrium positions.

The function ϕ^0 does not depend on the vibrational coordinate; thus, it is convenient to redefine a new Hamiltonian $\mathscr{H}_{\rm osc}$ given by

$$\mathscr{H}_{\rm osc} = \mathscr{H}'_{\rm osc} - \phi^0 = \sum_i p_i^2/2m_i + \tfrac{1}{2}\sum_{i,j} f_{\alpha\beta}(i,j)\xi_\alpha(i)\xi_\beta(j) + \cdots. \quad (6.23)$$

To simplify the subsequent discussion, it is convenient to introduce mass-reduced coordinates $u_\alpha(i)$ and mass-reduced generalized force constants $g_{\alpha\beta}(i, j)$ given, respectively, by

$$u_\alpha(i) = \xi_\alpha(i)(m_i)^{1/2}, \quad (6.24)$$

$$g_{\alpha\beta}(i,j) = (m_i m_j)^{-1/2} f_{\alpha\beta}(i,j) \quad (6.25)$$

and then express $\mathscr{H}_{\rm osc}$ in terms of $u_\alpha(i)$ and $g_{\alpha\beta}(i, j)$:

$$\mathscr{H}_{\rm osc} = \sum_i \tfrac{1}{2}P_i^2 + \tfrac{1}{2}\sum_{i,j} g_{\alpha\beta}(z,j)u_\alpha(i)u_\beta(j) + V_a, \quad (6.26)$$

where $P_\alpha(i) = \dot{u}_\alpha(i)$, $P_i^2 = \sum_\alpha P_\alpha^2(i)$, and V_a represents the anharmonic term, i.e., the sum of higher-order terms not written out explicitly. If V_a is neglected, then we have the harmonic approximation.

Equation (6.25) is a quadratic form in $u_\alpha(i)$, $u_\beta(j), \ldots$. It is possible to carry out a linear transformation from $\{u_\alpha(i)\}$ to a new set of coordinates q_j in such a way that $\mathscr{H}_{\rm osc}$ is expressed as a linear combination of only squares of the new coordinates, where the cross-product terms have been eliminated. This is called a canonical form; the new coordinates q_j are known as the normal coordinates.

To find the normal coordinates, consider the terms

$$F = \sum_{i,j} g_{\alpha\beta}(i,j)u_\alpha(i)u_\beta(j) = \mathbf{u}^{\rm T}\mathbf{g}\mathbf{u}, \quad (6.27)$$

where $\mathbf{u}$ represents the column matrix with $3n$ elements, $\mathbf{u}^{\rm T}$ is the transpose of $\mathbf{u}$ and $\mathbf{g}$ matrix with elements given by $g_{\alpha\beta}(ij)$.

Let the vector $\mathbf{u}$ be expressed in terms of $\mathbf{g}$ by the equation

$$\mathbf{u} = \mathbf{D}\mathbf{g}, \quad (6.28)$$

where the component of $\mathbf{g}$ is g_j, which has $3n$ elements ,and $\mathbf{D}$ is a square matrix of order $3n$ (a $3n \times 3n$ matrix). Introducing Eq. (6.28) into (6.27) then gives

$$F = (\mathbf{D}\mathbf{g})^{\rm T}\mathbf{g}(\mathbf{D}\mathbf{g}) = \mathbf{g}^{\rm T}(\mathbf{D}^{\rm T}\mathbf{g}\,\mathbf{D})\mathbf{g}. \quad (6.29)$$

We see that if F is to involve only squares of the variables g_i, then the matrix **D** must be so chosen that $\mathbf{D}^{\mathrm{T}}\,\mathbf{g}\,\mathbf{D}$ is a diagonal matrix; that is,

$$\mathbf{D}^{\mathrm{T}}\mathbf{g}\mathbf{D} = \begin{bmatrix} \lambda_1 & & & 0 \\ & \lambda_2 & & \\ & & \ddots & \\ 0 & & & \lambda_{3n} \end{bmatrix}. \tag{6.30}$$

The matrix **D** is constructed by solving the eigenvalue problem of the **g** matrix. Let $\hat{e}_i$ be the normalized eigenvector of **g** with the eigenvalue λ_i. Thus, we have the relations

$$\mathbf{g}\hat{e}_1 = \lambda_1 \hat{e}_1, \qquad \mathbf{g}\hat{e}_2 = \lambda_2 \hat{e}_2, \ldots, \mathbf{g}\hat{e}_{3n} = \lambda_{3n} \hat{e}_{3n}. \tag{6.31}$$

The matrix **D** is constructed in such a way that the elements of the unit vectors $\hat{e}_1, \hat{e}_2, \ldots, \hat{e}_{3n}$ form the elements of the successive columns of **D**:

$$\mathbf{D} = (\hat{e}_1, \hat{e}_2, \ldots, \hat{e}_{3n}). \tag{6.32}$$

Since

$$\mathbf{g}\mathbf{D} = (\lambda_1 \hat{e}_1, \lambda_2 \hat{e}_2, \ldots, \lambda_{3n} \hat{e}_{3n}) = \mathbf{D} \begin{bmatrix} \lambda_1 & & & \\ & \lambda_2 & 0 & \\ & 0 & \ddots & \\ & & & \lambda_{3n} \end{bmatrix}, \tag{6.33}$$

we have

$$\mathbf{D}^{\mathrm{T}}\mathbf{g}\mathbf{D} = \mathbf{D}^{\mathrm{T}}\mathbf{D} \begin{bmatrix} \lambda_1 & & & \\ & \lambda_2 & 0 & \\ & 0 & \ddots & \\ & & & \lambda_{3n} \end{bmatrix}, \tag{6.34}$$

$$= \begin{bmatrix} \lambda_1 & & & \\ & \lambda_2 & 0 & \\ & 0 & \ddots & \\ & & & \lambda_{3n} \end{bmatrix}. \tag{6.35}$$

The last step follows because $D^{\mathrm{T}} = D^{-1}$. Using Eq. (6.29), F is reduced to the form

$$F = \lambda_1 q_1^2 + \lambda_2 q_2^2 + \cdots + \lambda_{3n} q_{3n}^2, \tag{6.36}$$

where the numbers λ_i are the eigenvalues of **g**.

The kinetic-energy term can also be expressed as the sum of diagonal terms:

$$T = \sum_i \tfrac{1}{2} P_i^2 = \tfrac{1}{2}\mathbf{P}^{\mathrm{T}}\mathbf{P} = \tfrac{1}{2}\dot{\mathbf{u}}^{\mathrm{T}}\dot{\mathbf{u}} = \tfrac{1}{2}(\mathbf{D}\dot{\mathbf{g}})^{\mathrm{T}}(\mathbf{D}\mathbf{g})$$
$$= \tfrac{1}{2}\dot{\mathbf{g}}^{\mathrm{T}}(\mathbf{D}^{\mathrm{T}}\mathbf{D})\dot{\mathbf{g}} = \tfrac{1}{2}(\dot{\mathbf{g}}^{\mathrm{T}}\dot{\mathbf{g}}) = \tfrac{1}{2}(\dot{q}_1^2 + \dot{q}_2^2 + \cdots + \dot{q}_{3n}^2). \tag{6.37}$$

Combining Eqs. (6.36) and (6.37), we can rewrite Eq. (6.27) in terms of the normal coordinates (in the harmonic approximation) as:

$$\mathscr{H}_{\mathrm{osc}} = \sum_k (\tfrac{1}{2}\pi_k^2 + \tfrac{1}{2}\omega_k^2 q_k^2), \tag{6.38}$$

where $\lambda_k = \omega_k^2$ and $\pi_k = \dot{q}_k$. Equation (6.38) corresponds to a system of noninteracting harmonic oscillators; π_α is the momentum conjugate to q_α. They obey the commutation relations

$$[q_k, \pi_l] = i\hbar\delta_{kl}. \tag{6.39}$$

Here it is convenient to introduce annihilation and creation operators, which are denoted as a_k^- and a_k^+, given respectively by

$$a_k^+ = (1/\sqrt{2\hbar\omega_k})(\pi_k + i\omega_k q_k),$$
$$a_k^- = (1/\sqrt{2\hbar\omega_k})(\pi_k - i\omega_k q_k). \tag{6.40}$$

It is easy to verify that a_k^- and a_k^+ satisfy the commutation rules

$$[a_k^-, a_l^+] = \delta_{kl}. \tag{6.41}$$

In terms of the creation and annihilation operators, the Hamiltonian can be written as

$$\mathscr{H}_{\mathrm{osc}} = \sum_k \hbar\omega_k(a_k^+ a_k^- + \tfrac{1}{2}) = \sum_k \mathscr{H}_k. \tag{6.42}$$

Let $|n_k\rangle$ be the eigenstate of $\mathscr{H}_k$ with the eigenvalue E_k. Simple calculations show that $E_k = \hbar\omega_k(n_k + \frac{1}{2})$; i.e.,

$$\mathscr{H}_k|n_k\rangle = \hbar\omega_k(n_k + \tfrac{1}{2})|n_k\rangle. \tag{6.43}$$

It follows that $a_k^\pm|n_k\rangle$ is also an eigenstate of $\mathscr{H}_k$:

$$\mathscr{H}_k a_k^\pm|n_k\rangle = \hbar\omega_k(n_k \pm 1)\, a_k^\pm|n_k\rangle. \tag{6.44}$$

Clearly, from the foregoing,

$$a_k^-|n_k\rangle = C|n_k - 1\rangle. \tag{6.45}$$

The constant C is determined by the normalization requirement

$$\langle n_k - 1|n_k - 1\rangle = 1. \tag{6.46}$$

This implies that

$$C^2 = \langle n_k | a_k^+ a_k^- | n_k \rangle = \langle n_k | \left(\frac{1}{\hbar\omega_k} \mathscr{H}_k - \tfrac{1}{2} \right) | n_k \rangle = n_k.$$

Therefore,

$$a_k^- | n_k \rangle = \sqrt{n_k} | n_k - 1 \rangle. \tag{6.47}$$

Similarly, we have

$$a_k^+ | n_k \rangle = \sqrt{n_k + 1} | n_k + 1 \rangle. \tag{6.48}$$

The creation operator creates a state one quantum unit higher than the original state. Conversely, the annihilation operator annihilates one quantum unit. The material presented here will further the discussion of vibrational spectroscopy.

6.4 CORRELATION FUNCTIONS OF ISOTROPIC RAMAN POLARIZABILITY IN DENSE MOLECULAR FLUIDS

Raman scattering has been widely used for the study of molecular motions in the liquid state. In general, the width of a Raman spectral line is determined by both vibration- and reorientation-relaxation processes. We shall consider in this section the effects of vibrational relaxation processes on the Raman spectral line shape.

Using Eq. (6.13), the spectral density for the scattered light associated with transition of the nuclear state $|a\rangle$ to $|b\rangle$, observed in the j polarization with the incident light also polarized in the k direction, is expressed as

$$I_{jk}(\omega) = C\overline{|\langle \alpha | \alpha_{jk} | b \rangle|^2} \delta(\omega - \omega_0 - \omega_{ab}), \tag{6.49}$$

where the quantity C is a constant equal to $I_0(\omega^4/4c^4)$. For brevity of notation, we have suppressed the superscripts a, b associated with the transition of the nuclear states. The overbar on the matrix element $|\langle a | \alpha_{jk} | b \rangle|^2$ indicates the ensemble average over states $|a\rangle$ and $|b\rangle$ of various molecules; i.e.,

$$I_{jk}(\omega) = C \sum_{a,b} \rho_a |\langle a | \alpha_{jk} | b \rangle|^2 \delta(\omega - \omega_0 - \omega_{ab}), \tag{6.50}$$

where $\rho_a = e^{-\beta\hbar\omega}/Z$ is the Boltzmann factor. The ensemble average is needed for a system of identical scattering molecules present in the thermal reservoir. These molecules remain in equilibrium because of thermal agitation in fluids or lattice vibrations in solids.

We can now manipulate Eq. (6.50) in much the same manner as we did to derive $\chi''(\omega)$ given in Eq. (1.34). We introduce the Fourier transform representing $\delta(\omega - \omega_0 - \omega_{ab})$, using the expressions ($\hbar = 1$)

$$e^{-i\mathscr{H}t}|a\rangle = e^{-iE_a t}|a\rangle, \qquad e^{i\mathscr{H}t}|b\rangle = e^{iE_b t}|b\rangle, \tag{6.51}$$

eliminating the summation over b by using the closure relation, and finally obtaining

$$\hat{I}_{jk}(\omega) = \frac{1}{C} I_{jk}(\omega) = \frac{1}{2\pi}\int_{-\infty}^{\infty} dt\, e^{-i\Delta\omega t}\langle\alpha_{jk}(t)\alpha_{jk}^{+}\rangle, \tag{6.52}$$

where $\alpha_{jk}(t)$ is the Heisenberg operator given by $\alpha_{jk}(t) = e^{i\mathscr{H}t}\alpha_{jk}\, e^{-i\mathscr{H}t}$ and $\Delta\omega = \omega - \omega_0$ is the shift from the incident-laser frequency.

Equation (6.52) is the fundamental equation needed for the discussion of Raman line shape. As shown in Chapter 3 [see Eq. (3.98)], the isotropic component of the correlation function of Raman polarizability can be experimentally separated from the polarized Raman spectrum by the expression

$$G_{\text{iso}}(t) = G_{zz}(t) - \tfrac{4}{3}G_{zy}(t), \tag{6.53}$$

where $G_{jk}(t) = \langle\alpha_{jk}(t)\,\alpha_{jk}^{+}\rangle$; $G_{zz}(t)$ and $G_{zy}(t)$ are time-correlation functions that can be experimentally determined from polarized and depolarized Raman spectral density. The isotropic-polarizability time-correlation function is given by

$$G_{\text{iso}}(t) = \langle\alpha(t)\alpha(0)\rangle, \tag{6.54}$$

where $\alpha = \frac{1}{3}\,\text{Tr}\,\boldsymbol{\alpha} = \frac{1}{3}\sum_j \alpha_{jj}$, which is invariant with molecular rotation. The isotropic Raman tensor $\boldsymbol{\alpha}$ is a function of vibrational coordinates and, for scattering from a fundamental vibration, can be expanded in terms of the normal coordinate q (or the creation and annihilation operators)

$$\alpha = (\partial\alpha/\partial q)_0\, q = \mu q. \tag{6.55a}$$

Substituting Eq. (6.55a) into (6.54), we obtain

$$G_{\text{Iso}}(t) = \mu^2\langle q(t)q\rangle. \tag{6.55b}$$

Thus, the correlation function associated with the isotropic part of the spontaneous Raman-scattering process is proportional to the autocorrelation function of the normal vibrational mode.

According to the fluctuation–dissipation theorem [see Eqs. (1.150) and (1.163)], by identifying $A = B = q$ and taking the inverse Fourier transform, we obtain

$$\langle q(t)q\rangle = \frac{\hbar}{\pi}\int_{-\infty}^{\infty} dt\, e^{-i\omega t}\frac{\chi''(\omega)}{(1 - e^{-\beta\hbar\omega})\omega}, \tag{6.56}$$

where $\chi''(\omega)$ represents the imaginary part of the susceptibility associated with the vibrational mode subject to an external laser field. As will be shown in the next section, for a homogeneously broadened line the functional form for $\chi''(\omega)$ is given by

$$\chi''(\omega) \simeq \frac{2T_2^{-1}\omega}{(\omega^2 - \omega_0^2)^2 + (2\omega/T_2)^2}, \tag{6.57}$$

where T_2 denotes the dephasing time of the vibrational amplitude. This quantity will be defined exactly in the next section.

For the spectral line at the frequency $\hbar\omega \gg kT$, we can neglect the $e^{-\beta\hbar\omega}$ term in the denominator of Eq. (6.56). We next substitute Eq. (6.57) into (6.56) and integrate to obtain (for $\omega_0 \gg T_2^{-1}$)

$$\langle q(t)q \rangle \simeq (\cos \omega_0 t) \exp(-|t|/T_2). \tag{6.58}$$

Thus, the isotropic part of the spontaneous Raman-scattering spectrum from a homogeneously broadened line provides information about the dephasing time.

However, for the case of an inhomogeneously broadened Raman band, Eq. (6.58) is no longer valid, since the distribution of vibrational frequencies needs to be included in the calculation of the time-correlation function of the normal coordinate. The time-correlation function no longer decays exponentially; instead, it decays more rapidly because of the spread of vibrational frequencies. In other words, the linewidth of an inhomogeneously broadened line is larger than T_2^{-1}.

In the condensed phase, the interaction potential between molecules yields the mechanism for dephasing. The effect of intermolecular interaction on the Raman-scattering spectrum of the normal mode can be described by the interaction Hamiltonian V given by

$$V = V^0 + \tfrac{1}{2} \sum_{i,j} (V_{ij}^{(1)} q_i + V_{ij}^{(2)} q_i q_j), \tag{6.59}$$

where V^0 does not depend on the normal coordinate and q_i is the normal coordinate associated with molecule i. The interaction potential given here is a function of normal coordinates because of molecular vibration that modulates the intermolecular potential. It is of the form applicable to the intermolecular dipole–dipole interaction, quadrupole–quadrupole interaction, hydrogen-bonding interaction, and other types of interaction. The $V^{(1)}$ and $V^{(2)}$ terms provide pathways for the exchange of energy among the harmonic oscillators, as well as between the system of oscillators and the thermal bath. The $V^{(2)}$ term allows the resonant energy to transfer from one oscillator to another and is an important mechanism for the dephasing process.

However, for the $V^{(1)}$ term to be effective, the fluctuating motion of the thermal bath must be fast enough to simultaneously pick up or dissipate one unit of the oscillator energy. Since the frequency of an internal vibration is on the order of 10^{13} to 10^{14} Hz, the power density for the fluctuating motion of the thermal bath at this frequency range is very small. Nevertheless, the $V^{(1)}$ term is an important mechanism for determining the lifetime of an excited vibration state.

To illustrate the discussion presented here, we consider a system of identical molecules inside the scattering volume. The normal mode of molecule i is q_i.

The time-correlation function that yields the isotropic part of the spontaneous Raman-scattering spectrum is

$$G_{\mathrm{iso}}(t) = \mu^2 \left\langle \left[\sum_i q_i(t)\right]\left[\sum_j q_j\right]\right\rangle = \sum_i \langle q_i(t) q_i\rangle + \sum_{i \neq j} \langle q_i(t) q_j\rangle. \quad (6.60)$$

It is often assumed that Raman scattering is an incoherent process, and only the single-particle term is probed by Raman scattering. In this approximation, the two-particle pair-correlated terms (the cross terms) do not appear. However, in the presence of intermolecular interaction, the phases of vibrational wave functions of the interacting pair cannot be considered independent, and the pair-correlation effect can also contribute to the Raman-scattering spectrum of interacting molecules in liquids. We can include both self- and pair-correlation terms in the calculation of $G_{\mathrm{iso}}(t)$.

Expressing the normal coordinate in terms of the creation and annihilation operators and neglecting the $\langle a_i^+(t)\, a_j^+\rangle$ and $\langle a_i^-(t)\, a_j^-\rangle$ terms (which require that the thermal bath makes simultaneous transitions of two vibrational quanta, a very unlikely process), we obtain

$$G_{\mathrm{iso}}(t) = \frac{\mu^2 \hbar}{\omega_0} \sum_{i,j} \langle a_j^+(t)\, a_l^- + a_j^-(t)\, a_l^+\rangle = \frac{\mu^2 \hbar}{\omega_0} \sum_{jl} [g_{jl}(t) + g_{jl}^*(t)], \quad (6.61)$$

where

$$g_{jl}(t) = \langle a_j^+(t)\, a_l^-\rangle,$$
$$a_j^+(t) = e^{i\mathscr{H}t}(a_j^+)^{-i\mathscr{H}t}. \quad (6.62)$$

The total Hamiltonian $\mathscr{H}$ is the sum of $\mathscr{H}_{\mathrm{osc}}$, $\mathscr{H}_{\mathrm{L}}$, and $\mathscr{H}'$. Here, $\mathscr{H}_{\mathrm{osc}}$ represents the Hamiltonian of the system of identical noninteracting harmonic oscillators; $\mathscr{H}_{\mathrm{L}}$ represents the Hamiltonian of the lattice. (Here the word lattice is not to be confused with the crystal lattice; it simply represents the nonvibrational degrees of freedom.) Finally, $\mathscr{H}'$ is the coupling Hamiltonian, given by

$$\mathscr{H}' = V - V^0 = \tfrac{1}{2}\sum_{i,j}(V_{ij}^{(1)}q_i + V_{ij}^{(2)}q_iq_j). \tag{6.63}$$

We can derive a kinetic equation for the correlation function $g_{ij}(t)$ using the technique discussed at length in Chapter 4.

Using the dynamic variable **A** represented by the column matrix

$$\mathbf{A} = \begin{bmatrix} a_1^+ \\ a_2^+ \\ \vdots \\ a_N^+ \end{bmatrix}, \tag{6.64}$$

where i denotes the ith molecule (not to be confused with the ith normal coordinate of a molecule discussed in the previous section).

Identifying A_j to q_j and following the steps leading to Eq. (4.60), we obtain a generalized Langevin equation for the dynamic variable a_j^+ as

$$da_j^+(t)/dt = i\Omega_{jk}a_k^+(t) - \int_0^t d\tau\, K_{jk}(\tau)a_k^+(t-\tau) + F_j(t), \tag{6.65}$$

where the repeated index indicates summation over that index.

The correlation-function matrix $g_{jl}(t)$ follows the equation

$$dg_{jl}(t)/dt = i\Omega_{jk}g_{kl}(t) - \int_0^t d\tau\, K_{jk}(\tau)g_{kl}(t-\tau). \tag{6.66}$$

The frequency matrix Ω_{jk} is evaluated using the recipe given in Eq. (4.61) and found to be

$$\Omega_{jk} = \omega_0\delta_{jk} + (1/2\omega_0)\langle V_{jk}^{(2)}\rangle_{\mathrm{L}} = \omega_0\delta_{jk} + \Delta_{jk}^{(0)}, \tag{6.67}$$

where $\Delta_{jk}^{(0)}$ is the induced frequency shift owing to the $V^{(2)}$ term. The calculation of the memory matrix is more involved. Using Eq. (4.62), we have found, after considerable algebra, that

$$K_{jk}(\tau) = \sum_l\sum_{l'}\{(1/2\omega_0 n)\langle V_{jl}^{(1)}(\tau)V_{kl'}^{(1)}\rangle_{\mathrm{L}} + (1/4\omega_0^2)[\langle\delta V_{lj}^{(2)}(\tau)\, V_{l'k}^{(2)}\rangle_{\mathrm{L}}\delta_{ll'}e^{i\omega_0\tau}$$
$$+ (1 + 1/n)\langle\delta V_{lj}^{(2)}(\tau)V_{l'k}^{(2)}\rangle_{\mathrm{L}}\delta_{ll'}e^{-i\omega_0\tau}\} = \Gamma_{jk}(\tau) + i\Delta_{jk}(\tau), \tag{6.68}$$

where n is the average occupation number of oscillators with frequency ω_0, and the subscript L indicates an average over the lattice variables. The quantity $\Gamma_{jk}(\tau)$ is responsible for the decay of the polarizability-correlation function $G_{\mathrm{iso}}(t)$, and $\Delta_{jk}(\tau)$ is the dynamically induced frequency shift.

Combining Eqs. (6.66) and (6.67) with (6.65) gives

$$dg_{jl}(t)/dt = i\omega_0 g_{jl}(t) + i\Delta_{jk}^{(0)} g_{kl}(t) + i\int_0^t d\tau\, \Delta_{jk}(\tau)\, g_{kl}(t-\tau) - \int_0^t d\tau\, \Gamma_{jk}(\tau) g_{kl}(t-\tau). \quad (6.69)$$

Equation (6.69), in combination with Eqs. (6.67) and (6.68), is the basic result of the kinetic equation of the correlation function $g_{jl}(t)$. The equation contains statically and dynamically induced shifts, which appear only through the $V^{(2)}$ terms. Note that both $V^{(1)}$ and $V^{(2)}$ contribute to the damping of the correlation function. Detailed calculations using an intermolecular dipole–dipole interaction mechanism showed that the $V^{(2)}$ terms dominate the decay of $g_{jl}(t)$ (Wang, 1977). It should be useful to apply this general theory to evaluate Γ_{jk} using an explicit potential function and then to compare the result with the experimentally measured value of the dephasing time. Current research activities in analyzing the spontaneous Raman line-shape function are directed in this direction.

6.5 DEPHASING AND POPULATION RELAXATION

In the preceding section we discussed possible mechanisms that lead to dephasing. We will now further discuss relaxations of the normal mode $\langle q(t) \rangle$ and the excess vibrational population in the excited-state molecule given by $\langle n(t) \rangle$. Relaxation of the amplitude of the vibrational normal mode is known as dephasing, chracterized by a time constant T_2. The relaxation of the excess vibrational energy is characterized by another constant T_1.

Relaxation processes result from the stochastic nature of the condensed-phase system. It should be kept in mind that the description of damping behavior in terms of relaxation times T_1 and T_2 does not hold for very short time intervals $\Delta t \simeq \tau_c$, where τ_c is the correlation time associated with intermolecular interactions in the condensed phase, which is on the order of 10^{-13} sec.

In a number of experiments in which T_1 and T_2 were determined for the same normal mode involved, a considerable difference was observed for the two time constants. For instance, a ratio of $T_1/T_2 \simeq 10^{11}$ was found for the fundamental vibrational mode in liquid N_2 (Brueck and Osgood, 1976; Calaway and Ewing, 1975; Laubereau, 1974). On the other hand, the T_1/T_2 ratio for the symmetric stretching mode of CH_3 in CH_3CCl_3 is equal to 2.4 (Laubereau *et al.*, 1972). Values obtained for the dephasing time for most molecules in the condensed phase are on the order of 10^{-10} to 10^{-12} sec.

The relaxation behavior of $\langle q(t)\rangle$ and $\langle n(t)\rangle$ can be treated as follows. Consider a situation where the external electromagnetic field of the laser does not contain a resonant component at the transition frequency ω_0. Assume that the coupling of the laser radiation to the normal mode q of the molecular system leads to the Hamiltonian $\mathscr{H}'$ and the unperturbed Hamiltonian of the material system is $\mathscr{H}_0$. The time evolution of the density matrix describing an ensemble of the material system is given by ($\hbar = 1$)

$$\partial\rho(t)/\partial t = -i[\mathscr{H}, \rho(t)], \tag{6.70}$$

where

$$\mathscr{H} = \mathscr{H}_0 + \mathscr{H}'. \tag{6.71}$$

For the normal mode q connecting two nondegenerate states $|a\rangle$ and $|b\rangle$, the expection value of q is given by

$$\langle q(t)\rangle = \mathrm{Tr}[\rho(t)q] = \rho_{ab}q_{ba} + \rho_{ba}q_{ab}, \tag{6.72}$$

where the time dependence in ρ_{ab} and ρ_{ba} is not written out explicitly. Since the states $|a\rangle$ and $|b\rangle$ are nondegenerate, it is convenient to choose the phases of the eigenfunctions of these states to be real, such that $q_{ab} = q_{ba}$. Thus, Eq. (6.71) can be written as

$$\langle q(t)\rangle = q_{ab}R_1, \tag{6.73}$$

where

$$R_1 = \rho_{ab} + \rho_{ba}. \tag{6.74}$$

To describe the behavior of R_1, it is convenient to define R_2 and R_3 as

$$R_2 = i(\rho_{ab} - \rho_{ba}), \tag{6.75}$$

$$R_3 = \rho_{bb} - \rho_{aa}. \tag{6.76}$$

Note that while ρ_{ab} and ρ_{ba} are complex, R_1 and R_2 are real. The R_3 component is a measure of the excess excited-state population. Neglecting the effect of relaxation, Feynman *et al.* (1957) showed that the equations of motion for R_1, R_2, and R_3 are given by

$$\begin{aligned}\partial R_1/\partial t &= \omega_0 R_2 + iR_3(\mathscr{H}'_{ab} - \mathscr{H}'_{ba}),\\ \partial R_2/\partial t &= -\omega_0 R_1 - R_3(\mathscr{H}'_{ab} + \mathscr{H}'_{ba}),\\ \partial R_3/\partial t &= -i(\mathscr{H}'_{ab} - \mathscr{H}'_{ba})R_1 + (\mathscr{H}'_{ab} + \mathscr{H}'_{ba})R_2,\end{aligned} \tag{6.77}$$

where $(\mathscr{H}_0)_{ab} = (\mathscr{H}_0)_{ba} = 0$ and $\omega_0 = (\mathscr{H}_{bb} - \mathscr{H}_{aa})$. Note that Eqs. (6.77) are similar to the well-known spin-vector model for the magnetic dipole vector $\mathbf{R} = (R_1, R_2, R_3)$ subject to an external magnetic field given by $\mathbf{H} = [-(\mathscr{H}'_{ab} + \mathscr{H}'_{ba}), i(\mathscr{H}'_{ab} - \mathscr{H}'_{ba}), \omega_0]$ [see Eq. (5.22)].

However, Eqs. (6.77) represent an ideal situation. Any material system at finite temperature is subject to unavoidable perturbation of a stochastic nature. On the average, random stochastic perturbations will lead to relaxation effects. The physical origin of random perturbations are thermal agitation in fluids or lattice vibrations in crystals. As shown in Chapter 5, in magnetic resonance spectroscopy for times greater than the correlation time τ_c, Eqs. (6.77) can be generalized to include relaxation effects. From Eq. (5.218), we have

$$d\rho_{\alpha\alpha'}/dt = -i(\mathscr{H}, \rho)_{\alpha\alpha'} + \sum_{\beta\beta'} R_{\alpha\alpha'\,\beta\beta'}(\rho_{\beta\beta'} - \rho^0_{\beta\beta'}), \tag{6.78}$$

where $\rho^0_{\beta\beta'}$ denotes the equilibrium value of the element $\rho^0_{\beta\beta'}$. For times great compared with the correlation time τ_c, $R_{\alpha\alpha'\,\beta\beta'}$ are constant factors. The most significant terms of the tensor **R** are those associated with $\alpha - \alpha' = \beta - \beta'$.

For a two-level system, symmetry properties require that

$$R_{aa,\,bb} = R_{bb,\,aa} = -T_1^{-1}, \qquad R_{ab,\,ab} = R_{ba,\,ba} = -T_2^{-1}. \tag{6.79}$$

Thus, if the relaxation effects are incorporated, then instead of Eq. (6.76) we have the equations

$$\begin{aligned}
\partial R_1/\partial t &= \omega_0 R_2 + i(\mathscr{H}'_{ab} - \mathscr{H}'_{ba})R_3 - R_1/T_2, \\
\partial R_2/\partial t &= -\omega_0 R_1 - (\mathscr{H}'_{ab} + \mathscr{H}'_{ba})R_3 - R_2/T_2, \\
\partial R_3/\partial t &= -i(\mathscr{H}'_{ab} - \mathscr{H}'_{ba})R_1 + (\mathscr{H}'_{ab} + \mathscr{H}'_{ba})R_2 - (R_3 - \overline{R}_3)/T_1.
\end{aligned} \tag{6.80}$$

This set of equations is very useful for deriving the kinetic equations associated with relaxation of the normal mode $\langle q(t)\rangle$ and the excess excited-state population $\langle n(t)\rangle$.

Consider that the interaction Hamiltonian $\mathscr{H}'$ is caused by coupling of the electromagnetic field with the Raman polarizability. Then, $\mathscr{H}'$ is given by

$$\mathscr{H}' = -\sum_{i,j} (\partial\alpha_{ij}/\partial q)_0 E_i E_j q. \tag{6.81}$$

Hence,

$$\mathscr{H}'_{ab} = \mathscr{H}'_{ba} = -\sum_{i,j} (\partial\alpha_{ij}/\partial q)_0 E_i E_j q_{ab} = \tfrac{1}{2} F q_{ab}. \tag{6.82}$$

Substituting Eq. (6.82) into (6.80), we obtain

$$\begin{aligned}\partial R_1/\partial t &= \omega_0 R_2 - R_1/T_2,\\ \partial R_2/\partial t &= -\omega_0 R_1 - Fq_{ab}R_3 - R_2/T_2,\\ \partial R_3/\partial t &= Fq_{ab}R_2 - (R_3 - R_3)/T_1.\end{aligned} \tag{6.83}$$

Recalling that $\langle q(t)\rangle = q_{ab}R_1$ and defining $\langle n(t)\rangle = R_3 N$, N being the total number of molecules, we obtain from Eq. (6.83) two coupled equations for $\langle q(t)\rangle$ and $\langle n(t)\rangle$ by eliminating R_2 from Eq. (6.83) as

$$[\partial^2/\partial t^2 + 2T_2^{-1}(\partial/\partial t) + (\omega_0^2 + T_2^{-2})]\langle q(t)\rangle = (-\omega_0 Fq_{ab}^2/N)\langle n(t)\rangle, \tag{6.84}$$

$$(\partial/\partial t + T_1^{-1})(\langle n(t)\rangle - n^0) = (NFq_{ab}/\omega_0)[(\partial/\partial t)\langle q(t)\rangle + T_2^{-1}\langle q(t)\rangle]. \tag{6.85}$$

Solving the coupled Eqs. (6.84) and (6.85), we can show that the vibrational amplitute $\langle q(t)\rangle$ and the excess excited-state population $\langle n(t)\rangle$ decay exponentially with the time constants T_2 and T_1, respectively, after the excitation process has terminated.

Using Eq. (1.97) and setting $A = B = \Delta B = q$ and considering the right-hand side of Eq. (6.84) to be the generalized force $f(t)$, we can write

$$\langle q(t)\rangle = \int_0^t dt'\, \phi_{qq}(t - t')f(t'), \tag{6.86}$$

where

$$\phi_{qq}(t) = (i/\hbar)\langle[q(t), q]\rangle. \tag{6.87}$$

Recalling that $\chi(\omega)$ is related to $\phi_{qq}(t)$ by [Eq. (1.111)]

$$\chi(\omega) = \int_0^\infty dt\, e^{-i\omega t}\phi_{qq}(t), \tag{6.88}$$

where $\omega_0 \gg T_2^{-1}$, and also approximating $(\omega_0 Fq_{ab}/N)\ \langle n(t)\rangle$ to $f(t)$, we obtain from Eq. (6.84)

$$\chi(\omega) = \frac{\langle q(\omega)\rangle}{f(\omega)} = \frac{1}{(\omega^2 - \omega_0^2) + i2\omega/T_2} = \chi'(\omega) - i\chi''(\omega). \tag{6.89}$$

Therefore

$$\chi''(\omega) = \frac{2T_2^{-1}\omega}{(\omega^2 - \omega_0^2)^2 + (2\omega/T_2)^2}. \tag{6.90}$$

Equation (6.90) was previously introduced in Eq. (6.57).

6.6 RAMAN AND BRILLOUIN SCATTERING FROM SOFT MODES ASSOCIATED WITH PHASE TRANSITIONS IN SOLIDS

We shall next consider Raman and Brillouin spectroscopy of soft modes associated with phase transitions in solids. Light is scattered by fluctuations in the dielectric tensor of a material. Fluctuations of certain physical quantities are coupled to fluctuations in the dielectric tensor. Near phase transitions, the fluctuations of some of these quantities are enhanced. This makes light scattering an important tool for the study of phenomena of phase transitions.

Raman and Brillouin scattering techniques have proved very useful to study the effects of normal modes of vibrations in crystals that undergo phase transitions. There have been many experimental investigations [e.g., as reviewed by Scott (1974)] of the temperature dependence of spectra as phase transition is approached from above or below. Many of these experimental results for structural phase transitions have been interpreted in terms of the dynamic behavior of the quasi-harmonic soft modes, which display critical behavior near the phase-transition temperature.

We are interested in the scattering of light from solids in the vicinity of phase transitions. Consider that scattered light polarized in the direction $\hat{e}_s$ at frequency ω_s and at wave vector $\mathbf{K}_s$ propagating in the medium is induced by incident light polarized in the direction $\hat{e}_i$ at frequency ω_i and at wave vector $\mathbf{K}_i$. The scattered light with $\omega_s \neq \omega_i$ and $\mathbf{K}_s \neq \mathbf{K}_i$ is induced because of temporal and spatial fluctuations of the dielectric tensor of the medium. Instead of discussing polarizability, it is just as convenient to consider the dielectric tensor and to express the spatial and temporal dependence of the dielectric constant tensor $\varepsilon_{ij}(\mathbf{r}, t)$ as

$$\varepsilon_{ij}(\mathbf{r}, t) = \hat{e}_i \cdot \boldsymbol{\varepsilon}(\mathbf{r}, t) \cdot \hat{e}_j = \varepsilon_{ij}^0 + \delta\varepsilon_{ij}(\mathbf{r}, t), \tag{6.91}$$

where ε_{ij}^0 is the homogeneous part that is time and space independent and $\delta\varepsilon_{ij}(\mathbf{r}, t)$, the inhomogeneous part, is responsible for the light-scattering spectrum. Near the phase transition, it is assumed that fluctuations of some physical quantity known as the order parameter make the dominant contribution to light-scattering intensity.

It is useful to decompose $\delta\varepsilon_{ij}(\mathbf{r}, t)$ in terms of Fourier components in frequency Ω and in wave vector $\mathbf{q}$ according to

$$\begin{aligned}\delta\varepsilon_{ij}(\mathbf{q}, \Omega) &= \frac{1}{2\pi V} \iint d^3r\, dt\, \delta\varepsilon_{ij}(\mathbf{r}, t)\, e^{i\mathbf{q}\cdot\mathbf{r} - i\Omega t} \\ &= \frac{1}{2\pi} \int_{-\infty}^{\infty} dt\, \delta\varepsilon_{ij}(\mathbf{q}, t)\, e^{-i\Omega t}\end{aligned} \tag{6.92}$$

where V is the scattering volume, which is microscopically large (compared with the range of the spatial fluctuation).

The conservations of energy and momentum in the scattering process require that

$$\Omega = \omega_i - \omega_s, \tag{6.93}$$

$$\mathbf{q} = \mathbf{K}_i - \mathbf{K}_s. \tag{6.94}$$

For scattering with a small change of frequency such that

$$|\Omega| = |\omega_i - \omega_s| \ll \omega_i \simeq \omega_s,$$

it follows from Eq. (6.94) that the wavelength Λ of the fluctuation in the medium as selected by the scattering process is given by

$$\Lambda = 2\pi/|\mathbf{q}| = \lambda_0(n_i^2 + n_s^2 - 2n_i n_s \cos\theta)^{-1/2}, \tag{6.95}$$

where n_i and n_s are the refractive indexes of the scattering medium associated with the incident and the scattered light, respectively; θ is the scattering angle and λ_0 the wavelength of the incident light in a vacuum.

The spectral power density of the scattered light is obtained from Eq. (6.92) by [see Eq. (1.50)]

$$I_{ij}(\mathbf{q}, \Omega) \propto |\delta\varepsilon_{ij}(\mathbf{q}, \Omega)|^2 \propto \frac{1}{2\pi}\int_{-\infty}^{\infty} dt\, e^{-i\Omega t}\langle\delta\varepsilon_{ij}(\mathbf{q}, t)\, \delta\varepsilon_j^*(\mathbf{q}, 0)\rangle$$

$$\propto [n(\Omega) + 1]\, \mathrm{Im}\, \chi_{ij}^{(\varepsilon)}(\mathbf{q}, \Omega), \tag{6.96}$$

where the second identity is the result of the Wiener–Khintchine theorem and the angular brackets represent the ensemble average. The third identity is the fluctuation–dissipation theorem, $\chi_{ij}^{(\varepsilon)}$ the generalized susceptibility, and, $n(\Omega)$ the Bose–Einstein occupation number at Ω, given by

$$[\exp(\hbar\Omega/kT) - 1]^{-1}.$$

Equation (6.96) shows that the spectral power density of the scattered light is proportional to the Fourier transform of the time-correlation function of fluctuation of the dielectric tensor $\delta\varepsilon_{ij}(\mathbf{q}, t)$. Thus, the mechanism underlying the phase transition in solids is contained in the time-correlation function $\langle\delta\varepsilon_{ij}(\mathbf{q}, t)\, \delta\varepsilon_{ij}^*(\mathbf{q})\rangle$.

If the time-correlation function at phase transition is dominated by a physical quality η, known as the order parameter, then we can approximately characterize the order parameter by two quantities—a length ξ_c and a time τ_c. These two quantities are the coherence length ξ_c and the correlation time τ_c associated with the order parameter. The spectral line that follows the dynamic behavior of the order parameter is known as the soft mode. As previously mentioned, the light-scattering spectral density is affected by phase transition because, near the phase transition, $\delta\varepsilon_{ij}$ is dominated by the static and dynamic behavior of the soft mode. There may be several soft modes for the phase transition, and in that case the order parameter will become multidimensional. However, for simplicity, we assume here that η is one-dimensional.

An approximate equation of motion for the order parameter η is given by

$$m\ddot{\eta} + \gamma\dot{\eta} + \partial\phi/\partial\eta = f(t), \tag{6.97}$$

where m is the effective mass of the soft mode and γ the damping constant. Both m and γ are assumed to be insensitive to temperature. The function $f(t)$ is the sum of the random force and the external force conjugate to n, and ϕ is the thermodynamic potential that can be expanded in a power series of η and the gradient of η according to the Landau theory (Landau and Lifshitz, 1970). For small η, it can be shown that $\partial\phi/\partial\eta = (\partial^2\phi/\partial\eta^2)_0\,\eta = A\eta$, where the coefficient A is proportional to $|T - T_c|$, according to the Landau mean-field theory.

Solving Eq. (6.97) yields the susceptibility $\chi^{(\eta)}$ at frequency Ω as

$$\chi^{(\eta)} = f(\Omega)/\eta(\Omega) = (\Omega_0^2 - i\Omega\Gamma)^{-1}, \tag{6.98}$$

where $\Omega_0^2 = A/m$ and $\Gamma = \gamma/m$; $f(\Omega)$ and $\eta(\Omega)$ are the Fourier components of $f(t)$ and $\eta(t)$, respectively, at frequency Ω.

Since $\delta\varepsilon_{ij} = (\partial\varepsilon_{ij}/\partial\eta)_0\,\eta \equiv e_{ij}\eta$, it follows that $\chi_{ij}^{(\varepsilon)} = |e_{ij}|^2\chi^{(\eta)}$. Substituting Eq. (6.98) into (6.95), we obtain an expression for the spectral power density of the scattered light as

$$\begin{aligned} I_{ij}(\mathbf{q}, \Omega) &\propto [n(\Omega) + 1]|e_{ij}\eta|^2\,\mathrm{Im}(\Omega_0^2 - \Omega^2 - i\Omega\Gamma)^{-1} \\ &\propto [n(\Omega) + 1]\frac{P^2\Omega\Gamma}{(\Omega_0^2 - \Omega^2)^2 + \Omega^2\Gamma^2}. \end{aligned} \tag{6.99}$$

where $P = e_{ij}\eta$. If the soft mode is not Raman active in the high-temperature phase, simple one-phonon theory predicts that $P^2 = b^2(T_c - T)$, where b is a constant.

Equation (6.99) shows that the spectrum of the scattered light displays two maxima at the frequencies $\Omega = \pm(\Omega_0^2 - \Gamma^2/2)^{1/2}$ and a minimum at $\Omega = 0$. Since $\Omega_0^2 = A/m = a|T - T_c|$, on approaching the phase transition the peaks draw nearer to each other and depend on temperature as $|T - T_c|^{1/2}$ in the temperature range when $\Omega_0 > \Gamma$. When $\Omega_0^2 < \Gamma^2/2$, the two side peaks merge into a single peak centered at $\Omega = 0$. The spectra predicted by Eq. (6.99) are plotted in Fig. 6.1 for several values of the parameter $y_0 = \Omega_0/\Gamma$.

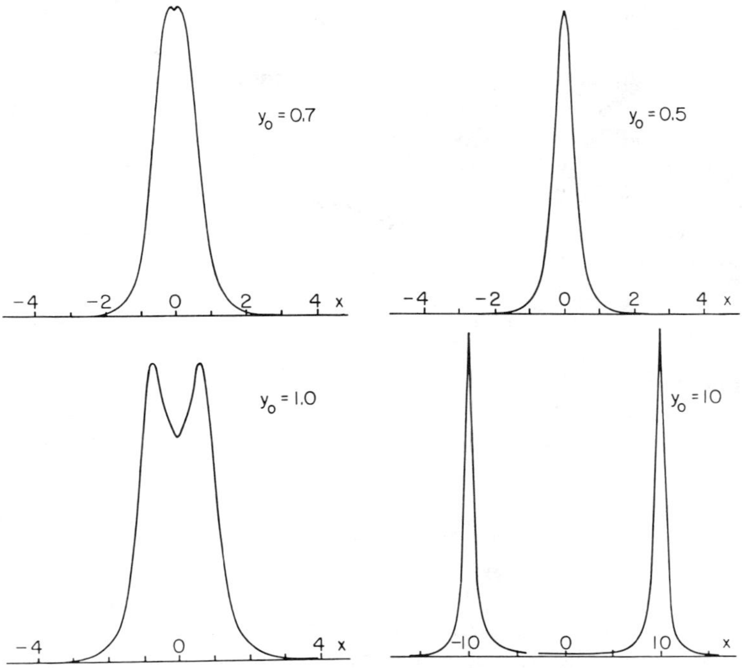

FIG. 6.1 Spectra calculated from Eq. (6.99) for several values of y_0 ($=\Omega_0/\Gamma$). The horizontal axis x is scaled in terms of Γ ($x = \Omega/\Gamma$). Temperature is equal to 500 Γ/k, k being the Boltzmann constant. Variation of temperature affects the height but not the shape of the spectra.

When the effective mass $m = 0$, which is appropriate for the order–disorder phase transition, Eq. (6.97) reduces to a Langevin equation. Solving the Langevin equation and substituting the result into Eq. (6.96) gives

$$I_{ij}(\mathbf{q}, \Omega) \propto P^2\tau/[1 + (\Omega\tau)^2], \tag{6.100}$$

where $\tau = \gamma/A = \gamma/a|T - T_c|$ is the relaxation time that diverges as the phase-transition point is approached. This phenomenon is known as critical slow down.

To obtain Eq. (6.100), it is assumed that the random force $f(t)$ does not correlate with the order parameter and that the correlation time of the random force is very short. Equation (6.100) predicts that the soft mode for transitions of the order–disorder type is of a purely relaxational character. On approaching the phase-transition point, the linewidth increasingly narrows. Because of the contribution from the elastic peak originating from defects that are difficult to remove, Eq. (6.100) has not been used as extensively as Eq. (6.99) for the quasi-harmonic soft mode in interpreting experimental results.

For the underdamped soft mode, the expression given in Eq. (6.99) is frequently used to fit experimental spectra. The fit gives values of the parameters P, Ω_0, and Γ at each temperature.

Several attempts have been made to provide quantitative tests of the soft-mode theory given in Eq. (6.99). One of the best candidates to test the theory is lead germanate, $Pb_5Ge_3O_{11}$ ($T_c = 450$ K). The Raman spectrum of this compound is not contaminated by Rayleigh scattering or the central peak. Careful Raman measurements were made by Hosea *et al.* (1979). Yocoby (1976) tried to fit the experimental spectrum of this compound to the single soft-mode theory. He showed that Ω_0^2 indeed decreases to zero according to $\Omega_0^2 = a(T_c - T)$ and that the linewidth is temperature independent, according to the prediction of the mean-field theory. However, a detailed comparison of the measured spectral shape with theory does not yield a quantitative agreement; it provides only a qualitative agreement with the experiment.

Although the soft-mode theory is unsatisfactory in detail, in several cases it is also found that the frequency of the soft mode Ω_0 does not decrease as $|T - T_c|^{1/2}$, which is predicted by the theory; the damping constant Γ increases rapidly and possibly even diverges as $T \to T_c$; and the experimental line shape is not accurately described by Eq. (6.99).

For molecular solids the interactions of the soft mode with other modes of motion, such as molecular reorientation or other phonons of the same branch, are responsible for the discrepancy. For lead germanate, $Pb_5Ge_3O_{11}$, the discrepancy results from failure to consider the two-phonon and one-phonon interference terms in the scattering expression. However, for a large number of systems, as long as T is not too close to T_c the experimental results are in reasonable agreement with the simple soft-mode theory.

6.7 NONLINEAR OPTICS AND COHERENT RAMAN SPECTROSCOPY

Third-order nonlinear optical spectroscopy has matured into a rich field for probing the condensed phase. As implied by the name, nonlinear optical spectroscopy derives information through the observation of resonance behavior associated with the mixing of high-intensity light sources by the nonlinear dielectric response of the material.

Generally, the nonlinear response of the material to incident light fields is written as the power series expansion of electric polarization in terms of the electric fields of input-laser radiations Bloembergen, 1965):

$$P_i = \chi_{ij}^{(1)} E_j + \chi_{ijk}^{(2)} E_j E_k + \chi_{ijkl}^{(3)} E_j E_k \, E_l + \cdots, \tag{6.101}$$

where the Einstein summation convention of the repeated index is used. The quantity $\chi^{(n)}$ is the susceptibility tensor of $n + 1$ rank that relates the input fields to the induced-polarization vector. For an isotropic medium or for systems with inversion symmetry, $\chi^{(2)}$ vanishes owing to symmetry. Thus, for liquids, amorphous solids, or glasses, the lowest-order nonlinear effect results from the third-order nonlinear term $\chi^{(3)}$, which is responsible for many new forms of nonlinear spectroscopy, including two-photon absorption, polarization spectroscopy, and numerous forms of coherent Raman spectroscopy. All materials make a nonvanishing contribution to $\chi^{(3)}$ as well as to $\chi^{(1)}$; $\chi^{(1)}$ is responsible for linear spectroscopy previously discussed. However, $\chi^{(3)}$ is a fourth-rank tensor characterized by four frequencies and four polarizations, which permits the mixing of three independent laser fields with known polarization, frequency, phase, and amplitude. Thus, we expect the study of $\chi^{(3)}$ to provide more information about the properties of materials than does spontaneous Raman spectroscopy.

In addition, since spontaneous Raman scattering is a very inefficient process (approximately one out of 10^6 photons is Raman scattered), the weak Raman signal may not be observed when the laser-induced fluorescence emission is also present. Nonlinear optical spectroscopy, however, is not handicapped by such a difficulty. This is because the various nonlinear processes associated with the $\chi^{(3)}$ term result in the generation of a coherent, well-defined Raman-type signal that has a specific frequency, propagation direction, and polarization.

In Fig. 6.2, we show a general four-wave mixing involving three quantum states. The quantum states $|a\rangle$, $|b\rangle$ are the rotation–vibrational states associated with the ground electronic state, and $|c\rangle$ is a rotation–vibration state in the first excited-electron state of a molecule. When three independent

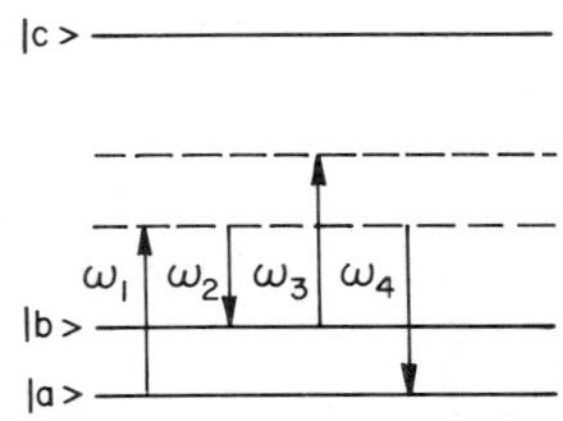

FIG. 6.2 Energy diagram for a nonlinear optical process in which two virtual states are introduced.

laser fields of frequencies ω_1, ω_2, and ω_3 are introduced to interact with the molecule, a fourth field at the frequency $\omega_4 = \omega_1 - \omega_2 + \omega_3$ will be generated by the optical mixing process associated with the $\chi^{(3)}$ term. However, when the frequency difference $\omega_1 - \omega_2$ is equal to the rotation–vibration frequency associated with the energy difference of states $|b\rangle$ and $|a\rangle$, the signal generated at ω_4 is greatly enhanced. Note that the signal at ω_4 is driven by three optical fields at ω_1, ω_2, and ω_3; the signal field is found to propagate in a specific direction $\mathbf{K}_4$, which is given by the vector relation

$$\mathbf{K}_4 = \mathbf{K}_1 - \mathbf{K}_2 + \mathbf{K}_3. \tag{6.102}$$

Equation (6.102) is the phase-matching condition for the mixing experiment involving four optical fields. To observe the signal field, the phases of the input fields must satisfy the matching condition. Although the phase-matching condition is cumbersome and may lead to loss of signal if not fulfilled, it provides a great advantage in discriminating against fluorescence and permits collection of all the signal field generated in a specific direction. Furthermore, frequency resolution of the nonlinear signal field is limited only by the inherent damping process of the medium and the intrinsic linewidths of the laser sources.

The study of nonlinear optical effects began in the early 1960s, shortly after the advent of Q-switched lasers. However, the field of coherent Raman spectroscopy made significant progress only after the development of tunable dye lasers in the early 1970s. Since then, coherent Raman spectroscopy has advanced rapidly and has found many scientific applications (Armstrong *et al.*, 1962; Bloembergen, 1965; Maker and Terhune, 1965). In the next section, we shall consider the application of the coherent Raman spectroscopic technique to the study of relaxation processes.

6.8 STIMULATED RAMAN SCATTERING

The stimulated Raman effect was observed shortly after the development of Q-switched lasers in the early 1960s (Woodbury and Ng, 1962). This phenomenon causes the gain (or loss) of a probe-laser pulse in the presence of a strong pump-laser pulse interacting via excitations (e.g., vibrational, rotational, diffusional) of the material. Maximum gains (or losses) occur when the pump and probe frequencies are separated by a characteristic frequency (e.g., vibrational) of the material. A Raman gain is produced when the probe frequency ω_2 is less than the pump frequency ω_1. A Raman loss is produced when $\omega_2 > \omega_1$. (This is known as the inverse Raman effect.)

The stimulated Raman effect provides an effective excitation process to induce the coherent and excess population of the excited vibrational state. After the passage of the first excitation pulse, a second weak probe pulse, properly delayed with respect to the first pulse, monitors the instantaneous state of the vibrationally excited system. From the time-dependent behavior of the intensity and shape of the probing pulse, information about dephasing time and lifetime can be obtained.

We shall first follow a semiclassical approach to the nonlinear optical effect and then focus on the stimulated Raman effect, treating the material system quantum mechanically and the light field classically. The semiclassical approach is accurate because the number of photons involved is large.

The Maxwell equations provide a connection between induced nonlinear polarization and the electric field. In a charge-free and current-free material system, these equations are similar to Eqs. (2.1)–(2.4), which in turn lead to the wave equation

$$\nabla^2\mathbf{E} - (1/c^2)(\partial^2/\partial t^2)(n^2\mathbf{E}) = (4\pi/c^2)(\partial^2\mathbf{P}/\partial t^2). \tag{6.103}$$

To obtain Eq. (6.103), we use $\nabla \cdot \mathbf{D} = 0$ in a charge-free region and $\nabla \cdot \mathbf{P} = 0$ in the plane-wave approximation. In this equation, n, the index of refraction of the medium, is assumed to be constant and isotropic, and the nonlinear polarization, $\mathbf{P} = \mathbf{P}^{(1)} + \mathbf{P}^{(3)}$, where $\mathbf{P}^{(1)} = \boldsymbol{\chi}^{(1)} \cdot \mathbf{E}(\omega, z, t)$. The polarization vector accounts for the coupling between the light fields and the molecular-vibration mode.

Assuming a plane-wave approximation with the wave propagating in the z-direction, we write

$$\mathbf{E}(z, t) = \tfrac{1}{2}\{\boldsymbol{\varepsilon}(\omega, z, t)\exp[i(\omega t - Kz)] + \text{c.c.}\}, \tag{6.104}$$

where $\boldsymbol{\varepsilon}(z, t)$ is slowly varying in both space and time. Thus,

$$\nabla^2\mathbf{E} = \tfrac{1}{2}[\partial^2\boldsymbol{\varepsilon}(\omega, z, t)/\partial z^2 - 2iK(\partial\boldsymbol{\varepsilon}(\omega, z, t)/\partial z - K^2\boldsymbol{\varepsilon}(\omega, z, t)]$$
$$\times e^{i(\omega t - Kz)} + \text{c.c.}, \tag{6.105}$$

$$\partial^2(n^2\mathbf{E})/\partial t^2 = n^2(\partial^2\mathbf{E}/\partial t^2 = \tfrac{1}{2}n^2[\partial^2\boldsymbol{\varepsilon}(\omega, z, t)/\partial t^2 - 2i\omega\,\partial\boldsymbol{\varepsilon}(\omega, z, t)/\partial t$$
$$- \omega^2\boldsymbol{\varepsilon}(\omega, z, t)\}\, e^{i(\omega t - Kz)} + \text{c.c.} \tag{6.106}$$

Since we assume that $\boldsymbol{\varepsilon}(\omega, z, t)$ is a slowly varying wave such that

$$|K\,\partial\boldsymbol{\varepsilon}(\omega, z, t)/\partial z \gg \partial^2\boldsymbol{\varepsilon}(\omega, z, t)/\partial z^2, \tag{6.107a}$$

$$|\omega\,\partial\boldsymbol{\varepsilon}(\omega, z, t)/\partial t| \gg \partial^2\boldsymbol{\varepsilon}(\omega, z, t)/\partial t^2, \tag{6.107b}$$

we can neglect the terms with second derivatives in both z and t and obtain

$$\partial \boldsymbol{\varepsilon}(\omega, z, t)/\partial z + (1/v)(\partial \boldsymbol{\varepsilon}(\omega, z, t)/\partial t) = -(4\pi i/c^2 k)\,(\partial^2 \mathbf{P}^{(3)}/\partial t^2\, e^{-i(\omega t - Kz)}, \tag{6.108}$$

where $v = c/n$ is the wave velocity in the medium. To obtain Eq. (6.108), we used the fact that

$$n^2(\omega) = 1 + 4\pi\chi^{(1)}(\omega). \tag{6.109}$$

We assume that $\boldsymbol{\varepsilon}(z, t)$ is time independent, that $P_i^{(3)}(\omega, z, t)$ has a sharp resonance at frequency ω, and that there is negligible depletion of the field amplitude over the interaction length so that

$$P_i^{(3)}(\omega, z, t) = \tfrac{1}{8} D\chi_{ijkl}^{(3)} \varepsilon_j(\omega_1)\varepsilon_k^*(\omega_2)\varepsilon_l(\omega_3) \times \exp\{i[(\omega_1 - \omega_2 + \omega_3)t + (K_1 - K_2 + K_3)z]\}, \tag{6.110}$$

where D is the combinational factor associated with the number of distinguishable combinations of the incident-field amplitudes. When all fields are distinguishable in frequency or polarization, $D = 6$, but $D = 3$ when the single field occurs twice in the interaction (Maker and Terhune, 1965). The term $\chi_{ijkl}^{(3)}$, a function of ω_1, ω_2, ω_3, and ω, is more appropriately written as $\chi_{ijkl}^{(3)}(-\omega, \omega_1, -\omega_2, \omega_3)$, but for brevity of notation the frequency dependence is not written out explicitly, except for situations where it is necessary.

Substituting Eq. (6.110) into (6.103), we obtain

$$\partial \varepsilon_i(\omega, z)/\partial z = (D\pi i\omega/2c)\, \chi_{ijkl}^{(3)} \varepsilon_j(\omega_1)\varepsilon_k^*(\omega_2)\varepsilon_l(\omega_3)\, e^{-i\Delta K z}, \tag{6.111}$$

where we let $\omega = \omega_1 - \omega_2 + \omega_3$. When $\Delta K \neq 0$, the z integration over length L yields

$$\varepsilon_i(\omega, z) = (D\pi i\omega L/2c)\, \chi_{ijkl}^{(3)} \varepsilon_j(\omega_1)\varepsilon_k^*(\omega_2)\varepsilon_l(\omega_3) f_{\mathrm{L}}(\Delta K), \tag{6.112}$$

where

$$f_{\mathrm{L}}(\Delta K) = \sin(\tfrac{1}{2}\Delta K L)/(\tfrac{1}{2}\Delta K L) \tag{6.113}$$

and

$$\Delta K = K_1 - K_2 + K_3 - K. \tag{6.114}$$

Since the intensity of the signal at frequency ω is proportional to $|\varepsilon_i|^2$, it is apparent that the intensity is proportional to $f_{\mathrm{L}}^2(\Delta K)$. Further, since the function $f_{\mathrm{L}}(\Delta K)$ diminishes dramatically for $\Delta K \neq 0$, Eq. (6.112) describes the coherent nature of the nonlinear process that requires a definite phase relationship between incident and generated photon fields. Equation (6.113) gives the phase-matching condition of the nonlinear interaction process and shows that if a portion of the active medium is out of phase with other portions, then destructive interference results and, consequently, no light field is emitted.

For a given value of ΔK, a coherent length L_c for the interaction can be defined when

$$f_L^2(\Delta K) = (2/\pi)^2 \tag{6.115}$$

or

$$L_c = \pi/\Delta K, \tag{6.116}$$

where L_c is an effective length. The efficiency of the nonlinear interaction increases with increasing L_c. Thus, a phase mismatch will result in a decreased correlation length.

We shall derive Eq. (6.112) by assuming that the fields propagate along the z axis. In three dimensions, the phase-matching condition $\mathbf{K} = \mathbf{K}_1 - \mathbf{K}_2 + \mathbf{K}_3$ is obtained.

We now return to the stimulated Raman effect. Woodbury and Ng (1962) first demonstrated that a ruby laser Q-switched with a nitrobenzene Kerr cell emitted a new spectral radiation at 7670 Å in addition to the normal ruby laser light at 6943 Å. This was correctly interpreted as a stimulated Raman effect (Eckardt *et al.*, 1962; Yajima and Takatsuji, 1964). Subsequent studies of this effect showed that only the frequencies belonging to the strongest spontaneous-Raman lines appear in the stimulated-Raman process. Occasionally, two additional lines appear. They belong to the totally symmetric vibration. Further increase of the pump-laser power only increases the intensity of the stimulated Stokes line and does not create other stimulated lines in the spontaneous spectrum. The Stokes intensity can build up to serve as the pump of the second Stokes lines, and so on. The effect is associated with the coupling of a light wave at the Stokes (or anti-Stokes) frequency ω_s to a vibrational wave at ω_v induced by a pump field at $\omega_L = \omega_s \pm \omega_v$. The nonlinear polarization density for stimulated Raman scattering can be either parallel or perpendicular to the pump-laser polarization. We assume that the pump field is at frequency ω_L and is polarized along the $\hat{x}$ axis and that the probe field is at frequency ω and is polarized along the $\hat{\imath}$ axis ($i = x$ or y). In this case, the third-order susceptibility for stimulated Raman scattering is $\chi_{iixx}(\omega, \omega, -\omega_L, \omega_L)$, the nonlinear polarization density at frequency ω is given by [see Eq. (6.110)]

$$P_i^{(3)}(\omega, z, t) = \tfrac{3}{4}\chi_{iixx}^{(3)}\varepsilon_i(z, \omega)|\varepsilon_x(\omega_L)|^2\, e^{i(\omega t - Kz)}, \tag{6.117}$$

and the equation for the complex probe field is given by

$$\partial\varepsilon_i(z, \omega)/\partial z = (3\pi i\omega/c)\chi_{iixx}^{(3)}\varepsilon_i(\omega)|\varepsilon_x(\omega_L)|^2, \tag{6.118}$$

which, upon integration, gives

$$\varepsilon_i(L, \omega) = \exp[(3\pi i\omega L/c)\chi^{(3)}_{iixx}|\varepsilon_x(\omega_L)|^2]\varepsilon_i(0, \omega). \tag{6.119}$$

As will be shown in a later section, $\chi^{(3)}$ contains real and imaginary parts, i.e.,

$$\chi^{(3)} = \text{Re}\,\chi^{(3)} + i\,\text{Im}\,\chi^{(3)}. \tag{6.120}$$

The intensity of the signal wave at frequency ω is given by

$$I_i(L, \omega) = I_i(0, \omega)\exp[-(6\pi\omega L/c)\,\text{Im}\,\chi^{(3)}_{iixx}|\varepsilon_x(\omega_L)|^2]. \tag{6.121}$$

Thus, depending on the sign of Im $\chi^{(3)}$, the signal wave is expected to experience exponential gain or loss. In a later section we shall show that if $\omega > \omega_L$, then Im $\chi^{(3)}$ is positive, indicating stimulated anti-Stokes loss. On the other hand, if $\omega < \omega_L$, then Im $\chi^{(3)}$ is negative, and the probe wave is amplified at the expense of the pump wave.

Stimulated-Raman gain spectroscopy restricts itself to the low-gain regime, well below the threshold for oscillation. In the weak-signal limit, the exponent in Eq. (6.121) can be expanded in a power series such that

$$I_i(L, \omega) \simeq I_i(0, \omega_s)[1 - (6\pi\omega L/c)\,\text{Im}\,\chi_{iixx}|\varepsilon_x(\omega_2)|^2]. \tag{6.122}$$

In this limit, effective signal gain or loss is proportional to the pump-laser intensity through the $|\varepsilon_x(\omega_L)|^2$ term.

As shown here, stimulated Stokes scattering by a strong pulse produces an excess population of the excited vibrational state. Subsequently, a properly time-delayed weak pulse of different frequency probes the instantaneous vibrational excitation by spontaneous anti-Stokes–Raman scattering. The scattered intensity that can be observed at a 90° scattering angle is proportional to the population of the excited vibrational state [and to the pump-laser intensity, as shown in Eq. (6.122)].

Experimentally, the time-integrated, incoherent anti-Stokes signal produced by the excess population $\langle n_i(t)\rangle$ of a vibrational state i is given by

$$S^{\text{inc}}(t_D) \simeq n\,(d\sigma_i/d\Omega)\,\Delta\Omega \int_{-L/2}^{L/2} dz \int_{-\infty}^{\infty} dt\,\langle n_i(z, t)\rangle I_P(t - t_D), \tag{6.123}$$

where n is the number density of molecules in the scattering volume, $\langle n_i(z, t)\rangle$ the excess population at time t and position z in the interaction path, $I_P(t - t_D)$ the probe-laser intensity time delayed by t_D; $(d\sigma_i/d\Omega)$, the Raman-scattering section of the observed vibrational mode i, and $\Delta\Omega$ the solid angle of collection optics in the detection system.

Time- and frequency-resolved investigations can be carried out with probing pulses of several picosecond duration. In reality, after excess population is produced in one vibrational mode by the pump-laser pulse, the excess population can find passages to other vibrational states by various energy-transfer processes. Measurement of $S_i^{\text{inc}}(t_D)$ as a function of frequency provides direct information on the instantaneous population distribution among different vibrational modes. Time-resolved data on the instantaneous population were provided by Laubereau and Kaiser (1978) for a number of molecular systems. They showed that the parameters characterizing the interaction of a probe pulse with excess population are the ratios of time constant t_P/T_2 and T_1/T_2, where t_P is the time duration of the probe pulse. Using Eq. (6.122) and the Bloch-type equations previously given, they showed that the signal curves display exponential tails. This is in agreement with the experimental results observed in several molecular liquids. The slope of the exponentially decaying tail is a direct measurement of T_1.

The evolution of the shape of the probe pulse in the stimulated Raman experiment can be investigated in a quantitative manner. Since we chose the z axis to be the propagation direction of the pump-laser field, which is assumed to be polarized in the x direction, the Stokes (or anti-Stokes) field will have x and y polarizations, unless the polarizability tensor is isotropic; in that case, the Stokes field has only the x polarization component. The total electric field in the active medium will be the sum of two fields, given by

$$\mathbf{E}(z, t) = \tfrac{1}{2}\hat{x}\{\varepsilon_L(\omega_L, z, t)\exp[i(K_L z - \omega_L t)] + \varepsilon_x(\omega, z, t)\exp[i(Kz - \omega t)]\} + \tfrac{1}{2}\hat{y}\varepsilon_y(\omega, z, t)\exp[i(Kz - \omega t)] + \text{c.c.} \quad (6.124)$$

Consistent with the notation introduced here, the wave vector and the frequency of the pump-laser field are given by K_L and ω_L, respectively, and those of the Stokes field by K and ω. Higher-order Stokes and anti-Stokes contributions are omitted in Eq. (6.124). This is justified since we consider only a small conversion of the laser to Stokes radiation.

For a semiclassical description, we assume that the nonlinear polarization $\mathbf{P}^{(3)}$ is given by the sum of the linear superposition of the induced dipole moments of all molecules in the interaction volume:

$$\mathbf{P}^{(3)}(z, t) = (\hat{x}/V)\left\{E_x(z, t)\sum_j \langle\alpha'_{xx}(j)q_j(z, t)\rangle + E_y(z, t)\sum_j \langle\alpha'_{xy}(j)q_j(z, t)\rangle\right\} + (\hat{y}/V)\left\{E_x(z, t)\sum_j \langle\alpha'_{xy}(j)q_j(z, t)\rangle + E_y(z, t)\sum_j \langle\alpha'_{yy}(j)q_j(z, t)\rangle\right\}, \quad (6.125)$$

where $\alpha'_{kl}(j)$ denotes the Raman polarizability tensor of molecule j, equal to $(\partial\alpha_{kl}^{(J)}/\partial q)$; $q_j(z, t)$ is the normal mode of molecule j and V the effective interaction volume. We assume that the vibrational normal coordinate $q_j(z, t)$ is given by

$$q_j(z, t) = (i/2)Q_j(z, t)\exp(iK_j z - i\omega_j t) + \text{c.c.}, \quad (6.126)$$

where K_j and ω_j represent the wave vector and the vibrational frequency of molecule j, respectively, and $Q_j(z, t)$ is the slowly varying amplitude.

Substituting Eqs. (6.125) into (6.108) and evaluating the $\partial^2\mathbf{P}^{(3)}(z, t)/\partial t^2$ term with the help of Eqs. (6.124) and (6.126), we obtain two equations for the two components of the Stokes field:

$$\left(\frac{\partial}{\partial z} + \frac{1}{v}\frac{\partial}{\partial t}\right)\varepsilon_x = \kappa\varepsilon_L \frac{1}{v}\sum_j \langle\alpha'_{xx}(j)Q_j^*\rangle \exp(-i\Delta\omega_j t), \tag{6.127}$$

$$\left(\frac{\partial}{\partial z} + \frac{1}{v}\frac{\partial}{\partial t}\right)\varepsilon_y = \kappa\varepsilon_L \frac{1}{v}\sum_j \langle\alpha'_{xy}(j)Q_j^*\rangle \exp(-i\Delta\omega_j t), \tag{6.128}$$

$$\kappa = \pi\omega^2/c^2k, \tag{6.129}$$

$$\Delta\omega_j = \omega_L - \omega_s - \omega_j. \tag{6.130}$$

To obtain Eqs. (6.127) and (6.128), we used the phase-matching condition $K_L - K - K_j = 0$. The Stokes field is neglected in the calculation of the right-hand sides of Eqs. (6.127) and (6.128).

Similarly, for each molecule j, Eqs. (6.84) and (6.85) are simplified to

$$(\partial/\partial t + T_2^{-1})Q_j = \kappa'\varepsilon_L[\varepsilon_x^*\alpha'_{xx}(j) + \varepsilon_y^*\alpha'_{xy}(j)], \tag{6.131}$$

$$(\partial/\partial t + T_1^{-1})\,\delta n_j = \tfrac{1}{2}N\varepsilon_L[\varepsilon_x^*\langle\alpha'_{xx}(j)Q_j^*\rangle + \varepsilon_y^*\langle\alpha'_{xy}(j)Q_j^*\rangle]\exp(-i\Delta\omega_j t) + \text{c.c.}, \tag{6.132}$$

where $\delta n_j = \langle n_j(t)\rangle - n_j$ and $\kappa' = q_{ab}^2/4V$.

It should be emphasized that several approximations are used in obtaining Eqs. (6.127), (6.128), (6.131), and (6.132): (1) the calculation is restricted to a small conversion of the pump-laser field into Stokes light and thus depletion of the pump-laser pulse is not considered; (2) second-order derivatives in Q_j (as well as in ε_x and ε_y) are neglected, and (3) in obtaining Eq. (6.131), we replace $\langle n_j\rangle$ by n_j on the right-hand side of the equation. Given an initial condition, Eq. (6.127), (6.128), (6.131), and (6.132) form a set of equations from which we can, in principle, derive the time-dependent behavior of the Stokes field, the vibrational amplitude, and the excess population. However, the actual calculation for the general case is very complex. It is therefore useful to discuss the essential features of the phenomena by considering a simplified situation in which we assume that the Raman tensor is isotropic and the Raman spectral line is homogeneously broadened. Fortunately, most molecular systems in which the stimulated process has been observed are associated with vibrational modes with very small depolarization ratios. In other words, for these systems the Raman tensor is practically isotropic. As a result of these approximations, we have $\alpha'_{xy} = 0$, $\alpha_{xx} = \alpha$, and $\Delta\omega_j = 0$; the Stokes component polarized perpendicular to the pump-laser field vanishes, and the Stokes field is not a function of molecular orientation. The subscript j labeling the molecules can be dropped, and Eqs. (6.127)–(6.132) are simplified to

$$[\partial/\partial z + (1/v\ \partial/\partial t]\varepsilon_x = \kappa\alpha n\varepsilon_L Q, \tag{6.133}$$

$$(\partial/\partial t + T_2^{-1})Q = \kappa'\alpha'\varepsilon_L\, \varepsilon_x^*, \tag{6.134}$$

$$(\partial/\partial t + T_1^{-1})\, \delta n_i = \tfrac{1}{2}N(\varepsilon_L\, \varepsilon_x^* Q^* + \varepsilon_L^*\, \varepsilon_x Q). \tag{6.135}$$

Solutions of Eqs. (6.133)–(6.135) were obtained in terms of the Bessel functions I_0 and I_1 of complex arguments. Using these results the Stokes field, the excitation amplitude, and the excess population were found to depend in a very important manner on the input shape of the probe pulse and T_2. It was found that the rise times of the excess population, the coherent vibrational excitation $\langle |Q|^2 \rangle$, and the intensity of the stimulated Stokes pulse depend on t_p (the half-power pulse width of the pump laser) but are quite independent of the t_P/T_2 ratio; the trail edge of these quantities, however, is very significantly dependent on the parameter t_P/T_2.

Although the detailed shape of the probe pulse is complicated, it is possible to collect data points along the exponential slope (the trailing edge) and to determine the dephasing time with an accuracy of $\sim 10\%$. It is found that the T_2 value obtained for liquid N_2 is considerably greater than that for other molecules. The measured T_2 values using the transient-pulse technique are also compared with the spontaneous Raman-Scattering results, whenever available. The values obtained by the two techniques are in fair agreement. However, since values obtained from the pulse technique utilize phase-matching conditions, these results are free of the effect of inhomogeneous broadening and are thus more accurate.

Many molecular liquids are composed of several isotopic species. The presence of vibrational modes with neighboring isotope components causes difficulty in determining the dephasing time by the spontaneous Raman-scattering technique, because the multiplicity of the vibrational modes owing to different isotopic species cannot be resolved by the Raman spectrometer. However, this difficulty is circumvented in the pulse technique by utilizing the phase-matching condition, which allows the signal of a particular isotopic species to be selected. When no phase matching is utilized, the presence of different isotopic species results in beats in the coherent signal. However, the envelope of the beat signal again gives information about the dephasing time.

The situation for determining dephasing times of the inhomogeneously broadened lines is similar to the case of vibrational modes with neighboring isotopic components. In this case, determining dephasing time by spontaneous Raman scattering is impossible; however, use of the transient technique carried out in a highly selective collinear K-matching geometry will allow the determination of the dephasing time. This was discussed in detail in a review article by Laubereau and Kaiser (1978).

The population lifetimes T_1 of well-defined vibrational modes were determined by measuring the time-delayed spontaneous anti-Stokes signal in accordance with Eq. (6.123) for a number of molecules. An example dealing with the totally symmetric CH_3 vibration mode at $\bar{\nu} = 2939\ \text{cm}^{-1}$ of CH_3CCl_3 in CCl_4 was investigated at several concentrations. For the neat liquid, T_1 is found to be 5.2 psec. but it increases as the CCl_4 concentration is increased. The physical processes leading to population relaxation result from inter- and intramolecular mechanisms. These mechanisms cause energy transfer and redistribution of population among various vibrational states (Laubereau and Kaiser, 1978).

Note that it is impossible to deduce the population lifetime from measurement of the spontaneous-Raman linewidth. The transient technique is the only way to determine T_1.

6.9 COHERENT ANTI-STOKES–RAMAN SPECTROSCOPY

Although stimulated Raman scattering was historically the first demonstrated coherent Raman effect, coherent anti-Stokes–Raman scattering (CARS) has attracted much greater attention in application. This technique is based on the mixing of two lasers at frequencies ω_1 and ω_2 to generate a third wave at frequency $\omega_3 = 2\omega_1 - \omega_2$. But CARS did not become a practical spectroscopic tool to probe condensed phases until tunable dye lasers were developed to enable continuous scanning of the CARS spectrum (Levenson, 1972; Levenson *et al.*, 1972; Wynne, 1972). The power of the CARS spectroscopic technique is its ability to enhance the Raman signal in the sample with a large fluorescent background. A review of CARS and numerous applications of this technique are given by Tolles *et al.* (1977) and Nibler and Knighten (1978).

Adapting Eq. (6.110) to CARS, we consider the nonlinear polarization given by

$$P_x(\omega, z)e^{i\omega t} = \tfrac{3}{4}\chi_{xxxx}(-\omega, \omega_1, -\omega_2, \omega_1)\varepsilon_x^*(\omega_1)\varepsilon_x(\omega_2)\varepsilon_x(\omega_1) \times \exp i[(2\omega_1 - \omega_2)t - (2K_1 - K_2)z], \tag{6.136}$$

where $\omega_1 > \omega_2$. This equation shows that the optical field is detected at $\omega = 2\omega_1 - \omega_2$. Generally, the technique of CARS employs two lasers with parallel polarizations at frequencies ω_1 and ω_2 (if $\omega_1 < \omega_2$, we have coherent Stokes–Raman scattering). As shown in Fig. 6.2, two virtual states are needed.

We shall now show that the third-order susceptibility is a function of the frequency difference $\omega_1 - \omega_2$. When $\omega_1 - \omega_2$ is approximately equal to ω_v, the molecular-vibration frequency (or optical-phonon frequency in a solid), the third-order susceptibility becomes large; hence, the signal field generated by the polarization at ω is enhanced.

Before providing a quantum theory for the third-order susceptibility, we shall first give a simplified description of the frequency dependence of $\chi^{(3)}$. Consider a simple two-level material system. The energy difference of the two levels is the vibrational frequency ω_v. The material system is subject to two laser fields at frequencies ω_1 and ω_2. The influence of the electronic states is included by considering that it is coupled to the fields through the Raman polarizability tensor; none of the field frequencies is in resonance with the electronic transition or the vibrational frequency. The pertinent part of the Hamiltonian describing the Raman type of interaction between matter and the fields is given by [see Eq. (6.81)]

$$\mathscr{H}'(t) = -(q/4)\alpha^{\mathrm{R}}_{jk}\varepsilon_j(\omega_1)\varepsilon^*_k(\omega_2)e^{i(\omega_1 - \omega_2)t} + \cdots, \tag{6.137}$$

where $\alpha^{\mathrm{R}}_{jk} = \partial\alpha_{jk}/\partial q$ is the polarizability tensor for spontaneous Raman scattering and q the normal mode that connects the two energy levels. Other terms not written out explicitly are the complex conjugate term and those varying with $e^{\pm i(\omega_1 + \omega_2)t}$.

We are interested in calculating nonlinear polarization owing to the normal mode q. In the presence of an electric field at frequency ω_3, the dielectric polarization density associated with the normal mode is given by

$$P^{(q)}_i(t) = N\alpha^{\mathrm{R}}_{il}\langle q(t)\rangle\varepsilon_l(\omega_3)e^{i\omega_3 t}. \tag{6.138}$$

Thus, we are interested in obtaining an expression for $\langle q(t)\rangle$. For the two-level system, the equation of motion for $\langle q(t)\rangle$ is given by (Eq. 6.84). Using the interaction Hamiltonian $\mathscr{H}'(t)$ previously given, Eq. (6.84) can be rewritten as

$$\left(\frac{\partial^2}{\partial t^2} + \frac{2}{T_2}\frac{\partial}{\partial t} + \omega_q^2\right)\langle q(t)\rangle = \frac{\bar{n}\alpha^{(\mathrm{R})}_{jk}\omega_v q^2_{ab}}{2N}\,\varepsilon_j(\omega_1)\varepsilon^*_k(\omega_2)e^{i(\omega_1 - \omega_2)t}, \tag{6.139}$$

where $\omega_q^2 = \omega_v^2 + T_2^{-2}$. Further, we replace $\langle n(t)\rangle$ by $\bar{n}$ because we are interested in the solution of $\langle q(t)\rangle$ accurate to the second order in laser fields.

We want to find the steady-state solution of Eq. (6.139). We assume

$$\langle q(t)\rangle = q(\omega_1 - \omega_2)e^{i(\omega_1 - \omega_2)t} + \text{c.c.} \tag{6.140}$$

Substituting Eq. (6.140) into (6.139), we obtain

$$q(\omega_1 - \omega_2) = \frac{\bar{n}\omega_v q_{ab}^2 \alpha_{jk}^{(R)} \varepsilon_j(\omega_1)\varepsilon_k^*(\omega_2)}{2N[\omega_q^2 - (\omega_1 - \omega_2)^2 + 2iT_2^{-1}(\omega_1 - \omega_2)]}. \tag{6.141}$$

Substituting Eqs. (6.140) and (6.141) into (6.138) gives

$$\begin{aligned} P^{(q)}(t) = \tfrac{1}{2}[&P_i(\omega_1 - \omega_2 + \omega_3)\exp[i(\omega_1 - \omega_2 + \omega_3)t] \\ &+ P_i(-\omega_1 + \omega_2 + \omega_3)\exp[i(-\omega_1 + \omega_2 + \omega_3)t] + \text{c.c.}], \end{aligned} \tag{6.142}$$

where the amplitudes of the Fourier components can be expressed in terms of the third-order nonlinear susceptibility tensor,

$$P_i(\omega_1 + \omega_2 + \omega_3) = D\chi_{ijkl}(\omega, \omega_1, -\omega_2, \omega_3)\varepsilon_j(\omega_1)\varepsilon_k^*(\omega_2)\varepsilon_l(\omega_3), \tag{6.143}$$

where D is the combination factor. The four frequencies must satisfy the relation

$$\omega = \omega_1 - \omega_2 + \omega_3. \tag{6.144}$$

The explicit form for the contribution to the nonlinear susceptibility resonant at $|\omega_1 - \omega_2| = \omega_q$ is

$$\chi_{jki}^{(q)}(\omega, \omega_1, -\omega_2, \omega_3) = \frac{\bar{n}\omega_0 q_{ab}^2 \alpha_{il}^{(R)} \alpha_{ik}^{(R)}}{[\omega_q^2 - (\omega_1 - \omega_2)^2 + 2iT_2^{-1}(\omega_1 - \omega_2)]}. \tag{6.145}$$

Similar contributions occur when $|\omega_3 - \omega_1| = \omega_q$, etc. Note that $\chi_{ijkl}^{(q)}$ is complex. When $|\omega_1 - \omega_2| = \omega_q$, $\chi_{ijkl}^{(q)}$ becomes very large and is purely imaginary. For CARS, we require $\omega_3 = \omega_1$, and two contributions occur in pairs with respect to the polarization subscript l and j:

$$\chi_{ijkl}^{(q)}(\omega, \omega_1, -\omega_2, \omega_1) = \frac{\bar{n}\omega_0 q_{ab}^2 [\alpha_{il}^{(R)}\alpha_{jk}^{(R)} + \alpha_{ij}^{(R)}\alpha_{lk}^{(R)}]}{[\omega_q^2 - (\omega_1 - \omega_2)^2 + 2iT_2^{-1}(\omega_1 - \omega_2)]}. \tag{6.146}$$

Equations (6.145) and (6.146) describe all the coherent Raman spectroscopy resonances in transparent media. The symmetry property of $\chi^{(3)}$ will be discussed in a later section.

In addition to the resonant contribution, the total nonlinear susceptibility tensor includes a nonresonant contribution resulting from virtual transitions. Thus

$$\chi_{ijkl}^{(3)} = \chi^{\text{nr}} + \chi^{\text{r}} = \chi^{\text{nr}} + \sum_q \chi^{(q)}. \tag{6.147}$$

Generally, the nonresonant term is independent of $(\omega_1 - \omega_2)$ and is a real constant.

We shall now calculate the field amplitude by substituting Eq. (6.136) into (6.111). The result is

$$\varepsilon_x(\omega, L) = (23\pi i\omega L/c)\chi^{(3)}_{xxxx}(-\omega, \omega_1, -\omega_2, \omega_1)\varepsilon_\chi^2(\omega_1)\varepsilon_\chi^*(\omega_2)f_{\mathrm{L}}(\Delta K), \quad (6.148)$$

where the phase-matching condition is

$$\Delta\mathbf{K} = 2\mathbf{K}_1 - \mathbf{K}_2 - \mathbf{K}. \quad (6.149)$$

The signal intensity $I_K(\omega)$ is given by

$$I_K(\omega) \simeq |\varepsilon_x(\omega, z)|^2 \simeq |\chi^{(3)}_{xxxx}|^2 I^2(\omega_1)I(\omega_2)f_{\mathrm{L}}^2(\Delta K). \quad (6.150)$$

Thus, the CARS signal has a quadratic dependence on laser intensity at frequency ω_1 and a linear dependence on laser intensity at ω_2. This result encourages the use of a high-peak-power laser at ω_1, which is often referred to as the pump laser. However, too much pump power is not entirely useful. Generally, 100 kW or less is sufficient for CARS experiments in liquids and solids; greater power can lead to self-focusing, stimulated Raman scattering, and damage to the sample.

The spectral shape of the CARS signal is determined by the third-order susceptibility $\chi^{(3)}$. We can write $\chi^{(3)}_{xxxx}$ as

$$\chi^{(3)}_{xxxx} = \chi^{\mathrm{nr}}_{xxxx} + \chi'_{xxxx} - i\chi''_{xxxx}, \quad (6.151)$$

where χ^{nr} is real and represents the nonresonant contribution to $\chi^{(3)}$, as pointed out earlier, and χ' and χ'' are the real and the imaginary parts, respectively, of the resonant term. Thus, four contributions are present in the CARS signal: $(\chi')^2$, $(\chi'')^2$, $(\chi^{\mathrm{nr}})^2$, and the cross (interference) term between χ' and χ^{nr} (the cross term between χ^{nr} and χ'' does not appear). Obviously, the CARS signal will have a complex line shape as ω_2 is tuned through the resonance. Furthermore, it is often found that χ^{nr} is the dominant term, and meaningful CARS signals can become difficult to obtain for materials with large nonresonant contributions.

The phase-matching angle can be calculated from Eq. (6.149). For small dispersion, we can set $\Delta\mathbf{K} = 0$. If the angle between $\mathbf{K}_1$ and $\mathbf{K}_2$ is θ (see Fig. 6.3), then

$$\cos\theta = \frac{(4K_1^2 + K_2^2 - K^2)}{4K_1K_2} = \frac{4\omega_1\omega_2 n^2 - 4\omega_1^2(n^2 - n_1^2) - \omega_2^2(n^2 - n_2^2)}{4\omega_1\omega_2 n_1 n_2}. \quad (6.152)$$

In condensed media, the coherence length is shorter than the interaction length of cross beams, and it is advantageous to fulfill the phase-matching condition. Note that if $n_1 = n_2 = n$, then we have $\theta = 0°$, i.e., a collinear

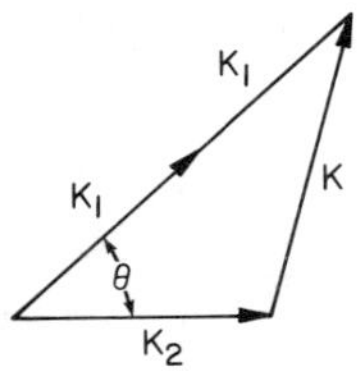

FIG. 6.3 Diagram showing the phase-matching condition.

direction. In general, θ is a very small angle ($\sim 1°$) and increases with increasing Raman shift.

In condensed phases, the nonresonant contribution to the third-order nonlinear susceptibility can be relatively large, and the CARS line shape reflects the destructive interference caused by the cross term between χ^{nr} and χ'. Various modifications of the CARS technique have been proposed to eliminate or suppress the interference term. We shall now describe two methods to accomplish this objective.

6.10 RAMAN-INDUCED KERR-EFFECT SPECTROSCOPY

A form of coherent Raman spectroscopy, known as Raman-induced Kerr-effect spectroscopy (RIKES), utilizes two lasers: a strong pump laser that can be circularly polarized at frequency ω_1 and a weak probe laser that is linearly polarized at frequency ω_2. The strong pump-laser field propagating through a medium induces an intensity-dependent birefringence that alters the polarization of the transmitted weak probe beam. The change in the polarization of the probe beam is detected as an increase in the intensity transmitted through a cross polarizer. Molecular vibration, in addition to reorientation of molecules in the medium, contribute to nonlinear susceptibility that produces this optical effect. The coherent Raman term makes a major contribution to the Kerr effect when the pump and probe frequencies differ by a vibrational frequency (Easley, 1981; Heiman *et al.*, 1976; Maker and Terhune, 1965). Consider that the two-laser field propagates in the z direction. The polarization of these fields can be written as

$$\boldsymbol{\varepsilon}(\omega_2) = \varepsilon_y(\omega_2)\hat{y} \tag{6.153}$$

and

$$\boldsymbol{\varepsilon}(\omega_1) = \varepsilon_x(\omega_1)\hat{x} + i\varepsilon_y(\omega_1)\hat{y}, \tag{6.154}$$

with

$$|\varepsilon_x(\omega_1)| = |\varepsilon_y(\omega_1)| = |\boldsymbol{\varepsilon}(\omega_1)|/\sqrt{2}.$$

We now want to detect the cross-polarized component at frequency ω; i.e., we want to find the vector $\mathbf{P}^{(3)}(z, t)$, which is given by

$$\begin{aligned}\mathbf{P}_x^{(3)}(z, t) &= \tfrac{1}{8}D\chi_{xjkl}^{(3)}(-\omega, \omega_2, -\omega_1, \omega_1)\varepsilon_j(\omega_2)\varepsilon_k(\omega_1)\varepsilon_l(\omega_1)e^{i(\omega_2 t - K_2 z)}\\ &= \tfrac{3}{8}[\chi_{xyxy}^{(3)}(-\omega, \omega_2, -\omega_1, \omega_1)\varepsilon_y(\omega_2)|\varepsilon_x(\omega_1)|^2\\ &\quad - \chi_{xyyx}^{(3)}(-\omega, \omega_2, -\omega_1, \omega_1)\varepsilon_y(\omega_2)\varepsilon_y^*(\omega_1)\varepsilon_x(\omega_1)]e^{i(\omega_2 t - K_2 z)},\end{aligned}$$

or

$$\mathbf{P}_x^{(3)}(z, t) = \tfrac{3}{16}(\chi_{xyyx}^{(3)} - \chi_{xyyx}^{(3)})\varepsilon_y(\omega_2)|\varepsilon(\omega_1)|^2 e^{i(\omega_2 t - K_2 z)}, \tag{6.155}$$

where we have used the result $\chi_{xyxx} = \chi_{xyyy} = 0$ for isotropic media. Thus, the third-order polarization in the RIKES configuration is proportional to $\chi_{xyxy}^{(3)} - \chi_{xyyx}^{(3)}$. Since the nonresonant contributions to $\chi_{xyxy}^{(3)}$ and $\chi_{xyyx}^{(3)}$ are equivalent [Kleinman symmetry (Kleinman, 1962)], the RIKES experiment removes the nonresonant background.

Note further that in RIKES the signal field is proportional to the pump intensity and is detected at the probe-laser frequency ω_2. The phase-matching condition is automatically fulfilled, regardless of the angle of overlap of the probe and pump lasers. This permits multiplexing the RIKES signals. In multiplex RIKES, the signal is produced by a broadband probe laser. Because the RIKES output is linear in the probe field, the RIKES signal from each frequency component of the broadband laser can be separated by a spectrograph and detected with an optical multichannel analyzer.

However, stress-induced birefringence in the sample and optics produces a troublesome background signal in the RIKES experiment; this affects the sensitivity of the RIKES technique. To improve sensitivity, optical heterodyne-detected RIKES (OHD-RIKES) has been introduced.

6.11 OPTICAL HETERODYNED RAMAN-INDUCED KERR EFFECT

The OHD-RIKES technique significantly reduces birefringence background signals and has demonstrated greater sensitivity than any other coherent Raman technique. One form of OHD-RIKES uses a pulsed or modulated pump laser and a stable continuous wave (CW) laser as a probe. An optical local-oscillator field $\boldsymbol{\varepsilon}_{\mathrm{LO}}$ polarized parallel to the RIKES field $\boldsymbol{\varepsilon}_{\mathrm{R}}$ is purposely injected at the detector. The net intensity at the detector is

$$I_{\mathrm{D}} \simeq |\boldsymbol{\varepsilon}_{\mathrm{LO}} + \boldsymbol{\varepsilon}_{\mathrm{R}}|^2 = |\varepsilon_{\mathrm{LO}}|^2 + |\varepsilon_{\mathrm{R}}|^2 + 2\,\mathrm{Re}|\boldsymbol{\varepsilon}_{\mathrm{LO}}^* \cdot \boldsymbol{\varepsilon}_{\mathrm{R}}|, \tag{6.156}$$

which is detected photoelectrically.

Since $\varepsilon_{\mathrm{LO}} \gg \varepsilon_{\mathrm{R}}$, the heterodyne cross-term

$$I_{\mathrm{H}} \simeq \mathrm{Re}|\boldsymbol{\varepsilon}_{\mathrm{LO}}^* \cdot \boldsymbol{\varepsilon}_{\mathrm{R}}| \tag{6.157}$$

makes a dominant ac contribution to the electrical signal; it can be detected by a boxcar integrator and then plotted as a function of the pump-laser frequency.

If the local-oscillator field at the detector is in phase with the probe field

$$\varepsilon_{\mathrm{LO}}(\omega_2) = \eta\varepsilon_y(\omega_2), \tag{6.158}$$

where $\eta \ll 1$ is the portion of the probe field injected into the detector, then the heterodyne signal becomes

$$I_{\mathrm{H}} \simeq \mathrm{Re}[\chi_{\mathrm{eff}}(-\omega_2, \omega_2, -\omega_1, \omega_1)]I(\omega_1)I(\omega_2), \tag{6.159}$$

where, according to Eq. (6.155),

$$\chi_{\mathrm{eff}} = \chi_{xyxy}^{(3)} - \chi_{xyyx}^{(3)}. \tag{6.160}$$

On the other hand, if the local-oscillator field is in phase quadrature with the probe field,

$$\varepsilon_{\mathrm{LO}}(\omega_2) = i\eta\varepsilon_y(\omega_2),$$

then

$$I_{\mathrm{H}} \simeq \mathrm{Im}[\chi_{\mathrm{eff}}(-\omega_2, \omega_2, -\omega_1, \omega_1)]I(\omega_1)I(\omega_2). \tag{6.161}$$

Equation (6.159) relates the dispersion line shape of the Raman mode to the heterodyne RIKES signal. Conversely, Eq. (6.161) relates the signal to the resonance line shape. As will be shown in the next section, χ_{eff} is related to the spontaneous Raman cross section, thus, for an imaginary local oscillator, the heterodyned RIKES spectrum reproduces the spontaneous Raman-scattering spectrum with a considerably higher sensitivity, resolution, and fluorescence rejection. Note also that the OHD-RIKES signal varies as the product of the pump-laser intensity and the probe-laser intensity and is thus better suited than CARS for line-shape study using stable CW lasers.

The OHD-RIKES technique has been demonstrated to be one of the most applicable methods of coherent Raman spectroscopy. The use of this technique to study the dephasing mechanism in solids in which optical static-stain birefringences are present was demonstrated by Levenson and Song (1980). The application of this technique in future research in the dynamics of the condensed phase is very promising.

6.12 QUANTUM THEORY OF NONLINEAR SUSCEPTIBILITY

There have been several excellent quantum-mechanical formulations of the nonlinear susceptibility tensor (Armstrong *et al.*, 1962; Bloembergen and Shen, 1964; Friedman and Wilson-Gordon, 1978a, b; Orr and Ward, 1971; Ward, 1965). Of course, since the electronic wave functions introduced in the formulation cannot be evaluated, the exact quantum formulation is useful only for exploring symmetry properties and for correlating the susceptibility coefficients with each other and with parameters describing other effects. Following a presentation similar to that in the excellent review article by Hellwarth (1977), we shall first provide a formal formulation of nonlinear susceptibility in the electric-dipole approximation and then introduce models and approximations to obtain simplified results.

We shall consider a system of N interacting molecules subject to an external time-dependent optical field $\mathbf{E}(t)$. We assume it is possible to choose a subvolume V in the medium in such a way that within V there are many molecules, so that we may describe the molecular property statistically; yet the dimension of V is small compared with the wavelength of the optical wave, so that all the molecules in V experience nearly the same field amplitude.

Before the application of the external optical field, the Hamiltonian for the system is given by $\mathscr{H}$. The external field couples to the electric dipole moment operator $\mathbf{u}$ by $\mathscr{H}_{\text{coh}}(t) = -\mathbf{u} \cdot \mathbf{E}(t)$, where $\mathbf{u} = \sum_\alpha e_\alpha \mathbf{r}_\alpha$, e_α and $\mathbf{r}_\alpha$ being the charge and position, respectively, of particle α (electron and nucleus).

The total polarization density (linear and nonlinear) $\mathbf{P}$ that acts as the source term in the Maxwell equation can be calculated according to

$$\mathbf{P}(t) = (N/V)\,\mathrm{Tr}[\boldsymbol{\mu}\rho(t)], \tag{6.162}$$

where N is the number of molecules in subvolume V.

The evolution of the density operator $\rho(t)$ follows the equation of motion given by ($\hbar = 1$)

$$\partial\rho(t)/\partial t = -i[\mathscr{H} + \mathscr{H}_{\text{coh}}(t), \rho(t)] \tag{6.163}$$

or

$$\partial\rho(t)/\partial t = -i[\mathscr{L} + \mathscr{L}_1(t)]\rho(t), \tag{6.164}$$

following the notation of Chapter 1 [Eq. (1.82)].

It is useful to transform Eq. (6.164) into the interaction representation

$$\rho(t) = e^{-i\mathscr{L}t}\rho^*(t) = e^{-i\mathscr{H}t}\rho^*(t)e^{i\mathscr{H}t}. \tag{6.165}$$

The equation for $\rho^*(t)$ is obtained from Eq. (6.164) as

$$\partial\rho^*(t)/\partial t = -i[\mathscr{H}^*_{\text{coh}}(t), \rho^*(t)] = -i\mathscr{L}_1^*(t)\rho^*(t), \tag{6.166}$$

where

$$\mathscr{H}^*_{\text{coh}}(t) = e^{i\mathscr{H}t}\mathscr{H}_{\text{coh}}(t)e^{-i\mathscr{H}t}. \tag{6.167}$$

Converting Eq. (6.166) into an integral equation

$$\rho^*(t) = \rho(0) - i\int_0^t dt'\,[\mathscr{H}^*_{\text{coh}}(t'), \rho^*(t')] = \rho(0) - i\int_0^t dt_1\,\mathscr{L}_1^*(t_1)\rho^*(t'). \tag{6.168}$$

This integral equation can be solved by iteration:

$$\begin{aligned}\rho^*(t) = \rho(0) &+ (-i)\int_0^t dt_1\,\mathscr{L}_1^*(t_1)\rho(0)\\ &+ (-i)^2\int_0^t dt_1\int_0^{t_1} dt_2\,\mathscr{L}_1^*(t_1)\mathscr{L}_1^*(t_2)\rho(0)\\ &+ (-i)^3\int_0^t dt_1\int_0^{t_1} dt_2\int_0^{t_2} dt_3\,\mathscr{L}_1^*(t_1)\mathscr{L}_1^*(t_2)\mathscr{L}_1^*(t_3)\rho(0) + \cdots.\end{aligned} \tag{6.169}$$

Transforming $\rho^*(t)$ back to $\rho(t)$ and then taking the trace with respect to μ_i, we obtain

$$\begin{aligned}\langle\mu_i(t)\rangle &= \text{Tr}[\mu_i\rho(t)]\\ &= \langle\mu_i\rangle_0 + i\int_0^t dt_1\,\langle[\mu_i(t-t_1), \mu_j]\rangle E_j(t_1)\\ &\quad + (i)^2\int_0^t dt_1\int_0^{t_1} dt_2\,\langle[[\mu_i(t-t_2), \mu_j(t_1-t_2)], \mu_k]\rangle E_j(t_1)E_k(t_2)\\ &\quad + (i)^3\int_0^t dt_1\int_0^{t_1} dt_2\int_0^{t_2} dt_3\\ &\quad \times \langle[[[\mu_i(t-t_3), \mu_j(t_1-t_3)], \mu_k(t_2-t_3)], \mu_1]\rangle\\ &\quad \times E_j(t_1)E_k(t_2)E_1(t_3),\end{aligned} \tag{6.170}$$

where $\langle A\rangle = \text{Tr}[\rho(0)A]$ and $\langle\mu_i\rangle_0$ is the equilibrium average value of μ_i.

We can define the $(n+1)$th-order response function as

$$\phi^{(n+1)}_{a_0a_i\cdots a_n}(t, t_1, \ldots, t_n) = \langle[\ldots, [\mu_{a_0}(t), \mu_{a_1}(t_1)], [\mu_{a_2}(t_2)], \ldots, \mu_{a_n}(t_n)]\rangle. \tag{6.171}$$

Note that the function $\phi^{(n+1)}_{a_0 a_1 \cdots a_n}$ is defined in terms of the multitime commutator and is invariant to time translation (meaning $\phi^{(n+1)}$ are functions only of differences among their time arguments).

Let

$$E_i(t) = 1/2\pi \int_{-\infty}^{\infty} d\omega_\alpha\, e^{+i\omega_\alpha t} E_i(\omega_\alpha); \tag{6.172}$$

then

$$P_i^{(n)}(t) = \int_{-\infty}^{\infty} d\omega_1 \cdots \int_{-\infty}^{\infty} d\omega_n\, \chi^{(n+1)}_{ia_1 \cdots a_n}(-\omega, \omega_1, \ldots, \omega_n) \times E_{a_1}(\omega_1) \cdots E_{a_n}(\omega_n) e^{i\omega t}, \tag{6.173}$$

where $\omega = -\sum_{\alpha=1}^{n} \omega_a$. The kernels $\chi^{(n+1)}_{ijk\cdots}$ $(-\omega, \omega_1, \ldots, \omega_n)$ are the Fourier transforms of $\phi^{(n+1)}_{ijk\cdots}$ and are known as the nonlinear susceptibility-tensor coefficients.

Since $P_i^{(n)}(t)$ is invariant to the permutation of the electric fields, the nonlinear susceptibility coefficient $\chi^{(n+1)}_{ia_1 \cdots a_n}(-\omega, \omega_1, \ldots, \omega_n)$ is unchanged if (a_j, ω_j) is exchanged with (a_k, ω_k). This is the general symmetry property of susceptibility.

We shall now examine the approximations that were introduced in the calculation of nonlinear susceptibilities.

The Steady-State Approximation

The steady-state approximation was used extensively by Bloembergen and Shen (1964). It begins by considering that the density matrix follows the equation of motion [see Eq. (5.215)]

$$\partial\rho/\partial t = -i[\mathscr{H}, \rho] - i[\mathscr{H}_{\text{coh}}(t), \rho] + R\rho, \tag{6.174}$$

where $\mathscr{H}$ is the Hamiltonian of the system in the absence of the external field; $\mathscr{H}_{\text{coh}}(t)$ is the time-dependent Hamiltonian representing the coherent interaction of the externally applied laser field with the medium; and R is the time-independent operator representing the damping effect caused by coupling of the system with the thermal reservoir. When $\mathscr{H}_0$ is diagonal, Eq. (6.174) can be written as

$$\partial\rho_{kk}/\partial t = -i[\mathscr{H}_{\text{coh}}(t), \rho]_{kk} + \sum_n W_{kn}\,\rho_{kn} - \sum_n W_{nK}\,\rho_{kk} \tag{6.175}$$

for the diagonal matrix element, where W_{kn} is the transition probability given by Eq. (5.217), and

$$\partial \rho_{kl}/\partial t = -i\omega_{kl}\rho_{kl} - i[\mathscr{H}_{\text{coh}}(t), \rho]_{kl} - R_{kl}\rho_{kl} \tag{6.176}$$

for the off-diagonal matrix element.

We now define the Fourier components of the incident field by

$$\mathbf{E}(t) = \tfrac{1}{2}\sum_j [\boldsymbol{\varepsilon}(\omega_j)e^{-i\omega_j t} + \boldsymbol{\varepsilon}^*(\omega_j)e^{i\omega_j t}]. \tag{6.177}$$

Here we also include in the expression the much smaller output-signal amplitude given by

$$\mathbf{E}_s(t) = \tfrac{1}{2}(\boldsymbol{\varepsilon}e^{-i\omega t} + \boldsymbol{\varepsilon}^* e^{i\omega t}). \tag{6.178}$$

The total matter–field interaction Hamiltonian is given by

$$\mathscr{H}_{\text{coh}}(t) = -\boldsymbol{\mu}\cdot\mathbf{E}(t). \tag{6.179}$$

We are interested in the steady-state solution of $\rho(t)$ in terms of a power series of field amplitude. It is convenient to convert Eqs. (6.175) and (6.176) into algebraic equations. In the first-order approximations, the equation for the off-diagonal density-matrix element is

$$-i\omega\rho_{kl}^{(1)} = -i\omega_{kl}\rho_{kl}^{(1)} - R_{kl}\rho_{kl}^{(1)} - i[\mathscr{H}_{\text{coh}}(\omega), \rho^{(0)}]_{kl}, \tag{6.180}$$

where $\rho(0)$ is the steady-state density matrix in the absence of coherent perturbation. We assume $\rho(0)$ to be given by the density operator for the Boltzmann distribution. Further, assuming that the electric dipole moment is to have only real off-diagonal elements $\boldsymbol{\mu}_{gm} = \boldsymbol{\mu}_{mg}$, $\boldsymbol{\mu}_{gg} = \boldsymbol{\mu}_{mm} = 0$, we obtain from Eq. (6.180) (for a single-frequency input)

$$\rho_{mg}^{(1)}(\omega) = \frac{\boldsymbol{\mu}_{mg}\cdot\boldsymbol{\varepsilon}e^{-i\omega t}}{2(\omega - \omega_{mg} + i\Gamma_{mg})}(\rho_{mm}^{(0)} - \rho_{gg}^{(0)}). \tag{6.181}$$

A similar expression for $\rho_{ng}^{(1)}(-\omega)$ is obtained by replacing ω with $-\omega$ and $\boldsymbol{\varepsilon}$ with $\boldsymbol{\varepsilon}^*$.

The linear electric polarization is

$$\begin{aligned}\langle \mathbf{P}^{(1)}(\omega)\rangle &= \frac{N}{V}\sum_g\sum_m \boldsymbol{\mu}_{gm}\rho_{mg}^{(1)}(\omega) = \frac{N}{V}\sum_g\sum_m \frac{\boldsymbol{\mu}_{gm}\boldsymbol{\mu}_{mg}\cdot(\boldsymbol{\varepsilon}e^{-i\omega t})}{2(\omega - \omega_{ng} + i\Gamma_{ng})}(\rho_{mm}^{(0)} - \rho_{gg}^{(0)}) \\ &= \frac{N}{V}\sum_j\sum_m \frac{\omega_{mg}\boldsymbol{\mu}_{gm}\boldsymbol{\mu}_{mg}\cdot(\boldsymbol{\varepsilon}e^{-i\omega t})}{\omega_{mg}^2 - (\omega + i\Gamma_{ng})^2}\rho_{gg}^{(0)}.\end{aligned} \tag{6.182}$$

If no distinction is made between the macroscopic field and the applied field, then we obtain

$$\langle P_i^{(1)}(\omega)\rangle = \tfrac{1}{2}\chi_{ij}(\omega)\varepsilon_j e^{-i\omega t}, \tag{6.183}$$

where

$$\chi_{ij}(\omega) = \frac{N}{V}\sum_{ng}\frac{2(\mu_{ng})_i(\mu_{ng})_j\omega_{mg}}{(\Omega_{mg}^2 - \omega^2) - 2i\omega\Gamma_{mg}}\rho_{gg}. \tag{6.184}$$

If we write

$$\chi_{ij}(\omega) = \chi'_{ij}(\omega) + i\chi''_{ij}(\omega), \tag{6.185}$$

where $\Omega_{mg}^2 = \omega_{mg}^2 + \Gamma_{mg}^2$, then

$$\chi'_{ij}(\omega) = n_0\sum_m\frac{2(\mu_{mg})_i(\mu_{gm})_j(\Omega_{mg}^2 - \omega^2)\omega_{mg}}{(\Omega_{mg}^2 - \omega^2)^2 + 4\omega^2\Gamma_{mg}^2}, \tag{6.186}$$

$$\chi''_{ij}(\omega) = n_0\sum_m\frac{4(\mu_{mg})_i(\mu_{gm})_j\omega\Gamma_{mg}\omega_{mg}}{(\Omega_{mg}^2 - \omega^2)^2 + 4\omega^2\Gamma_{mg}^2}, \tag{6.187}$$

where n_0 is the number density of the molecules in the ground state $|g\rangle$. For negligible damping, $\Omega_{mg} \to \omega_{mg}$ and

$$\chi'_{ij}(\omega) = n_0\sum_m\frac{2(\mu_{mg})_i(\mu_{gm})_j\omega_{mg}}{\omega_{mg}^2 - \omega^2}, \tag{6.188}$$

which is the Kramers–Heisenberg dispersion formula.

We next consider the lowest-order nonlinear susceptibility. The second-order equation for the off-diagonal element of the density matrix at frequency 2ω is

$$-i2\omega\rho_{kl}^{(2)} = -i\omega_{kl}\rho_{kl}^{(2)} - iR_{kl}\rho_{kl}^{(2)} - i[\mathscr{H}_{\text{coh}}(\omega), \rho^{(1)}]_{kl}. \tag{6.189}$$

Solving for $\rho_{m'm}^{(2)}$ and again assuming one single-input frequency, we obtain

$$\begin{aligned}\rho_{m'm}^{(2)} &= \frac{[\mathscr{H}_{\text{coh}}(\omega), \rho^{(1)}]_{m'm}}{2\omega - \omega_{m'm} + iR_{m'm}}\\ &= \frac{1}{4}\sum_{m''}\frac{(\mu_{m''m'})_j(\mu_{m'm''})_k(\rho_{m''m''}^{(0)} - \rho_{m'm'}^{(0)})}{(2\omega - \omega_{m'm} + iR_{m'm})(\omega - \omega_{m'm''} + i\Gamma_{m'm''})}\varepsilon_j\varepsilon_k e^{-2i\omega t}.\end{aligned} \tag{6.190}$$

There are also other terms with zero frequency and frequency at 2ω. The 2ω term can be obtained from Eq. (6.178) by replacing 2ω with -2ω and replacing ε with ε^*. The zero-frequency term corresponds to saturation and is of no interest in the present calculation at the second-harmonic frequency.

The second-harmonic polarization density is

$$\langle P_i^{(2)}(2\omega)\rangle = \chi_{ijk}(2\omega)\varepsilon_j\varepsilon_k e^{-2i\omega t} = (N/V)\sum_m\sum_{m'}(\mu_{mm'})_i\rho_{m'm}^{(2)}. \tag{6.191}$$

Since the $\rho^{(2)}$ term is quadratic in the dipole-moment matrix element, in the $P_i^{(2)}$ expression the dipole-moment matrix element occurs three times. Thus, $\chi_{ijk}(2\omega)$ vanishes for systems having inversion symmetry. Furthermore, if the ground electronic state of the system lacks degeneracy, the second-order nonlinear susceptibility also vanishes. Clearly, χ_{ijk} vanishes for isotropic systems.

THE TWO-STATE APPROXIMATION

We can now calculate the third-order susceptibility. The calculation is straightforward but is very tedius, especially for the case where two or several periodic electric fields are applied (Bloembergen and Shen, 1964). Because of the complexity, we shall not undertake the whole calculation for the third-order susceptibility using this method. Instead, we shall consider the case with two fields at frequencies ω_1 and ω_2 applied to the system, and select from the many terms only those that have a small resonant or near-resonant denominator when $|\omega_1 - \omega_2| = \omega_v$, the vibrational frequency of the system. We shall further assume that neither ω_1 nor ω_2 is in resonance with any of the electronic transitions of the material system; thus, the adiabatic approximation introduced previously in the Placzek theory of nonresonant Raman polarizability is applicable. In other words, we consider here only a simplified model for a molecule that exhibits Raman effects.

In this model, we consider two vibrational states $|a\rangle$ and $|b\rangle$ in the ground electronic state with energy separation given by $\hbar\omega_v$. The excited electronic states $|m\rangle$ are far removed from the applied optical frequencies ω_1 and ω_2. To shorten the effort in deriving an expression for the effective Hamiltonian appropriate for the vibrational states $|a\rangle$ and $|b\rangle$, we shall use the projection operator method introduced by Friedman and Wilson-Gordon (1978).

Following the method of quantum mechanics, we can show that if ψ is the wave function describing the state of the system subject to the coherent matter–field interaction Hamiltonian $\mathscr{H}_{\text{coh}}(t)$, then the equation of motion for ψ in the interaction representation is given by ($\hbar = 1$)

$$\mathscr{H}_{\text{I}}(t)\psi_{\text{I}}(t) = i\partial\psi_{\text{I}}(t)/\partial t, \tag{6.192}$$

where ψ_{I} is related to ψ by $\psi_{\text{I}} = e^{i\mathscr{H}_{\text{o}}t}\psi$, and

$$\mathscr{H}_{\text{I}}(t) = e^{i\mathscr{H}_{\text{o}}t}\mathscr{H}_{\text{coh}}(t)e^{-i\mathscr{H}_{\text{o}}t}. \tag{6.193}$$

To obtain the effective Hamiltonian, we shall cast Eq. (6.192) into an equation defined only by two states $|a\rangle$ and $|b\rangle$. To accomplish this, we introduce two projection operators P and Q given by

$$P = |a\rangle\langle a| + |b\rangle\langle b|, \tag{6.194}$$

$$Q = \sum_m |m\rangle\langle m|. \tag{6.195}$$

Clearly,

$$P + Q = I, \tag{6.196}$$

where I is the identity operator.

Let

$$f = P\psi_{\mathrm{I}}, \tag{6.197}$$

$$g = Q\psi_{\mathrm{I}}. \tag{6.198}$$

In accordance with the terminology of Chapter 4, f is the relevant wave function and g the irrelevant one. From Eq. (6.192), we obtain coupled equations for f and g as follows:

$$i\,\partial f/\partial t = P\mathscr{H}_{\mathrm{I}} f + P\mathscr{H}_{\mathrm{I}} g, \tag{6.199}$$

$$i\,\partial g/\partial t = Q\mathscr{H}_{\mathrm{I}} f + Q\mathscr{H}_{\mathrm{I}} g. \tag{6.200}$$

To eliminate g from Eq. (6.199), we need to solve g in terms of f from Eq. (6.200) and then substitute the result into Eq. (6.199). Because $\mathscr{H}_{\mathrm{I}}$ is a time dependent operator function, the solution is not straightforward. Assuming that $S(t, t')$ is a solution to the equation

$$\partial S/\partial t = -iQ\mathscr{H}_{\mathrm{I}} S, \tag{6.201}$$

with $S(t', t') = I$, then the solution for $S(t, t')$ is

$$\begin{aligned} S(t, t') &= I - i\int_{t'}^{t} dt_1\, Q\mathscr{H}_{\mathrm{I}}(t_1) - \int_{t'}^{t}\int_{t'}^{t_1} dt_1\, dt_2\, Q\mathscr{H}_{\mathrm{I}}(t_1)Q\mathscr{H}_{\mathrm{I}}(t_2) + \cdots \\ &= T\exp\left[-i\int_{t'}^{t} dt''\, Q\mathscr{H}_{\mathrm{I}}(t'')\right], \end{aligned} \tag{6.202}$$

where T defines the time-ordering operator, which chronologically orders the operator function in accordance with Eq. (6.202).

In terms of S, the solution of g can be written formally as

$$g(t) = -i\int_{t'}^{t} dt''\, S(t, t')S^{-1}(t'', t')Q\mathscr{H}_{\mathrm{I}}(t'')f(t''). \tag{6.203}$$

Substituting Eq. (6.191) into (6.187) gives

$$i\,\partial f/\partial t = \int_{t'}^{t} dt''\, [P\mathscr{H}_{\mathrm{I}}(t'')\,\delta(t - t'') - iS(t, t')S^{-1}(t'', t')Q\mathscr{H}_{\mathrm{I}}(t'')]f(t''). \tag{6.204}$$

Clearly, we can find an expression for the effective Hamiltonian only if the time argument t'' in $f(t'')$ can be replaced by t. This is possible, provided that the wave function $f(t)$ does not change appreciably during the time interval between t and t'', a situation that is expected if the optical frequencies are not in resonance with the electronic-transition frequencies. Mathematically, the approximation corresponds to the adiabatic condition given by $\partial\varepsilon(\omega_j)/\partial t \ll \varepsilon(\omega_j)(\omega_{ma} - \omega_j)$, $j = 1, 2$. Thus, if the adiabatic condition is satisfied, then Eq. (6.192) can be approximated as (Friedman and Wilson-Gordon, 1978)

$$i\,\partial f(t)/\partial t = \mathscr{H}_{\text{eff}}(t) f(t), \tag{6.205}$$

where the effective two-level Hamiltonian in the interaction representation $\mathscr{H}_{\text{eff}}$ is given by

$$\mathscr{H}_{\text{eff}}(t) = P\mathscr{H}_{\text{I}}(t) - iP\mathscr{H}_{\text{I}}(t)S(t, -\infty)\int_{-\infty}^{t} dt_1\, S^{-1}(t_1, -\infty)Q\mathscr{H}_{\text{I}})(t_1), \tag{6.206}$$

where we set the lower limit in the integration $t' = -\infty$.

The diagonal matrix elements of $\mathscr{H}_{\text{eff}}(t)$ correspond to optical Stark shifts in states $|a\rangle$ and $|b\rangle$. Note that the lowest-order Stark shifts are quadratic in the applied fields. Terms of the type $\varepsilon(\omega_1)\varepsilon(\omega_2)$ and $\varepsilon(\omega_1)\varepsilon^*(\omega_2)$ also contribute to the optical Stark shifts. The off-diagonal elements are associated with transition moments. Since $|a\rangle$ and $|b\rangle$ have the same parity and are separated by a frequency nearly equal to the difference of the two incident frequencies ω_1 and ω_2, the first term on the right-hand side of Eq. (6.206) vanishes identically; the off-diagonal matrix element of $\mathscr{H}_{\text{eff}}(t)$ is

$$\langle b|\mathscr{H}_{\text{eff}}(t)|a\rangle = -i\langle b|\mathscr{H}_{\text{I}}(t)S(t, -\infty)\int_{-\infty}^{t} dt_1\, S^{-1}(t_1, -\infty)\, Q\mathscr{H}_{\text{I}}(t_1)|a\rangle \tag{6.207}$$

Note that the lowest-order nonvanishing interaction in this equation is bilinear in the incident-field amplitudes. Thus, for the CARS experiment with two incident frequencies ω_1 and ω_2, Eq. (6.207) reduces to

$$\begin{aligned}\langle b|\mathscr{H}_{\text{eff}}(t)|a\rangle &= -\frac{1}{4}e^{i\omega_{ba}t}\left\{\frac{[\mu_{bm}\cdot\boldsymbol{\varepsilon}(\omega_1)][\mu_{ma}\cdot\boldsymbol{\varepsilon}^*(\omega_2)]}{\omega_{ma}+\omega_2}\right.\\ &\quad\left.+\frac{[\mu_{bm}\cdot\boldsymbol{\varepsilon}^*(\omega_2)][\mu_{ma}\cdot\boldsymbol{\varepsilon}(\omega_1)]}{\omega_{ma}-\omega_1}\right\}e^{-i(\omega_1-\omega_2)t}\\ &= \frac{1}{4}e^{i\omega_{ba}t}\alpha_{jk}(-\omega_1, \omega_2)\varepsilon_j(\omega_1)\varepsilon_k^*(\omega_2)e^{-i(\omega_1-\omega_2)t} = e^{i\omega_{ba}t}\mathscr{H}'_{ba}(t),\end{aligned} \tag{6.208}$$

where the phase factor $e^{i\omega_{ba}t}$ occurs because $\mathscr{H}_{\text{eff}}(t)$ is expressed in terms of the interaction representation. The quantity $\alpha_{jk}(-\omega_1, \omega_2)$ is given by

$$\alpha_{jk}(-\omega_1, \omega_2) = \left[\frac{(\mu_{bm})_j(\mu_{ma})_k}{\omega_{ma} - \omega_2} + \frac{(\mu_{bm})_k(\mu_{ma})_j}{\omega_{ma} + \omega_1}\right]. \tag{6.209}$$

The expression given by Eq. (6.209) was first introduced by Maker and Terhune (1965) and justifies the expression for the interaction Hamiltonian $\mathscr{H}'$ given by Eq. (6.137). In Eq. (6.137), α_{jk} is written as $(\partial\alpha_{jk}/\partial q)_0\, q$ because α_{jk} is a function of vibrational coordinates. Further, in Eq. (6.137) the frequency dependence in α_{jk} is suppressed, which is consistent with the assumption that ω_1 and ω_2 are not in resonance with the electronic transition. Thus, α_{jk} is only weakly dependent on ω_1 and ω_2.

From the expression of $\mathscr{H}'_{ba}$, we can obtain the transition dipole-matrix element μ_{ab} associated with the signal field at frequency ω by taking the functional derivative of $\mathscr{H}'_{ba}(t)$ with respect to $\frac{1}{2}\varepsilon^* e^{i\omega t}$ as

$$\begin{aligned}\boldsymbol{\mu}_{ab} = \frac{\partial\mathscr{H}'_{ba}(t)}{\partial[(\boldsymbol{\varepsilon}^* e^{i\omega t})/2]} &= -\frac{1}{2}\left\{\frac{[\boldsymbol{\mu}_{bm}\cdot\boldsymbol{\varepsilon}(\omega_1)]\boldsymbol{\mu}_{ma}}{\omega_{ma} + \omega} + \frac{\boldsymbol{\mu}_{bm}(\boldsymbol{\mu}_{ma}\cdot\boldsymbol{\varepsilon}(\omega_1))}{\omega_{ma} - \omega_1}\right\} e^{-i\omega_1 t} \\ &= -\frac{\boldsymbol{\alpha}(-\omega_1, \omega)}{2}\cdot\boldsymbol{\varepsilon}(\omega_1) e^{-i\omega_1 t}.\end{aligned} \tag{6.210}$$

Note that only the Fourier component at ω_1 contributes to the output wave at ω.

Having obtained the expression of $\mathscr{H}'_{ab}$, we can then follow Eq. (6.137) to calculate the nonlinear susceptibility at output frequency ω using the two-level model. The result was given in Eq. (6.146).

The Born–Oppenheimer Approximation

We shall now more fully develop the theory of nonlinear susceptibility in the Born–Oppenheimer approximation (BOA). This approximation is applicable to the transparent medium in which the electronic-transition frequencies of the material are very much greater than the frequencies of the optical field present in the medium. In this situation, the electrons follow nearly adiabatically with respect to the optical field and the fluctuating fields of the nuclei.

As mentioned earlier, in the presence of the optical field, the Hamiltonian of the system becomes

$$\mathscr{H}(t) = \mathscr{H}_0 + \mathscr{H}_{\text{coh}}(t), \tag{6.211}$$

where $\mathscr{H}_{\text{coh}}(t) = -\boldsymbol{\mu}\cdot\mathbf{E}(t) = -\mu_i E_i(t)$.

If the energy of the ground Born–Oppenheimer state is ε_{00} at a given nuclear configuration, then in the presence of the matter–field interaction we can use the standard Rayleigh–Schrödinger perturbation scheme to calculate the correction to the ground-state energy in terms of a power series of $\mathbf{E}(t)$. The result under the adiabatic approximation can be written as

$$\varepsilon_0(\mathbf{E}) = \varepsilon_{00} - \{m_i E_i + \tfrac{1}{2}\alpha_{ij} E_i E_j + \tfrac{1}{3}\beta_{ijk} E_i E_j E_k + \tfrac{1}{6}\gamma_{ijkl} E_i E_j E_k E_l + \cdots\}, \tag{6.212}$$

where m_i, α_{ij}, β_{ijk}, etc., are no longer dependent on the electronic wave functions. It should be noted that since ε_0 is calculated using the unperturbed Born–Oppenheimer ground-state wave function in a given nuclear configuration, the quantities ε_{00}, m_i, α_{ij}, β_{ijk}, etc., depend explicitly on the nuclear coordinates.

The energy ε_0 and the electric fields E_i are real quantities; it thus follows that m_i, α_{ij}, β_{ijk}, etc., are real coefficients. Furthermore, because the energy is invariant to permutation of the electric-field amplitudes, it also follows that α_{ij}, β_{ijk}, etc., are symmetric under exchange of the space indices, i.e., $\alpha_{ij} = \alpha_{ji}$ and $\beta_{ijk} = \beta_{jik} = \beta_{ikj} = \cdots$. This is known as Kleinman symmetry (Kleinman, 1962).

Similarly, using perturbation theory, the Born–Oppenheimer ground-state wave function can also be corrected by taking the optical field into account. Thus, for a given configuration of nuclei and the optical field, the ground electronic state is represented by $|0, (\xi, \mathbf{E})\rangle$. The total dipole moment associated with the ground electronic state can then be calculated by

$$M_i(\xi, E) = \langle 0, (\xi, \mathbf{E})| -e \sum_{\alpha} \chi_{i\alpha} |0, (\xi, \mathbf{E})\rangle + \sum_{\alpha} e_\alpha r_{i\alpha}, \tag{6.213}$$

where $\chi_{i\alpha}$ and $r_{i\alpha}$ are the electron and nuclear coordinates, respectively. As is well known, M_i given here can also be calculated by taking the functional derivative of ε_0 with respect to $E_i(t)$:

$$M_i = -[\partial\varepsilon_0(E)/\partial E_i(t)] = m_i + \alpha_{ij} E_j + \beta_{ijk} E_j E_k + \gamma_{ijkl} E_j E_k E_l + \cdots. \tag{6.214}$$

In Eqs. (6.212) and (6.214), the electronic coordinates were eliminated from the problem. Thus, for calculating the effect of nuclear motion, the energy ε_0 and dipole moment M_i can be considered operators. Thus, in the BOA scheme, we shall replace Eq. (6.212) by

$$\mathscr{H} = \mathscr{H}_0(\xi) + V(\xi, \mathbf{E}), \tag{6.215}$$

where the expectation value of $\mathscr{H}_0$ with respect to the Born–Oppenheimer ground-state electronic wave function is ε_{00}, and

$$V(\xi, E) = -m_i E_i + \tfrac{1}{2}\alpha_{ij} E_i E_j + \tfrac{1}{3}\beta_{ijk} E_i E_j E_k = V_1 + V_2 + V_3 + \cdots,$$

where V_1, V_2, and V_3 are linear, quadratic, and cubic, respectively, in the electric field.

The coupling of M_i with the external electric field is responsible for all the phenomena observed in vibrational spectroscopy. To analyze the vibrational spectra, it is customary to expand m_i, α_{ij}, etc., in a power series of the normal (or phonon) coordinates q_l as

$$m_i = m_i(0) + (\partial m_i/\partial q_l)_0 \, q_l + \cdots, \tag{6.216}$$

$$\alpha_{ij} = \alpha_{ij}(0) + (\partial \alpha_{ij}/\partial q_l)_0 q_l + \cdots. \tag{6.217}$$

The term $(\partial m_i/\partial q_l)_0 q_l$ will account for the infrared absorption spectrum resulting from the normal mode q_l. Similarly, the term $(\partial \alpha_{ij}/\partial q_l)_{0\,ql}$ is responsible for the "one-phonon" Raman-scattering spectrum. Higher-order terms not written out explicitly result in overtone, two-phonon, or higher-order absorption or scattering processes.

The term associated with β_{ijk} gives light scattered at a frequency near twice the incident frequency. This is known as hyper-Raman scattering (Hellwarth, 1977; Terhune *et al.*, 1965). Because β_{ijk} is a tensor of higher rank than α_{ij}, hyper-Raman scattering can in principle provide more information about the scattering medium than can be obtained from infrared absorption and the usual Raman-scattering spectra. Hyper-Raman scattering can be analyzed by expanding β_{ijk} as a power series of the normal mode analogous to the manner of m_i and α_{ij} given earlier. Similarly, we expect that scattering from the γ_{ijkl} term will have a frequency near the third harmonic of the incident frequency. This high-order effect however, has, not been studied.

Having obtained Eqs. (6.214) and (6.215) under the BOA, we can now proceed to calculate the expectation value of M_i in ascending powers of the external field. The calculation will only involve vibrational wave functions associated with the ground electronic state.

The polarization density $P_i(t)$ is

$$\begin{aligned} P_i(t) &= (N/V)\,\mathrm{Tr}[M_i(t)\rho(t)] = n\,\mathrm{Tr}[m_i\rho(t)] + n\,\mathrm{Tr}[\alpha_{ij}\rho(t)]E_j(t) \\ &\quad + n\,\mathrm{Tr}[\beta_{ijk}\rho(t)]E_j(t)E_k(t) + n\,\mathrm{Tr}[\gamma_{ijkl}\rho(t)]\,E_j(t)E_k(t)E_l(t) + \cdots. \end{aligned} \tag{6.218}$$

We need to find the solution of $\rho(t)$ in a power series of $\mathbf{E}$ and then collect terms to obtain $P_i(t)$ in ascending powers of $\mathbf{E}$. To accomplish this, we first write the equation of motion for ρ as

$$\partial\rho/\partial t = -i[\mathscr{H}_0 + V_1 + V_2 + V_3, \rho]. \tag{6.219}$$

Let the solution of ρ be given by

$$\rho(t) = \rho^{(0)} + \rho^{(1)} + \rho^{(2)} + \rho^{(3)} + \cdots. \tag{6.220}$$

Equating terms with the same order in the electric field in Eq. (6.229), we obtain

$$\partial\rho^{(0)}/\partial t = -i[\mathscr{H}_0, \rho^{(0)}], \tag{6.221}$$

$$\partial\rho^{(1)}/\partial t = -i[\mathscr{H}_0, \rho^{(1)}] - i[V_1, \rho^{(0)}], \tag{6.222}$$

$$\partial\rho^{(2)}/\partial t = -i[\mathscr{H}_0, \rho^{(2)}] - i[V_1, \rho^{(1)}] - i[V_2, \rho^{(0)}], \tag{6.223}$$

$$\partial\rho^{(3)}/\partial t = -i[\mathscr{H}_0, \rho^{(3)}] - i[V_1, \rho^{(2)}] - i[V_2, \rho^{(1)}] - i[V_3, \rho^{(0)}]. \tag{6.224}$$

Solving Eqs. (6.231)–(6.234) for the density operators in different order and then substituting the results into Eq. (6.218), we obtain

$$P_i(t) = P_i^{(1)}(t) + P_i^{(2)}(t) + P_i^{(3)}(t) + \cdots. \tag{6.225}$$

The linear term is found to be

$$P_i^{(1)}(t) = n\langle\alpha_{ij}\rangle E_j(t) + in\int_0^t dt_1\,\langle[m_i(t - t_1), m_j]\rangle E_j(t_1), \tag{6.226}$$

where $m_i(t) = e^{i\mathscr{H}_0 t} m_i e^{-i\mathscr{H}_0 t}$.

From Eq. (6.226), we immediately find the expression for the linear susceptibility $\chi_{ij}^{(1)}(\omega)$ as

$$\chi_{ij}^{(1)}(\omega) = n\langle\alpha_{ij}\rangle + in\int_0^\infty d\tau\,\langle[m_i(\tau), m_j]\rangle e^{-i\omega\tau}. \tag{6.227}$$

The first term is the electronic contribution to the susceptibility. Note that the electronic part responds instantaneously to the external field. The second term is the nuclear contribution owing to vibration. The nuclear term does not respond to the external field instantaneously and is responsible for the frequency dependence of $\chi_{ij}^{(1)}(\omega)$. At the optical frequency, the nuclear contribution is negligible.

If we denote the vibrational eigenstates of $\mathscr{H}_0$ having energies ε_l by $|l\rangle$ and use the fact that $|l\rangle$ is a real function of the nuclear coordinates, then Eq. (6.227) can be written as

$$\chi_{ij}^{(1)}(\omega) = n\langle\alpha_{ij}\rangle + 2\sum_{l,l'} \frac{n_l\langle l|m_i|l'\rangle\langle l'|m_j|l\rangle\omega_{ll'}}{\omega_{ll'}^2 - \omega^2} \tag{6.228}$$

where n_l is the equilibrium population of molecules in $|l\rangle$. This form is similar to that given by Eq. (6.188). The term $\chi_{ij}^{(1)}$ is symmetric with respect to its indices and contains, at most, six independent elements. However, for cubic crystals, fluids, or glass, $\chi_{ij}^{(1)}$ has only one element, of the form $\chi\,\delta_{ij}$.

The second-order polarization density is next calculated. In accordance with Eqs. (6.218) and (6.220), $P_i^{(2)}$ can be written as

$$P_i^{(2)}(t) = n\,\mathrm{Tr}[m_i\rho^{(2)}(t)] + n\,\mathrm{Tr}[\alpha_{ij}\rho^{(1)}(t)]E_j(t) + n\langle\beta_{ijk}\rangle E_j(t)E_k(t). \tag{6.229}$$

The last term of the right-hand side of this equation is the electronic term. The response owing to this term is instantaneous. From Eqs. (6.232) and (6.233), the first two terms of Eq. (6.229) are found to be

$$\begin{aligned} &n\,\mathrm{Tr}[m_i\rho^{(2)}(t)] + n\,\mathrm{Tr}[\alpha_{ij}\rho^{(1)}(t)]E_j(t) \\ &\quad = -n\int_{-\infty}^{t} dt_1 \int_{-\infty}^{t_1} dt_2 \langle\{m_i(t-t_2), m_j(t_1-t_2)], m_k\}\rangle E_j(t_1)E_k(t_2) \\ &\quad + in\int_{-\infty}^{t} dt_1\, \langle[m_i(t-t_1)_1\, \tfrac{1}{2}\alpha_{jk}]\rangle E_j(t_1)E_k(t_1) \\ &\quad + in\int_{-\infty}^{t} dt_1\, \langle[\alpha_{ij}(t-t_1), m_k]\rangle E_k(t_1)E_j(t). \end{aligned} \tag{6.230}$$

The last term in Eq. (6.230) gives the electrooptic (Pockels) effect. Evidently, none of these second-order terms exists unless the medium lacks a center-of-inversion symmetry. The symmetry properties of crystalline solids and the implications of these symmetries for all important second-order nonlinear effects have been presented by Yariv (1971), and we shall not further discuss them here.

We next proceed to find the third-order susceptibility-tensor coefficients. In accordance with Eq. (6.217), $P_i^{(3)}$ can be written as

$$\begin{aligned} P_i^{(3)}(t) &= n\,\mathrm{Tr}[m_i\rho^{(3)}] + n\,\mathrm{Tr}[\alpha_{ij}\rho^{(2)}]E_j(t) + n\,\mathrm{Tr}[\beta_{ijk}\rho^{(1)}]E_j(t)E_k(t) \\ &\quad + n\langle\gamma_{ijkl}\rangle E_j(t)E_k(t)E_l(t). \end{aligned} \tag{6.231}$$

The last term is instantaneous and a purely electronic contribution. It exists in all materials.

The term next to the last is given by

$$\begin{aligned} &n\,\mathrm{Tr}[\beta_{ijk}\rho^{(1)}]E_j(t)E_k(t) \\ &\qquad = in\int_{0}^{t} dt_1\, \langle[\beta_{ijk}(t-t_1), m_l]\rangle E_l(t_1)E_j(t)E_k(t); \end{aligned} \tag{6.232}$$

the first two terms yield six terms:

$$n(i)^3 \int_0^t dt_1 \int_0^{t_1} dt_2 \int_0^{t_2} dt_3 \langle[[[m_i(t), m_j(t_1)], m_k(t_2)], m_l(t_3)]\rangle \times E_j(t_1)E_k(t_2)E_l(t_3), \tag{6.233}$$

$$n(i)^2 \int_0^t dt_1 \int_0^{t_1} dt_2 \langle[[m_i(t), \tfrac{1}{2}\alpha_{jk}(t_1)], m_l(t_2)]\rangle E_j(t_1)E_k(t_1)E_l(t_2), \tag{6.234}$$

$$n(i)^2 \int_0^t dt_1 \int_0^{t_1} dt_2 \langle[[m_i(t), m_j(t_1)], \tfrac{1}{2}\alpha_{kl}(t_2)]\rangle E_j(t_1)E_k(t_2)E_l(t_2), \tag{6.235}$$

$$n(i)^2 \int_0^t dt_1 \int_0^{t_1} dt_2 \langle[[\tfrac{1}{2}\alpha_{ij}(t), m_k(t_1)], m_l(t_2)]\rangle E_j(t)E_k(t_1)E_l(t_2), \tag{6.236}$$

$$n(i) \int_0^t dt_1 \langle[\tfrac{1}{2}\alpha_{ij}(t), \tfrac{1}{2}\alpha_{kl}(t_1)]\rangle E_j(t)E_k(t_1)E_l(t_1), \tag{6.237}$$

$$n(i) \int_0^t dt_1 \langle[m_i(t), \tfrac{1}{3}\beta_{jkl}(t_1)]\rangle E_j(t_1)E_k(t_1)E_l(t_1). \tag{6.238}$$

The total third-order polarization density is the sum of all terms given in Eqs. (6.231)–(6.238), eight altogether. However, the eight terms are not of equal importance. When the frequencies of the external field are in the optical region, i.e., when the frequencies in E_i and P_i are well above any of the frequencies of multiple-time correlations associated with nuclear motion, only the term given in Eq. (6.237) and the instantaneously responding electronic term are of importance; the other terms are small because the integrals for each of these terms oscillate at the optical frequency and hence nearly cancel each other when the time integration is carried out.

Therefore, for purely optical nonlinear effects, the third-order polarization density has the simple form

$$P_i^{(3)}(t) = n\langle\gamma_{ijkl}\rangle E_j(t)E_k(t)E_l(t) + E_j(t)\int_0^t dt_1\, D_{ijkl}(t - t_1)E_k(t_1)E_l(t_1), \tag{6.239}$$

where

$$D_{ijkl}(t - t_1) = (ni/4)\langle[\alpha_{ij}(t - t_1), \alpha_{kl}]\rangle = (ni/4)\langle[\alpha_{ij}(t), \alpha_{kl}(t_1)]\rangle. \tag{6.240}$$

The first term on the right of Eq. (6.239) is the electronic contribution and the second term the nuclear contribution. Clearly, the result obtained with the BOA is much simpler than the completely general expression for the third-order susceptibility.

We next consider symmetry properties of the third-order polarization density given by Eq. (6.239). Because the electronic susceptibility operator is symmetric in its two indices ($\alpha_{ij} = \alpha_{ji}$), D_{ijkl} is invariant with respect to the interchange of either the first or the last two indices:

$$D_{ijkl} = D_{jikl} = D_{ijlk} = D_{jilk}. \tag{6.241}$$

Furthermore, since α_{ij} (or α_{kl}) are real functions of the nuclear coordinates, we have

$$\langle[\alpha_{ij}(t), \alpha_{kl}]\rangle = \langle[\alpha_{kl}(t), \alpha_{ij}]\rangle. \tag{6.242}$$

Thus, D_{ijkl} is invariant to an exchange of its first pair of indices with its last:

$$D_{ijkl} = D_{klij}. \tag{6.243}$$

Strictly, the foregoing symmetry properties are valid only for optically inactive materials. However, the error is negligible even if the medium is optically active (Hellwarth, 1977).

Additional symmetry considerations come from the general transformation property of a fourth-rank tensor. Since both γ_{ijkl} and D_{ijkl} transform like the product of vector components $r_i r_j r_k r_l$ under all-space transformation, this transformation property in combination with Eqs. (6.241) and (6.243) reduces the number of the independent D_{ijkl} to 21 for materials with lowest symmetry. The results are similar to those for the elastic (stiffness) constants and were derived by Nye (1964) and Bhagavantam (1964). For isotropic materials, D_{ijkl} has only two independent components, in contrast to three for a general third-order susceptibility tensor.

Obviously, the form of $\langle\gamma_{ijkl}\rangle$ can be obtained from that of D_{ijkl} by symmetrizing D_{ijkl} among its indices. (This is possible because of Kleinman symmetry.) Thus, the maximum number of independent elements for $\langle\gamma_{ijkl}\rangle$ is 12 and for the isotropic medium is 1.

To sum up, in accordance with the results of Nye (1964) and Bhagavantam (1964), in isotropic media the general form of D_{ijkl} under BOA is

$$D_{ijkl} = A\,\delta_{ij}\,\delta_{kl} + \tfrac{1}{2}B(\delta_{il}\,\delta_{jk} + \delta_{ik}\,\delta_{jl}), \tag{6.244}$$

where A and B are complex functions of time; and

$$\langle\gamma_{ijkl}\rangle = \gamma(\delta_{ij}\,\delta_{kl} + \delta_{ik}\,\delta_{jl} + \delta_{il}\,\delta_{jk}), \tag{6.245}$$

where γ is a real constant.

We next focus on CARS by considering that two optical waves with frequencies ω_1 and ω_2 interact with a transparent medium. The medium mixes the two optical waves and radiates a new wave at frequency $\omega = 2\omega_1 - \omega_2$. This CARS (or three-wave mixing) effect can be described by substituting

$$\mathbf{E}(t) = \tfrac{1}{2} \sum_{j=1}^{2} \{\boldsymbol{\varepsilon}(\omega_j) \exp[i(\omega_j t - K_j z)] + \boldsymbol{\varepsilon}^*(\omega_j) \exp[-i(\omega_j t - K_j z)]\} \tag{6.246}$$

into Eq. (6.239) and then isolating the part of $P_i^{(3)}(t)$ that varies with frequency $2\omega_1 - \omega_2$. We find for this case

$$P_i^{(3)}(t) = \tfrac{3}{8}\chi_{ijkl}^{(3)}\varepsilon_j(\omega_1)\varepsilon_k(\omega_1)\varepsilon_l^*(\omega_2) \exp\{-i[(2\omega_1 - \omega_2)t - (2K_1 - K_2)z]\}, \tag{6.247}$$

where

$$\chi_{ijkl}^{(3)} = \tfrac{1}{4}n \langle \gamma_{ijkl} \rangle + \tfrac{1}{6} \int_0^\infty d\tau\, D_{ijkl}(\tau) \exp[i(\omega_1 - \omega_2)\tau], \tag{6.248}$$

and the upper limit has been extended to ∞.

The electronic part gives rise to the nonresonant contribution. Since it is independent of frequency, it contributes to the spectrum as background. Other parts of the nonresonant contribution result from some of the terms not included in Eq. (6.239). However, $n\langle\gamma_{ijkl}\rangle$ gives the major contribution to the nonresonant third-order susceptibility. The second term on the right-hand side of Eq. (6.248) is the resonant contribution to the CARS susceptibility. It is interesting to note that under the BOA, the nonlinear effect is expressed in terms of a time-correlation function, a quantity that we have shown to be associated with linear spectroscopy.

6.13 RELATION OF CARS TO SPONTANEOUS RAMAN SCATTERING

There is a very useful relation between CARS spectroscopy and ordinary Raman scattering. To show the relation, it is useful to consider $\chi_{ijkl}^{(r)}(\Omega)$, defined as

$$\chi_{ijkl}^{(r)}(\Omega) = \int_0^\infty d\tau\, e^{-i\Omega\tau} D_{ijkl}(\tau) = \chi'_{ijkl}(\Omega) - i\chi''_{ijkl}(\Omega), \tag{6.249}$$

where the superscript r indicates the resonant part. Following the development in Chapter 1, we can show that χ'_{ijkl} and χ''_{ijkl} satisfy the Kramers–Kronig relations:

$$\chi''_{ijkl}(\Omega) = \frac{1}{\pi} \mathrm{P} \int_{-\infty}^{\infty} dx \frac{\chi'_{ijkl}(\chi) - \chi'_{ijkl}(0)}{x - \Omega} \tag{6.250}$$

and

$$\chi'_{ijkl}(\Omega) - \chi'_{ijkl}(0) = \frac{1}{\pi} \mathrm{P} \int_{-\infty}^{\infty} dx \frac{\chi''_{ijkl}(x)}{\Omega - x}. \tag{6.251}$$

In Chapter 1, we showed that the imaginary part $\chi''_{ijkl}(\Omega)$ is related to the Fourier transform of the time-correlation function [compare with the fluctuation–dissipation theorem expressed in Eqs. (1.131) and (1.150)] as

$$\chi''_{ijkl}(\Omega) = (1/2\hbar)(1 - e^{-\beta\hbar\Omega}) \int_{-\infty}^{\infty} d\tau \, \langle \alpha_{ij}(\tau)\alpha_{kl} \rangle e^{-i\Omega\tau}. \tag{6.252}$$

We next show that $\chi''_{ijkl}(\Omega)$ is related to the spontaneous Raman spectrum. As was given previously in Eq. (6.96), spectral power density of the Raman-scattered light obtained with the incident light polarized along $\hat{n}_a$ and the scattered light polarized along $\hat{n}_b$ is given by

$$\begin{aligned} I_{ab}(\mathbf{q}, \Omega) &= \frac{1}{2\pi} \int_{-\infty}^{\infty} dt \, e^{-i\Omega t} \langle \delta\varepsilon_{ab}(\mathbf{q}, t) \, \delta\varepsilon^*_{ab}(\mathbf{q}, 0) \rangle \\ &\propto \int_{-\infty}^{\infty} dt \, e^{-i\Omega t} \langle \alpha_{ab}(\mathbf{q}, t) \alpha^*_{ab}(\mathbf{q}) \rangle, \end{aligned} \tag{6.253}$$

where Ω is the shift from the incident-laser frequency, and $\alpha_{ab}(\mathbf{g}, t)$ is given by

$$\alpha_{ab}(\mathbf{q}, t) = 1/V \int d^3r \, \alpha_{ab}(\mathbf{r}, t) e^{i\mathbf{q} \cdot \mathbf{r}}. \tag{6.254}$$

Since $\alpha_{ab}(\mathbf{r}, t)$ is expressed in terms of the laboratory coordinate system, we can perform a transformation to another coordinate system and write

$$\alpha_{ab}(\mathbf{q}, t) = n_{ai} n_{bj} \alpha_{ij}(\mathbf{q}, t), \tag{6.255}$$

where n_{ai} and n_{bj} are transformation coefficients and summation over the repeated indices is implied. Note also that $n_{ai} n_{bi} = \delta_{ab}$. Since the wavelength of the incident light used in the Raman experiment is much longer than the molecular dimension or the dimension of the unit cell of the crystalline solid, the q dependence in the Raman polarizability expression can be neglected. As a result, Eq. (6.255) is approximated as

$$\alpha_{ab}(\mathbf{q}, t) \simeq \frac{1}{V} n_{ai} n_{bj} \int d^3r \, \alpha_{ij}(\mathbf{r}, t). \tag{6.256}$$

We can express Eq. (6.256) in terms of molecular polarizability if the Raman polarizability density is written as

$$\alpha_{ij}(\mathbf{r}, t) = \sum_c \delta[\mathbf{r} - \mathbf{r}_c(t)]\, \alpha_{ij}^{(c)}(t), \tag{6.257}$$

where $\alpha_{ij}^{(c)}(t)$ is the Raman polarizability of molecule c located at position $\mathbf{r}_c$. For an internal mode, the $\mathbf{q}$ dependence can be neglected, and we can write $\alpha_{ab}(\mathbf{q}, t)$ as

$$\alpha_{ab}(\mathbf{q}, t) = (N/V) n_{ai} n_{bj} \alpha_{ij}(t), \tag{6.258}$$

where $\alpha_{ij}(t)$ is the Raman polarizability of a representative molecule.

Substituting Eq. (6.258) into (6.253) gives

$$\begin{aligned} I_{ab}(\mathbf{q}, \Omega) &\propto n_{ai} n_{bj} n_{ak} n_{bl} \int_{-\infty}^{\infty} dt\, e^{i\Omega t} \langle \alpha_{ij}(t) \alpha_{kl}(0) \rangle \\ &\propto n_{ai} n_{bj} n_{ak} n_{bl} \chi''_{ijkl}(\Omega)/(1 - e^{-\beta\hbar\Omega}), \end{aligned} \tag{6.259}$$

where in Eq. (6.259) only the quantities relevant to the discussion are written out explicitly. The proportionality instant is easily evaluated but not needed, because its absolute value has little accuracy; calibration is always needed. We have thus established that the Raman-scattering spectrum is related to the imaginary part of the third-order susceptibility $\chi_{ijkl}^{(r)}(\Omega)$.

However, what is observed in CARS is power radiated by the polarization density $P_i^{(3)}(t)$ given in Eq. (6.247). The radiated intensity is proportional to $|\chi_{ijkl}^{(3}(-\omega, \omega_1, \omega_1, -\omega_2)|^2$. Since $\chi^{(3)}$ is equal to the sum of nonresonant and resonant parts, the CARS signal is superimposed by the dispersive intensity resulting from the $(\mathrm{Re}\, \chi_{ijkl}^{(r)})^2$ term and the nonresonant background, owing mainly to the electronic term $\langle \gamma_{ijkl} \rangle^2$. Additional interference owing to the cross term $2(\mathrm{Re}\, \chi_{ijkl}^{(r)})\langle \gamma_{ijkl} \rangle$ is also present. The quantitative study of the CARS signal is thus complicated without a knowledge of these quantities. To clarify the nature of the CARS signal, Levenson *et al.* (1972) showed how the nonresonant contributions in a single material can be used to calibrate the others by measuring the CARS signal as a function of $(\omega_1 - \omega_2)$. Similar studies of the interference effect between solute and solvent molecules in solutions were also made by Levenson and Bloembergen (1974).

As mentioned earlier, in fluids and glasses $\chi_{ijkl}^{(r)}(\Omega)$ has only two independent components:

$$\chi_{ijkl}^{(r)}(\Omega) = A(\Omega)\, \delta_{ij}\, \delta_{kl} + \tfrac{1}{2} B(\Omega)(\delta_{ik}\, \delta_{jl} + \delta_{il}\, \delta_{jk}), \tag{6.260}$$

where $A(\Omega)$ and $B(\Omega)$ are the Fourier transforms of $A(t)$ and $B(t)$, defined in Eq. (6.244). In accordance with Eqs. (6.259) and (6.260), we find

$$I_{\perp}(\Omega) = I_{zy}(\Omega) \propto [\chi''_{zyzy}(\Omega)/(1 - e^{-\beta\hbar\Omega})],$$

or

$$\tfrac{1}{2}\mathrm{Im}\, B(\Omega) = \eta(1 - e^{-\beta\hbar\Omega})I_{\perp}(\Omega); \tag{6.261}$$

and

$$I_{\parallel}(\Omega) = I_{zz}(\Omega) \propto [\chi''_{zzzz}(\Omega)/(1 - e^{-\beta\hbar\Omega})],$$

or

$$\mathrm{Im}[A(\Omega) + B(\Omega)] = \eta(1 - e^{-\beta\hbar\Omega})I_{\parallel}(\Omega), \tag{6.262}$$

where η is a proportionality constant that can easily be evaluated but is not necessary. Measurements of $I_{\perp}(\Omega)$ and $I_{\parallel}(\Omega)$ thus allow the direct determination of the imaginary part of $\chi^{(r)}_{ijkl}$. Knowing Im $\chi^{(r)}_{ijkl}$ as a function of frequency, we can determine the real part of $\chi^{(r)}_{ijkl}$ by performing the Kramers–Kronig integration using Eq. (6.251). Therefore, the spontaneous Raman-scattering spectrum uniquely determines the nonlinear susceptibility $\chi^{(r)}_{ijkl}$. Hellwarth *et al.* (1975) applied this procedure to determine the resonant part of the third-order optical susceptibility of glasses.

Finally, we shall consider the Raman-induced Kerr-effect. Keep in mind that the signal associated with the Raman–Kerr effect is detected at the probe-wave frequency ω_2 and not at $2\omega_1 - \omega_2$, as in the case of the CARS signal. As previously mentioned, to study this effect we assume the pump field to be circularly polarized in the xy plane and the probe field linearly polarized along the y axis. We further assume that the strength of the pump field (at frequency ω_1) is greater than that of the probe field (at frequency ω_2). The complex amplitude $P^{(3)}_x$ of the third-order nonlinear polarization density at probe frequency ω_2 was given in Eq. (6.155). To find the expression

$$[\chi^{(3)}_{xyxy}(-\omega_2, \omega_2, -\omega_1, \omega_1) - \chi^{(3)}_{xyyx}(-\omega_2, \omega_2, -\omega_1, \omega_1)],$$

we substitute Eq. (6.246) into (6.239) and obtain

$$P^{(3)}_i(\omega_2) = \tfrac{3}{8}\chi^{(3)}_{ijkl}(-\omega_2, \omega_2, -\omega_1, \omega_1)\varepsilon_j(\omega_2)\varepsilon_k(\omega_1)\varepsilon^*_l(\omega_1), \tag{6.263}$$

where the third-order susceptibility is given by

$$\chi^{(3)}_{ijkl}(-\omega_2, \omega_2, -\omega_1, \omega_1) = n[\tfrac{1}{4}\langle\gamma_{ijkl}\rangle + \tfrac{1}{6}\chi^{(r)}_{ijkl}(0) + \tfrac{1}{6}\chi^{(r)}_{ijkl}(\omega_1 - \omega_2)]. \tag{6.264}$$

Using Eqs. (6.244) and (6.245), we thus obtain

$$\chi^{(3)}_{xyxy}(-\omega_2, \omega_2, -\omega_1, \omega_1) - \chi^{(3)}_{xyyx}(-\omega_2, \omega_2, -\omega_1, \omega_1)$$
$$= \tfrac{1}{6}[\chi^{(r)}_{xxyy}(\omega_1 - \omega_2) - \chi^{(r)}_{xyxy}(\omega_1 - \omega_2)]$$
$$= \tfrac{1}{6}[A(\omega_1 - \omega_2) - \tfrac{1}{2}B(\omega_1 - \omega_2)]. \tag{6.265}$$

Substituting Eq. (6.265) into (6.155) gives $P^{(3)}_x(z, t)$ as

$$P^{(3)}_x(z, t) = \tfrac{3}{96}[A(\omega_1 - \omega_2) - \tfrac{1}{2}B(\omega_1 - \omega_2)]\varepsilon_y(\omega_2)|\varepsilon(\omega_1)|^2 e^{i(\omega_2 t - K_2 z)}. \tag{6.266}$$

Clearly, in the Raman-induced Kerr-effect experiment, the nonresonant background owing to $\langle\gamma_{ijkl}\rangle$ and $\chi^{(0)}_{ijkl}$, is eliminated, leaving the signal very nearly equal to the Raman spectrum. It is because of this ability to eliminate the nonresonant background that the effective Raman signal-to-noise ratio in the Raman-induced Kerr-effect experiment can be much greater, and achieved in a much shorter time, than in spontaneous Raman scattering. In addition, the fact that the signal resulting from the Raman-induced Kerr effect can be partially separated from other light, such as fluorescence, makes this technique a very promising tool for resolving small Raman shifts in a medium beset with fluorescent background.

Appendix: Linear-Response Theory and Fluctuation Dissipation Theorem Including the Spatial Dependence

In a continuum system the dynamic variable varies with space and time. In this appendix we include the spatial dependence in the development of linear-response theory and the fluctuation–dissipation theorem. The derivation is carried out quantum mechanically.

A.1 DERIVATION OF A GENERALIZED LINEAR-RESPONSE FUNCTION

Consider a set of dynamic variables $A_1, A_2, \ldots, A_n$. Each of these is coupled to external fields $F_1, F_2, \ldots, F_n$. The Hamiltonian describing the material system as well as the material-field interaction is given by

$$\mathscr{H}(t) = \mathscr{H}_0 + \mathscr{H}_{\text{ext}}(t), \tag{A.1}$$

where $\mathscr{H}_0$ is the Hamiltonian of the system in the absence of the external field and $\mathscr{H}_{\text{ext}}(t)$ describes the matter–field interaction and is given by

$$\mathscr{H}_{\text{ext}}(t) = -\sum_{j} \int d^3r\, A_j^+(\mathbf{r})F_j(\mathbf{r}, t). \tag{A.2}$$

The statistical property of the system in the presence of the external field is described by the density operator $\rho(t)$, which obeys the Heisenberg equation of motion:

$$\frac{\partial}{\partial t}\rho(t) = -\frac{i}{h}[\mathscr{H}(t), \rho(t)]$$

or

$$\frac{\partial}{\partial t}\rho(t) = -\frac{i}{\hbar}[\mathscr{H}_0, \rho(t)] - \frac{i}{h}[\mathscr{H}_{\text{ext}}(t), \rho(t)]. \tag{A.3}$$

We shall assume that the external fields are turned on at $t = 0$. For $t < 0$, the system is in thermal equilibrium described by the equilibrium density operator $\rho_e = e^{-\beta\mathscr{H}_0/Z}$, where $\beta = 1/kT$ and $Z = \text{Tr}\, e^{-\beta\mathscr{H}_0}$ is the canonical partition function. We further assume that the fields $F_1, F_2, \ldots, F_n \equiv \{F_j\}$ are small so that Eq. (A.3) may be solved by the power series expansion of $\{F_j\}$.

Clearly, if the external fields are absent, $\mathscr{H}_{\text{ext}}(t) = 0$. In this case, the solution of Eq. (A.3) is given by

$$\rho(t) = e^{-i\mathscr{H}_0 t/\hbar}\rho(t = 0)e^{i\mathscr{H}_0 t/\hbar} = \rho_e = \rho_0. \tag{A.4}$$

In the presence of the external fields, we assume that the solution to Eq. (A.3) is given by

$$\rho = \rho_0 + \rho_1 + \cdots, \tag{A.5}$$

where ρ_0 is the equilibrium density operator; ρ_1 depends linearly on the external fields; the terms not written out explicitly depend on second and higher powers of the external fields. They represent nonlinear responses and are neglected in the present calculation.

Substituting Eq. (A.5) into Eq. (A.4), we obtain

$$\frac{\partial}{\partial t}\{\rho_0 + \rho_1 + \cdots\}$$
$$= -\frac{i}{\hbar}[\mathscr{H}_0, \rho_0 + \rho_1 + \cdots] - \frac{i}{h}[\mathscr{H}_{\text{ext}}(t), \rho_0 + \rho_1 + \cdots]. \tag{A.6}$$

From Eq. (A.6) we obtain the zero-order equation as

$$\frac{\partial}{\partial t}\rho_0 = -\frac{i}{\hbar}[\mathscr{H}_0, \rho_0] \tag{A.7}$$

and the first-order equation as

$$\frac{\partial}{\partial t}\rho_1 = -\frac{i}{\hbar}[\mathscr{H}_0, \rho_1] - \frac{i}{\hbar}[\mathscr{H}_{\text{ext}}(t), \rho_0]. \tag{A.8}$$

Clearly, the solution to Eq. (A.7) is $\rho_0 = \rho_e$. It is easy to show that the solution to Eq. (A.8) is given by

$$\rho_1 = -\frac{i}{\hbar}\int_0^t dr' \exp[-i\mathscr{H}_0(t-t')/\hbar]/[\mathscr{H}_{\text{ext}}(t'), \rho_e]\exp[i\mathscr{H}_0(t-t')/\hbar]. \tag{A.9}$$

The nonequilibrium expectation value of the dynamic variable $A_i(\mathbf{r})$ is given by

$$\begin{aligned}\langle A_i(\mathbf{r}, t)\rangle_{\text{non}} &= \text{Tr}\{A_i(\mathbf{r})\rho(t)\} = \text{Tr}\{A_i(\mathbf{r})[\rho_0 + \rho_1 + \cdots]\} \\ &= \langle A_i(\mathbf{r})\rangle + \text{Tr}\{A_i(\mathbf{r})\rho_1\}.\end{aligned} \tag{A.10}$$

The fluctuation in A_i is thus given by

$$\begin{aligned}\langle \delta A_i(\mathbf{r}, t)\rangle &\equiv \langle A_i(\mathbf{r}, t)\rangle_{\text{non}} - \langle A_i(\mathbf{r})\rangle \\ &= -\frac{i}{\hbar}\int_0^t dt' \,\text{Tr}\{A_i(\mathbf{r})\exp[-i\mathscr{H}_0(t-t')/\hbar] \\ &\quad \times [\mathscr{H}_{\text{ext}}(t'), \rho_e]\exp[i\mathscr{H}_0(t-t')/\hbar]\}.\end{aligned} \tag{A.11}$$

Using the fact that $\text{Tr}(BC) = \text{Tr}(CB)$ and

$$\text{Tr}\, B[C, D] = \text{Tr}[B, C]D,$$

we can rewrite Eq. (A.11) as

$$\langle \delta A_i(\mathbf{r}, t)\rangle = \frac{i}{\hbar}\sum_j \int_0^t dr' \int d^3r' \langle [A_i(\mathbf{r}, t), A_j^+(\mathbf{r}', t')]\rangle F_j(\mathbf{r}', t'), \tag{A.12}$$

where $A_i(\mathbf{r}, t)$ is the Heisenberg operator and is given by

$$A_i(\mathbf{r}, t) = e^{i\mathscr{H}_0 t/\hbar} A_i(\mathbf{r}) e^{-i\mathscr{H}_0 t/\hbar}. \tag{A.13}$$

We define the response function matrix $\phi_{ij}(\mathbf{r}, t; \mathbf{r}', t')$ as

$$\phi_{ij}(\mathbf{r}, t; \mathbf{r}', t) = \frac{i}{\hbar}\langle [A_i(\mathbf{r}, t), A_j^+(\mathbf{r}', t')]\rangle \tag{A.14}$$

and rewrite Eq. (A.12) as

$$\langle \delta A_i(\mathbf{r}, t)\rangle = \sum_j \int_0^t dt' \int d^3r' \; \phi_{ij}(\mathbf{r}, t_i; \mathbf{r}', t')F_j(\mathbf{r}', t'). \tag{A.15}$$

For an equilibrium ensemble, $\phi_{ij}(\mathbf{r}, t; \mathbf{r}', t')$ is invariant in time translation, that is,

$$\phi_{ij}(\mathbf{r}, t; \mathbf{r}', t') = \phi_{ij}(\mathbf{r}, t - t'; \mathbf{r}', 0) \tag{A.16}$$

For spatially homogeneous systems, ϕ_{ij} does not depend on $\mathbf{r}$ and $\mathbf{r}'$ separately; rather it depends only on $\mathbf{r} - \mathbf{r}'$; thus, $\phi_{ij}(\mathbf{r}, t; \mathbf{r}', t') = \phi_{ij}(\mathbf{r} - \mathbf{r}', t - t')$. Therefore, Eq. (A.15) can be written as

$$\langle \delta A_i(\mathbf{r}, t)\rangle = \sum_j \int_0^t dt' \int d^3r' \; \phi_{ij}(\mathbf{r} - \mathbf{r}', t - t')F_j(\mathbf{r}', t'). \tag{A.17}$$

Equation (A.15), or Eq. (A.17) for spatially homogeneous system, is the basic result of linear-response theory. The linear response in A_i at position $\mathbf{r}$ and time t owing to the application of the external fields $F_1, F_2, \ldots, F_n$ at position $\mathbf{r}'$ and earlier time t' is found to be a convolution integral of time and space. Note that the response in A_i is dynamically coupled to other variables.

Equation (A.17) can be converted into $(\mathbf{q}, z)$ space by taking the Fourier transform in space and the Laplace transform in time. We define

$$\langle \delta A_i(\mathbf{q}, z)\rangle = \int d^3r \int_0^\infty dt \exp[i(\mathbf{q} \cdot \mathbf{r} - zt)] \; \delta\langle A_i(\mathbf{r}, t)\rangle \tag{A.18}$$

and rewrite Eq. (A.17) by taking its Fourier–Laplace transform

$$\langle \delta A_i(\mathbf{q}, z)\rangle = \sum_j \chi_{ij}(\mathbf{q}, z)F_j(\mathbf{q}, z), \tag{A.19}$$

where $\chi_{ij}(\mathbf{q}, z)$ is called the generalized susceptibility tensor and is given by

$$\chi_{ij}(\mathbf{q}, z) = \int d^3r \int_0^\infty dt \exp[i(\mathbf{q} \cdot \mathbf{r} - zt)]\phi_{ij}(\mathbf{r}, t). \tag{A.20}$$

Similarly,

$$F_j(\mathbf{q}, z) = \int d^3r \int_0^\infty dt \exp[i(\mathbf{q} \cdot \mathbf{r} - zt)]F_j(\mathbf{r}, t). \tag{A.21}$$

Equation (A.19) is obtained from Eq. (A.18) by using the convolution theorem.

The above calculation was carried out using quantum mechanics. Though the classical result can be obtained as a special case of the quantal calculation, a purely classical calculation can be carried out in a similar manner as given above provided that Tr is replaced by integration over the phase-space coordinates and that the commutator is replaced by the Poissan bracket. We shall not consider the classical formulation, since the formal result is identical to the quantal one.

A.2 GENERALIZED SUSCEPTIBILITY TENSOR $\chi_{ij}(\mathbf{q}, \omega)$

In Eq. (A.20) we defined the generalized susceptibility in terms of the complex Laplace transform. We now set $z = \omega$, the real frequency. Thus we have

$$\chi_{ij}(\mathbf{q}, \omega) = \int_0^\infty dt\, e^{-i\omega t}\phi_{ij}(\mathbf{q}, t). \tag{A.22}$$

Using Eq. (A.14) it is easy to show that $\phi_{ij}(\mathbf{q}, t)$ is real. One can also show that $\phi_{ij}(\ , t)$ decays to zero at large t. It then follows that $\chi_{ij}(\mathbf{q}, \omega)$ exists and is a complex function of ω. Thus, we can write

$$\chi_{ij}(\mathbf{q}, \omega) = \chi'_{ij}(\mathbf{q}, \omega) - i\chi''_{ij}(\mathbf{q}, \omega). \tag{A.23}$$

Clearly, from Eq. (A.22) we obtain

$$\chi^*_{ij}(\mathbf{q}, \omega) = \chi_{ij}(\mathbf{q}, -\omega) = \chi'_{ij}(\mathbf{q}, -\omega) - i\chi''_{ij}(\mathbf{q}, -\omega) \tag{A.24}$$

Using Eqs. (A.23) and (A.24), it follows that χ'_{ij} is even in ω,

$$\chi'_{ij}(\mathbf{q}, \omega) = \chi'_{ij}(\mathbf{q}, -\omega) \tag{A.25}$$

and χ'' is odd in ω,

$$\chi''_{ij}(\mathbf{q}, \omega) = -\chi''_{ij}(\mathbf{q}, -\omega). \tag{A.26}$$

It is useful to introduce a relaxation function $R_{ij}(\mathbf{q}, t)$ given by

$$\phi_{ij}(\mathbf{q}, t) = -\frac{\partial R_{ij}(\mathbf{q}, t)}{\partial t}. \tag{A.27}$$

As implied by the name, the relaxation function is finite at $t = 0$ and relaxes to zero at long time. Using the relaxation function we can rewrite Eq. (A.22) as

$$\chi_{ij}(\mathbf{q}, \omega) = -\int_0^\infty dt\, e^{-i\omega t}\frac{\partial R_{ij}(\mathbf{q}, t)}{\partial t}, \tag{A.28}$$

which upon integration by parts yields

$$\chi_{ij}(\mathbf{q}, \omega) = R_{ij}(\mathbf{q}, t = 0) - i\omega \int_0^\infty dt\, e^{-i\omega t} R_{ij}(\mathbf{q}, t). \tag{A.29}$$

The second term vanishes identically at $\omega = 0$; thus

$$\chi_{ij}(\mathbf{q}, \omega = 0) = R_{ij}(\mathbf{q}, t = 0).$$

Therefore, the initial value of the relaxation function represents the zero frequency susceptibility. We denote the zero frequency (static) susceptibility as $\chi_{ij}(\mathbf{q})$.

Using Eq. (A.29), we obtain

$$\chi'_{ij}(\mathbf{q}, \omega) = \chi_{ij}(\mathbf{q}) - \omega \int_0^\infty dt\, R_{ij}(\mathbf{q}, t) \sin \omega t \tag{A.30}$$

and

$$\chi''_{j}(\mathbf{q}, \omega) = \omega \int_0^\infty dt\, R_{ij}(\mathbf{q}, t) \cos \omega t. \tag{A.31}$$

Since $\chi''_{ij}(\mathbf{q}, \omega)$ is an odd function of ω, we can invert Eq. (A.31) and obtain

$$\frac{1}{4} \{R_{ij}(\mathbf{q}, t) + R_{ij}(\mathbf{q}, -t)\} = \frac{1}{2\pi} \int_{-\infty}^\infty d\omega\, e^{i\omega t} \frac{\chi''_{ij}(\mathbf{q}, \omega)}{\omega}. \tag{A.32}$$

In most cases $R_{ij}(\mathbf{q}, t)$ is an even function of time, and as such the relaxation function can be written as

$$R_{ij}(\mathbf{q}, t) = \frac{1}{\pi} \int_{-\infty}^\infty d\omega\, e^{i\omega t} \frac{\chi''_{ij}(\mathbf{q}, \omega)}{\omega}. \tag{A.33}$$

Using Eq. (A.27) we find that the response function is given by

$$\phi_{ij}(\mathbf{q}, t) = -\frac{i}{\pi} \int_{-\infty}^\infty d\omega\, e^{i\omega t} \chi''_{ij}(\mathbf{q}, \omega). \tag{A.34a}$$

The Fourier inversion of this equation is

$$\chi''_{ij}(\mathbf{q}, \omega) = \frac{i}{2} \int_{-\infty}^\infty dt\, e^{-i\omega t} \phi_{ij}(\mathbf{q}, t). \tag{A.34b}$$

According to Eq. (A.34a), the generalized susceptibility $\chi_{ij}(\mathbf{q}, z)$ can also be written as

$$\chi_{ij}(\mathbf{q}, z) = \int_0^\infty dt\, e^{-izt} \phi_{ij}(\mathbf{q}, t) = -\frac{i}{\pi} \int_0^\infty dt\, e^{-izt} \int_{-\infty}^\infty d\omega\, e^{i\omega t} \chi''_{j}(\mathbf{q}, \omega),$$

which on the interchange of the order of integration gives

$$\chi_{ij}(\mathbf{q}, z) = \frac{1}{\pi} \int_{-\infty}^{\infty} d\omega \frac{\chi''_{ij}(\mathbf{q}, \omega)}{\omega - z}. \tag{A.35}$$

From Eq. (A.35), we obtain a thermodynamic sum rule as

$$\chi_{ij}(\mathbf{q}) = \frac{1}{\pi} \int_{-\infty}^{\infty} d\omega \frac{\chi''_{ij}(\mathbf{q}, \omega)}{\omega}. \tag{A.36}$$

Setting $z = \omega - i\varepsilon$, we find that in the limit of infinitesimal ε, the Kramers–Kronig relation

$$\chi'_{ij}(\mathbf{q}, \omega) = \frac{1}{\pi} \mathrm{P} \int_{-\infty}^{\infty} d\omega' \frac{\chi''_{ij}(\mathbf{q}, \omega')}{\omega' - \omega}, \tag{A.37}$$

where P indicates taking the principal value.

A.3 LINEAR-RESPONSE THEORY IN TERMS OF OF AN INITIAL VALUE PROBLEM

In Eq. (A.17) or (A.19) we expressed the linear response of the dynamic variable as in terms of the externally applied fields $\{F_j\}$. For problems in which external fields are actually applied (such as in the NMR or dielectric relaxation techniques), Eqs. (A.17) and (A.19) are very convenient in analyzing the experimental result. For other problems in which the external fields are not explicitly applied and in which the external fields cannot be explicitly assessed, the application of Eqs. (A.17) and (A.19) is then difficult.

Thus, it is advantageous to eliminate the external field in Eq. (A.19) and to express the dynamic response in terms of an initial value problem.

This can be carried out by considering that the external field was adiabatically switched on at $t = -\infty$ and is switched off at $t = 0$. Namely, we have

$$F_j(\mathbf{r}, t) = \begin{cases} F_j(\mathbf{r})e^{\varepsilon t} & \text{for} \quad t < 0, \\ 0 & \text{for} \quad t \geq 0, \end{cases} \tag{A.38}$$

where ε is infinitesimal.

Using Eq. (A.18) and following the same procedure that leads to Eq. (A.17), we have found that the response in A_i is given by

$$\langle \delta A_i(\mathbf{q}, t) \rangle = \sum_j \int_{-\infty}^{0} dt' \, \phi_{ij}(\mathbf{q}, t - t') F_j(\mathbf{q}) e^{\varepsilon t'}, \tag{A.39}$$

which upon the change of the variable by letting $t - t' = \tau$ gives

$$\langle \delta A_i(\mathbf{q}, t) \rangle = \sum_j \int_t^\infty d\tau\, \phi_{ij}(\mathbf{q}, \tau) F_j(\mathbf{q}) e^{\varepsilon(t-\tau)}. \tag{A.40}$$

From Eq. (A.40) we find that the linear response at $t = 0$ is given by

$$\begin{aligned} \langle \delta A_i(\mathbf{q}, t = 0) \rangle &= \sum_j \int_0^\infty d\tau\, \phi_{ij}(\mathbf{q}, \tau) F_j(\mathbf{q}) e^{-\varepsilon\tau} \\ &= \sum_j \chi_{ij}(\mathbf{q}) F_j(\mathbf{q}) = [\boldsymbol{\chi}(\mathbf{q})\mathbf{F}]_i, \end{aligned} \tag{A.41}$$

where $\boldsymbol{\chi}$ is a square matrix denoting the static susceptibility tensor. $\mathbf{F}$ denotes a column vector representing the external forces.

We next rewrite Eq. (A.40) as

$$\begin{aligned} \langle \delta A_i(\mathbf{q}, t) \rangle = &- \sum_j \int_0^t dt\, \phi_{ij}(\mathbf{q}, \tau) F_j(\mathbf{q}) e^{\varepsilon(t-\tau)} \\ &+ \sum_j \int_0^\infty d\tau\, \phi_{ij}(\mathbf{q}, t) F_j(\mathbf{q}) e^{\varepsilon(t-\tau)}, \end{aligned} \tag{A.42}$$

which upon taking the complex Laplace transform of Eq. (A.42) yields

$$\begin{aligned} \langle \delta \hat{A}_i(\mathbf{q}, z) \rangle &= \int_0^\infty dt\, e^{-izt} \langle \delta A_i(\mathbf{q}, t) \rangle \\ &= -\frac{1}{iz} \sum_j \chi_{ij}(\mathbf{q}, z) F_j(\mathbf{q}) + \frac{1}{iz} \sum_j \chi_{ij}(\mathbf{q}) F_j(\mathbf{q}) \end{aligned} \tag{A.43}$$

or

$$\begin{aligned} \langle \delta \hat{A}_i(\mathbf{q}, z) \rangle &= \frac{1}{iz} \{ -\boldsymbol{\chi}(\mathbf{q}, z) \cdot \mathbf{F}(\mathbf{q}) + \boldsymbol{\chi}(\mathbf{q}) \cdot \mathbf{F}(\mathbf{q}) \}_i \\ &= \frac{1}{iz} \{ [-\boldsymbol{\chi}(\mathbf{q}, z) \cdot \boldsymbol{\chi}^{-1}(\mathbf{q}) + \mathbf{I}] \cdot \boldsymbol{\chi}(\mathbf{q}) \cdot \mathbf{F}(\mathbf{q}) \}_i. \end{aligned} \tag{A.44}$$

The external force can now be eliminated by using Eq. (A.41), thus Eq. (A.44) becomes

$$\langle \delta \hat{A}_i(\mathbf{q}, z) \rangle = \frac{1}{iz} \sum_j \{ [-\boldsymbol{\chi}(\mathbf{q}, z) \cdot \boldsymbol{\chi}^{-1}(\mathbf{q}) + \mathbf{I}]_{ij} \langle \delta A_j(\mathbf{q}, t = 0) \rangle \}. \tag{A.45}$$

In Eq. (A.45) we have succeeded in expressing the linear response in A_i in terms of an initial value problem.

As an example we consider the diffusion of macromolecules in a solvent consisting of small molecules. The concentration of the macromolecules is given by n. The concentration fluctuation follows the diffusion equation

$$\frac{\partial}{\partial t} \langle \delta n(\mathbf{r}, t) \rangle = D \nabla^2 \langle \delta n(\mathbf{r}, t) \rangle. \tag{A.46}$$

The Fourier–Laplace transform of Eq. (A.46) gives

$$iz\langle\delta\hat{n}(\mathbf{q}, z)\rangle - \langle\delta n(\mathbf{q}, t = 0)\rangle + Dq^2\langle\delta\hat{n}(\mathbf{q}, z)\rangle = 0$$

or

$$\langle\delta\hat{n}(\mathbf{q}, z)\rangle = \frac{1}{iz + Dq^2}\langle\delta n(\mathbf{q}, t = 0)\rangle. \tag{A.47}$$

Comparing Eq. (A.47) with Eq. (A.45) we find that

$$\frac{1}{iz}\{-\chi(\mathbf{q}, z)\chi^{-1}(\mathbf{q}) + 1\} = \frac{1}{iz + Dq^2}$$

or

$$\chi(\mathbf{q}, z) = \frac{Dq^2}{iz + Dq^2}\chi(\mathbf{q}) \Rightarrow \chi''(q, \omega) = \frac{Dq^2\omega}{\omega^2 + (Dq^2)^2}, \tag{A.48}$$

which is equivalent to Eq. (2.52) if $\chi(\mathbf{q}, \omega)$ is identified with $\omega S(\mathbf{q}, \omega)/N$.

A.4 FLUCTUATION–DISSIPATION THEOREM

We have so far considered the calculation of the linear response due to the interaction of a material system with an external field. Attention is now given to the calculation of the rate of change of the energy of the system due to the application of the external field. The application of the field to the system originally at equilibrium will do work on the system and cause the energy to be absorbed by the system. The energy gained by the system will be dissipated as heat because of the fluctuations of translational and rotational degrees of freedom owing to molecular motion that takes place within the material system. The fluctuation–dissipation theorem gives a quantitative description of the relation between the energy absorption in terms of the time-correlation function that describes the fluctuations. Following the procedure of Chapter 1, we first show that the energy absorption is related to χ''.

In the presence of an external field, the total energy of the system is equal to $\varepsilon(t) = \mathrm{Tr}\{\mathscr{H}(t)\rho(t)\}$. The rate of energy change of the system is equal to

$$\begin{aligned}\frac{\partial\varepsilon(t)}{\partial t} &= \mathrm{Tr}\left\{\frac{\partial\mathscr{H}(t)}{\partial t}\rho(t) + \mathscr{H}(t)\frac{\partial\rho(t)}{\partial t}\right\} \\ &= \mathrm{Tr}\left\{\frac{\partial\mathscr{H}(t)}{\partial t}\rho(t) + \frac{i}{\hbar}\mathscr{H}(t)[\mathscr{H}(t), \rho(t)]\right\}.\end{aligned} \tag{A.49}$$

The second term on the right-hand side of Eq. (A.49) vanishes identically because of the property of the trace. Furthermore, since

$$\mathscr{H}(t) = \mathscr{H}_0 - \sum_j \int d^3r\, A_j^+(\mathbf{r})F_j(\mathbf{r}, t)$$

$$= \mathscr{H}_0 - \sum_j \int d^3r\, A_j(\mathbf{r})F_j^*(\mathbf{r}, t), \tag{A.50}$$

we find

$$\frac{\partial \varepsilon(t)}{\partial t} = -\sum_j \int d^3r\, \mathrm{Tr}\{A_j(\mathbf{r})\rho(t)\} \frac{\partial F_j^*(\mathbf{r}, t)}{\partial t}$$

$$= -\sum_j \int d^3r\, \mathrm{Tr}\{A_j(\mathbf{r})[\rho_0 + \rho_1 + \cdots]\} \frac{\partial F_j^*(\mathbf{r}, t)}{\partial t}$$

$$= -\sum_j \int d^3r\, \{\langle A_j(\mathbf{r})\rangle + \langle \delta A_j(\mathbf{r}, t)\rangle\} \frac{\partial F_j^*(\mathbf{r}, t)}{\partial t}. \tag{A.51}$$

If the field is described by superposition of normal modes,

$$F_j(\mathbf{r}, t) = \sum_{\mathbf{q}, \omega} F_j(\mathbf{q}, \omega) \exp[-i(\mathbf{q} \cdot \mathbf{r} - \omega t)] \tag{A.52}$$

Then Eq. (A.51) becomes

$$\frac{\partial \varepsilon(t)}{\partial t} = i\omega \sum_j \sum_{\mathbf{q}, \omega} \{\langle A_j(\mathbf{q})\rangle + \langle \delta A_j(\mathbf{q}, t)\rangle\} F_j^*(\mathbf{q}, \omega) e^{-i\omega t}. \tag{A.53}$$

The average increase of the total energy of the system over a period of $2T$ is

$$\Delta\varepsilon = \int_{-T}^{T} dt\, \frac{\partial \varepsilon(t)}{\partial t} = i \sum_j \sum_{\mathbf{q}, \omega} \int_{-T}^{T} dt\, \omega \langle \delta A_j(\mathbf{q}, t)\rangle F_j^*(\mathbf{q}, \omega) e^{-i\omega t}, \tag{A.54}$$

where the first term drops out because of the average over t.

We assume that the period T is long compared with the material system's characteristic relaxation times, which are in turn longer than the period of the external field. In this case we can extend T to infinity. As a result, we obtain

$$\Delta\varepsilon = i \sum_j \sum_{\mathbf{q}, \omega} \sum_k \int_{-\infty}^{\infty} dt\, \omega e^{-i\omega t} \int_0^t dt\, \{\phi_{jk}(\mathbf{q}, t - \tau) F_k(\mathbf{q}, \tau)\} F_j^*(\mathbf{q}, \omega)$$

or

$$\Delta\varepsilon = 2 \sum_j \sum_k \sum_{\mathbf{q}, \omega} \omega F_j^*(\mathbf{q}, \omega) \chi''_{jk}(\mathbf{q}, \omega) F_k(\mathbf{q}, \omega). \tag{A.55}$$

To obtain Eq. (A.55), we have used Eq. (A.346) and the property of the convolution integral.

Note that the energy absorption is proportional to the imaginary part of the generalized susceptibility. We have thus succeeded in obtaining the first part required in the proof of the fluctuation–dissipation theorem. The second part of the proof deals with relating $\chi''_{jk}(\mathbf{q}, \omega)$ to the time-correlation function. The time-correlation function describes fluctuations of the dynamic variable in the system.

Since the response function $\phi_{jk}(\mathbf{q}, t)$ is given by [c.f. Eq. (A.13)]

$$\begin{aligned} \phi_{jk}(\mathbf{q}, t) &= \frac{i}{\hbar}\langle[A_j(\mathbf{q}, t), A_k^+(\mathbf{q})]\rangle \\ &= \frac{i}{\hbar}\langle\{A_j(\mathbf{q}t)A_k^+(\mathbf{q}) - A_k^+(\mathbf{q})A_j(\mathbf{q}t)\}\rangle, \end{aligned} \tag{A.56}$$

we can rewrite the second term in Eq. (A.56) as

$$\begin{aligned} \langle A_k^+(\mathbf{q})A_j(\mathbf{q}, t)\rangle &= \frac{1}{Z}\operatorname{Tr}\{e^{-\beta\mathscr{H}_0}A_k^+ e^{i\mathscr{H}_0 t/\hbar}A_j e^{-i\mathscr{H}_0 t/\hbar}\} \\ &= \frac{1}{Z}\operatorname{Tr}\{e^{-\beta\mathscr{H}_0}e^{i\mathscr{H}_0 t/\hbar}e^{\beta\mathscr{H}_0}A_j e^{-i\mathscr{H}_0 t/\hbar}e^{-\beta\mathscr{H}_0}A_k^+\} \\ &= \frac{1}{Z}\operatorname{Tr}\{e^{-\beta\mathscr{H}_0}A_j(t - i\beta\hbar)A_k^+\} = \langle A_j(\mathbf{q}, t - i\beta\hbar)A_k^+(\mathbf{q})\rangle. \end{aligned} \tag{A.57}$$

Setting $i = j$ and $j = k$, we first rewrite Eq. (A.34b) as

$$\chi''_{jk}(\mathbf{q}, -\omega) = -2i\int_{-\infty}^{\infty} dt\, e^{i\omega t}\phi_{jk}(\mathbf{q}, t). \tag{A.58}$$

Then substituting Eqs. (A.56) and (A.57) into Eq. (A.58), we obtain

$$\chi''_{jk}(\mathbf{q}, -\omega) = \frac{2}{h}\int_{-\infty}^{\infty} dt\, e^{i\omega t}\{\langle A_j(\mathbf{q}, t)A_k^+(\mathbf{q})\rangle - \langle A_j(\mathbf{q}, t - i\beta\hbar)A_k^+(\mathbf{q})\rangle\}. \tag{A.59}$$

We consider the Fourier transform of the second form of Eq. (A.59):

$$I = \int_{-\infty}^{\infty} dt\, e^{i\omega t}\langle A_j(\mathbf{q}, t - i\beta\hbar)A_k^+(\mathbf{q})\rangle.$$

Change the integration variable from t to $\tau = t - i\beta\hbar$ and obtain

$$I = \int_{-\infty + i\beta\hbar}^{\infty + i\beta\hbar} dt \exp[i\omega(\tau + i\beta\hbar)]\langle A_j(\mathbf{q}, \tau)A_k^+(\mathbf{q})\rangle$$

$$= e^{\beta\hbar\omega} \int_{-\infty}^{\infty} d\tau \, \langle A_j(\mathbf{q}, \tau)A_k^+(\mathbf{q})\rangle e^{i\omega\tau}. \tag{A.60}$$

Thus, Eq. (A.60) becomes

$$\chi''_{jk}(\mathbf{q}, -\omega) = \frac{2}{\hbar}(1 - e^{-\beta\hbar\omega}) \int_{-\infty}^{\infty} dt \, e^{i\omega t}\langle A_j(\mathbf{q}, t)A_k^+(\mathbf{q})\rangle. \tag{A.61}$$

Equation (A.61) is the fluctuation–dissipation theorem as it relates $\chi''_{jk}(\mathbf{q}, -\omega)$ to the time-correlation function. Using the fact that χ''_{jk} is an odd function of ω, we can express Eq. (A.61) in terms of other forms. One of them is

$$\chi''_{jk}(\mathbf{q}, \omega) = \frac{2\pi}{\hbar} \operatorname{tank}\left(\frac{\beta\hbar\omega}{2}\right)\hat{C}_{jk}(\mathbf{q}, \omega), \tag{A.62}$$

where

$$\hat{C}_{jk}(\mathbf{q}, \omega) = \frac{1}{2\pi} \int_{-\infty}^{\infty} dt \, e^{i\omega t}\langle \tfrac{1}{2}\{A_j(\mathbf{q}, t)A_k^+(\mathbf{q}) + A_k^+(\mathbf{q})A_j(\mathbf{q}, t)\}\rangle. \tag{A.63}$$

A.5 APPLICATION TO LINEAR VISCOELASTIC SYSTEMS

As stated in Chapter 2, in a linear viscoelastic system the most general constitutive equation for the stress–strain relation is given by

$$\sigma_{ij}(t) = -P\,\delta_{ij} + \int_0^t dt' \, \Gamma_{ijkl}(t - t')\dot{\gamma}_{kl}(t'), \tag{A.64}$$

where σ_{ij} represents the component of the symmetric stress tensor. $\dot{\gamma}_{kl}$ is the rate-of-strain tensor and can be expressed in terms of the velocity gradient tensor as $\dot{\gamma}_{ij} = (\partial V_i/\partial x_j + \partial V_j/\partial x_i)/2$. Γ_{ijkl} is the stress modulus tensor. P is the hydrostatic pressure.

Newton's equation of motion for continuum is

$$\rho_m \frac{dV_i}{dt} = \rho_m\left(\frac{\partial V_i}{\partial t} + V_i \frac{\partial V_i}{\partial x_i}\right) = \frac{\partial \sigma_{ij}}{\partial x_i}. \tag{A.65}$$

The equation of continuity for the particle flow is

$$\frac{\partial \rho_m}{\partial t} = -\frac{\partial(\rho_m V_i)}{\partial x_i}, \tag{A.66}$$

where ρ_m is the mass density, equal to $m\rho$. ρ is the number density.

Substituting Eq. (A.64) into Eq. (A.66), letting $\rho = \rho_0 + \delta\rho$, and linearizing Eqs. (A.65) and (A.66), we obtain a set of two coupled equations for isotropic media:

$$\frac{\partial}{\partial t}\,\delta\rho(\mathbf{r}, t) = -\rho_0 \nabla \cdot \mathbf{V}(\mathbf{r}, t), \tag{A.67}$$

$$\rho_m^0 \frac{\partial}{\partial t} V_i(\mathbf{r}, t) = -\frac{\partial P}{\partial x_i} + \int_0^t dt' \left\{ \left[K(t - t') + \frac{1}{3} G(t - t') \right] \times \frac{\partial}{\partial x_i} \nabla \cdot \mathbf{V}(t') + G(t - t')\nabla^2 V_i(t') \right\}. \tag{A.68}$$

Equations (A.67) and (A.68) can be converted into an initial value problem. We neglect the effect of temperature fluctuation on pressure and write

$$\frac{\partial P}{\partial x_i} = \left(\frac{\partial P}{\partial \rho}\right)_T \left(\frac{\partial \rho}{\partial x_i}\right) = \frac{1}{\rho_0 \chi_T} \left(\frac{\partial \rho}{\partial x_i}\right). \tag{A.69}$$

We next define

$$\langle \delta \hat{A}_1(\mathbf{q}, z)\rangle = \delta\hat{\rho}(\mathbf{q}, z) = \int d^3r \int_0^\infty dt \exp[i(\mathbf{q}\cdot\mathbf{r} - zt)]\, \delta\rho(\mathbf{r}, t), \tag{A.70a}$$

$$\langle \delta \hat{A}_2(\mathbf{q}, z)\rangle = \delta p(\mathbf{q}, z) = \int d^3r \int_0^\infty dt \exp[i(\mathbf{q}\cdot\mathbf{r} - zt)] p(\mathbf{r}, t), \tag{A.70b}$$

where $p(\mathbf{q}, z)$ is momentum density in the direction of $\mathbf{q}$.

Taking the Fourier–Laplace transform of Eqs. (A.67) and (A.68), we convert these equations into an initial value problem:

$$iz\langle \delta\hat{A}_1\rangle + \frac{iq}{m}\langle \delta\hat{A}_2\rangle = \langle \delta A_1(\mathbf{q})\rangle, \tag{A.71a}$$

$$\left(iz + \frac{q^2}{\rho_m}\hat{M}\right)\langle \delta\hat{A}_2\rangle + \frac{iq}{\rho_0 \chi_T}\langle \delta\hat{A}_1\rangle = \langle \delta A_2(\mathbf{q})\rangle, \tag{A.71b}$$

where $\hat{M}$ is the Laplace transform of the longitudinal modulus, given by

$$\hat{M}(z) = \int_0^\infty dt\, e^{-izt}\left\{K(t) + \frac{4}{3}G(t)\right\}. \tag{A.72}$$

In Eqs. (A.71), we have not written out the explicit $\mathbf{q}$ and z dependence in $\langle \delta\hat{A}_1\rangle$ and $\langle \delta\hat{A}_2\rangle$.

Equations (A.71a) and (A.71b) can now be inverted to yield

$$\begin{bmatrix} \langle \delta \hat{A}_1 \rangle \\ \langle \delta A_2 \rangle \end{bmatrix} = \frac{1}{\Delta} \begin{bmatrix} iz + \dfrac{q^2}{\rho_m^0} \hat{M} & -\dfrac{iq}{m} \\ -\dfrac{iq}{\rho_0 x_T} & iz \end{bmatrix} \begin{bmatrix} \langle \delta A_1(\mathbf{q}) \rangle \\ \langle \delta A_2(\mathbf{q}) \rangle \end{bmatrix} \tag{A.73}$$

where $\Delta = iz(iz + q^2\hat{M}/\rho_m^0) + \omega_T^2$; $\omega_T = q(\rho_m^0 x_T)^{-1/2}$ is the isothermal sound frequency.

Equation (A.73) can be compared with Eq. (A.45):

$$\frac{1}{iz}[-\boldsymbol{\chi}(\mathbf{q}z)\boldsymbol{\chi}^{-1}(\mathbf{q}) + \mathbf{I}] = \frac{1}{\Delta} \begin{bmatrix} iz + \dfrac{q^2}{\rho_m^0} \hat{M} & -\dfrac{iq}{M} \\ -\dfrac{iq}{\rho_0 x_T} & iz \end{bmatrix} \tag{A.74}$$

Using the fact that $\boldsymbol{\chi}(\mathbf{q})$ is diagonal

$$\boldsymbol{\chi}(\mathbf{q}) = \beta \begin{bmatrix} \langle |\delta\rho(\mathbf{q})|^2 \rangle & 0 \\ 0 & \langle |p(\mathbf{q})|^2 \rangle \end{bmatrix} \tag{A.75}$$

and also setting $z = \omega$, we have found that

$$-\frac{\chi_{11}(q, \omega)}{\chi_{11}(q)} + 1 = \frac{1}{\Delta}\left(-\omega^2 + \frac{q^2}{\rho_m^0} i\omega\hat{M}\right) \tag{A.76a}$$

$$-\frac{\chi_{12}(\mathbf{q}, \omega)}{\chi_{22}(\mathbf{q}, \omega)} = \frac{1}{\Delta}\left(\frac{\omega q}{m}\right) \tag{A.76b}$$

$$-\frac{\chi_{21}(\mathbf{q}, \omega)}{\chi_{11}(\mathbf{q})} = \frac{1}{\Delta}\left(\frac{\omega q}{\rho_0 x_T}\right), \tag{A.76c}$$

$$-\frac{\chi_{22}(\mathbf{q}, \omega)}{\chi_{22}(\mathbf{q})} + 1 = \frac{1}{\Delta}(-\omega^2). \tag{A.76d}$$

We can equate the real and imaginary parts in both sides of Eqs. (A.76). For example, taking the real and imaginary part of Eq. (A.76a), we obtain

$$-\frac{\chi'_{11}(\mathbf{q}, \omega)}{\chi_{11}(\mathbf{q})} + 1 = \mathrm{Re}\left\{\frac{-\omega^2 + (q^2/p_m^0)[M'(\omega) + iM''(\omega)]}{\omega_T^2 - \omega^2 + (q^2/\rho_m^0)[M'(\omega) + iM''(\omega)]}\right\} \tag{A.77a}$$

and

$$\frac{\chi''_{11}(\mathbf{q}, \omega)}{\chi_{11}(\mathbf{q})} = \mathrm{Im}\left\{\frac{-\omega^2 + (q^2/\rho_m^0)[M'(\omega) + iM''(\omega)]}{\omega_T^2 - \omega^2 + (q^2/\rho_m^0)[M'(\omega) + iM''(\omega)]}\right\}, \tag{A.77b}$$

where M' and M'' are the real and imaginary parts of the relaxation modulus. Using the fluctuation–dissipation theorem we can write Eq. (A.77b) in terms of the spectral power density. The result is

$$I(\mathbf{q}\omega) = \frac{1}{2\pi}\int_{-\infty}^{\infty} e^{-i\omega t}\langle \delta\rho(\mathbf{q}t)\, \delta\rho^*(\mathbf{q})\rangle$$

$$\sim \chi''_{11}(\mathbf{q}, \omega) = \chi_{11}(\mathbf{q}) \frac{q^2 M''(\omega)/\rho_m^0}{[\omega_T^2 - \omega^2 + (q^2/\rho_m^0)M'(\omega)]^2 + [(q^2/\rho_m^0)M''(\omega)]^2} \tag{A.78}$$

Inclusion of the temperature fluctuation in the calculation has been carried out recently. The result for $I(\mathbf{q}, \omega)$ has the same form as Eq. (A.78), provided that the relaxation moduli M' and M'' are modified to include the effect of thermal diffusion Wang and Fischer, 1980).

References

Abragam, A. (1961). "The Principles of Nuclear Magnetism." Oxford University Press (Clarendon), London and New York.

Alms, G. R., Bauer, D. R., Brauman, J. I., and Pecora, R. (1973). *J. Chem. Phys.* **58**, 5570.

Alms, G. R., Bauer, D. R., Brauman, J. I., and Pecora, R. (1974). *J. Chem. Phys.* **61**, 2255.

Andersen, H. C., and Pecora, R. (1971a). *J. Chem. Phys.* **54**, 2584.

Andersen, H. C., and Pecora, R. (1971b). *J. Chem. Phys.* **55**, 1496.

Armstrong, J. A., Bloembergon, N., Ducuing, J., and Pershan, P. S. (1962). *Phys. Rev.* **127**, 1918.

Bernard, W., and Callen, H. B. (1959). *Rev. Modern Phys.* **31**, 1017.

Berne, B. J., and Pecora, R. (1976). "Dynamic Light Scattering." Wiley, New York.

Bhagavantam, S. (1964). "Crystal Symmetry and Physical Properties." Academic Press, New York.

Bleich, H. E., and Redfield, A. G. (1971). *J. Chem. Phys.* **55**, 540.

Bloembergen, N. (1965). "Nonlinear Optics." Benjamin, New York.

Bloembergen, N., and Shen, Y. R. (1964). *Phys. Rev. A* **133**, 37.

Bloembergen, N., Purcell, E. M., and Pound, R. V. (1978). *Phys. Rev.* **73**, 679.

Born, M., and Huang, K. (1962). "Dynamical Theory of Crystal Lattice." Oxford Univ. Press (Clarendon), London and New York.

Brueck, S. R., and Osgood, R. M. (1976). *J. Chem. Phys. Lett.* **39**, 568.

Calaway, W. F., and Ewing, G. E. (1975). *J. Chem. Phys.* **6**, 2842.

Cavanaugh, D. B., and Wang, C. H. (1982). *J. Appl. Phys.* **53**, 2793.

Chandrasekhar, S. (1943). *Rev. Modern Phys.* **15**, 1.

Chappell, P. J., and Kivelson, D. (1981). *J. Chem. Phys.* **76**, 1742.

deGroot, S. R., and Mazur, P. (1962). "Nonequilibrium Thermodynamics." Wiley (Interscience) New York.

Demco, D. E., Tegenfeldt, J., and Waugh, J. S. (1975). *Phys. Rev. B* **11**, 4133.

Eckardt, G., Hellwarth, R. W., McClung, F. H., Schwartz, S. E., Weiner, D., and Wooburg, E. J. (1962). *Phys. Rev. Lett.* **9**, 455.

Edmonds, A. R. (1960). "Angular Momentum in Quantum Mechanics." Princeton Univ. Press, Princeton, New Jersey.

Eesley, G. L. (1981). "Coherent Raman Spectroscopy." Pergamon, Oxford.

Evans, W. A. B. (1968). *Ann. Physics.* **48**, 72.

Fablinski, J. L., Sabirov, L. M., and Starunov, V. S. (1969). *Phys. Lett. A* **29**, 414.

Farro, D. (1960). *Phys. Rev.* **119**, 53.

Ferry, J. D. (1980). "Viscoelastic Properties of Polymers. Wiley, New York.

Feynman, R. P., Vernon, F. C., and Hellwarth, R. W. (1957). *J. Appl. Phys.* **28**, 49.

Fixman, M., and Rider, K. (1969). *J. Chem. Phys.* **51**, 2425.

Forster, D. (1975). "Hydrodynamic Fluctuations, Broken Symmetry, and Correlation Functions." Benjamin, New York.

Friedman, H., and Wilson-Gordon, A. D. (1978a). *Opt. Commun.* **24**, 5.

Friedman, H., and Wilson-Gordon, A. D. (1978b). *Opt. Commun.* **26**, 193.

Goldman, M. (1970). "Spin Temperature and Nuclear Magnetic Resonance in Solids." Oxford Univ. Press (Clarendon), London and New York.

Gordon, R. G. (1966). *J. Chem. Phys.* **44**, 1830.

Gordon, R. G. (1968). *Adv. Magn. Reson.* **3**, 1.

Grannell, P. K., Mansfield, P., and Whitaker, M. A. B. (1973). *Phys. Rev. B* **8**, 4149.

Haeberlen, U., and Waugh, J. S. (1968). *Phys. Rev.* **175**, 453.

Hahn, E. L. (1950). *Phys. Rev.* **80**, 580.

Hartmann, S. R., and Hahn, E. L. (1962). *Phys. Rev.* **128**, 2042.

Heiman, D., Hellwarth, R. W., Levenson, M. D., and Martin, G. (1976). *Phys. Rev. Lett.* **34**, 189.

Hellwarth, R. W. (1977). *Prog. Quantum Electron.* **5**, 1.

Hellwarth, R. W., Cherlow, J., and Yang, T. T. (1975). *Phys. Rev. B* **11**, 964.

Herzfeld, K. F., and Litovitz, T. A. "Absorption and Dispersion of Ultrasonic Waves." Academic Press, New York (1959).

Hosea, T. J., Lockwood, D. J., and Taylor, W. (1979). *J. Phys. C* **12**, 387.

Hu, C. M., and Zwanzig, R. (1974). *J. Chem. Phys.* **60**, 4354.

Huntress, W. T., Jr., (1968). *J. Chem. Phys.* **48**, 3524.

Keyes, T., and Kivelson, D. (1972). *J. Chem. Phys.* **56**, 1057.

Kirkwood, J. G. (1946). *J. Chem. Phys.* **14**, 180.

Kivelson, D., and Madden, P. (1980). *Annu. Rev. Phys. Chem.* **31**, 523.

Kleinman, D. A. (1962). *Phys. Rev.* **129**, 1977.

Kubo, R. (1962). *In* "Fluctuations, Relaxation, and Resonance in Magnetic Systems," (D. T. Haar, ed.). Oliver and Boyd, London.

Kubo, R. (1957). *J. Phys. Soc. Japan* **12**, 570.

Kubo, R. (1959). "Some Aspects of the Statistical-Mechanical Theory of Irreversible Processes," (Brittin and Dunham, eds.) vol. 1. Wiley (Interscience), New York.

Kubo, R., and Tomita, K. (1957). *J. Phys. Soc. Japan* **9**, 888.

Landau, L. D., and Lifshitz, E. M. (1960). "Electrodynamics of Continuous Media." Addison-Wesley, Reading, Massachusetts.

Landau, L. D., and Lifshitz, E. M. (1970). "Statistical Physics." Addison-Wesley, Reading, Massachusetts.

Lascombe, J., and Huong, P. V. (1982). "Raman Spectroscopy (Linear and Nonlinear)." Wiley, New York.

Laubereau, A. (1974). *Chem. Phys. Lett.* **27**, 600.

Laubereau, A., and Kaiser, W. (1978). *Rev. Modern Phys.* **50**, 607.

Laubereau, A., von der Linde, D., and Kaiser, W. (1972). *Phys. Rev. Lett.* **28**, 1162.

Levenson, M. D. (1972). *IEEE Quantum Electron.* **QE-10**, 110.

Levenson, M. D., and Bloembergen, N. (1974). *J. Chem. Phys.* **60**, 1323.

Levenson, M. D., and Song, J. J. (1980). *In* "Coherent Nonlinear Optics," (M. S. Feld and V. S. Letokhov, eds.) Springer-Verlag, Berlin.

Levenson, M. D., Flytzanis, C., and Bloembergen, N. (1972). *Phys. Rev. B* **6**, 3962.

Lin, Y. H., and Wang, C. H. (1979). *J. Chem. Phys.* **70**, 681.

Lowe, I. J., and Norberg, R. E. (1957). *Phys. Rev.* **107**, 46.

Lurie, E. M., and Slichter, C. P. (1964). *Phys. Rev. A* **133**, 1108.

Madden, P. A. (1982). *In* "Raman Spectroscopy, Linear and Nonlinear," (J. Lascombe and P. V. Huong, ed.). Wiley, New York.

Magnus, W. (1954). *Comm. Pure Appl. Math.* **7**, 649.

Maker, P. D. (1970). *Phys. Rev. A* **1**, 923.

Maker, P. D., and Terhune, R. W. (1965). *Phys. Rev. A* **137**, 801.

Mansfield, P., and Ware, D. (1966). *Phys. Lett.* **22**, 133.

Martin, P. C., and Kadanoff, L. P. (1963). *Ann. Physics* **24**, 419.

McArthur, D. A., Hahn, E. L., and Walstedt, R. E. (1969). *Phys. Rev.* **188**, 609.

McClung, R. E. W. (1972). *J. Chem. Phys.* **57**, 5478.

Mehring, M. (1976). "High Resolution NMR Spectroscopy in Solids." Springer-Verlag, Berlin and New York.

Merzbacher, E. (1970). "Quantum Mechanics," 2nd ed., Wiley, New York.

Mori, H. (1965a). *Progr. Theoret. Phys.* **33**, 423.

Mori, H. (1965b). *Progr. Theoret. Phys.* **34**, 399.

Mountain, R. D. (1966). *J. Res. Nat. Bur. Standards* **70A**, 207.

Mountain, R. D. (1968). *J. Res. Nat. Bur. Standards* **72A**, 75.

Mountain, R. D. (1970). *Crit. Rev. Solid State Mater. Sci.* **1**, 5.

Mountain, R. D., and Zwanzig, R. (1965). *J. Chem. Phys.* **43**, 4464.

Nelson, D. F., and Lax, M. (1970). *Phys. Rev. Lett.* **24**, 389.

Nelson, D. F., and Lax, M. (1971). *Phys. Rev. B* **3**, 2778.

Nibler, J. W., and Knighton, G. V. (1978). *In* "Raman Spectroscopy of Gases and Liquids." (A. Webster, ed.) Springer-Verlag, Berlin.

Nye, J. F. (1960). "Physical Properties of Crystals." Oxford Univ. Press (Clarendon), London and New York.

Nyquist, H. (1928). *Phys. Rev.* **32**, 110.

O'Dell, J., and Berne, B. J. (1975). *J. Chem. Phys.* **63**, 2376.

Oppenheim, I., Shuler, K. E., and Weiss, G. H. (1977). "Stochastic Processes in Chemical Physics: The Master Equation." MIT Press, Cambridge, Massachusetts.

Orr, B. J., and Ward, J. F. (1971). *Molecular Phys.* **20**, 513.

O'Steen, B. L. (1981). Ph.D. Thesis. California Institute of Technology, Pasadena, California.

Ostroff, E. D., and Waugh, J. S. (1966). *Phys. Rev. Lett.* **22**, 133.

Pechukas, P., and Light, J. C. (1966). *J. Chem. Phys.* **44**, 3897.

Pecora, R. (1968). *J. Chem. Phys.* **49**, 1036.

Pines, A., Gibby, M. G., and Waugh, J. S. (1973). *J. Chem. Phys.* **59**, 569.

Placzek, G. (1939). "Handbuch der Radiologie," Vol. 6, Part 2, p. 205. Akad. Verlagsges, Frankfurt.

Redfield, A. G. (1955). *Phys. Rev.* **98**, 1787.

Redfield, H. G. (1957). *IBM J. Res. Develop.* **1**, 19.

Rhim, W. K., Pines, A., and Waugh, J. S. (1971). *Phys. Rev. B* **3**, 684.

Rose, M. E. (1957). "Elementary Theory of Angular Momentum." Wiley, New York.

Rotenberg, M., Bivins, R., Metropolis, N., and Wooten, J. K. (1959). "The 3-*j* and 6-*j* Symbols." Technology Press, MIT, Cambridge, Massachusetts.

Rytou, S. M. (1957). *Zh. Eksp. Teor. Fiz.* **33**, 166, 514, 671. [*Soviet Phys. JETP* **6**, 130, 401, 513 (1958).]

Satija, S. K., and Wang, C. H. (1978). *J. Chem. Phys.* **69**, 2739.

Schaefer, J. F., Stejskal, E. O., and Buchdahl, R. (1974). *Macromolecules* **8**, 291.

Schaefer, J. F., Stejskal, E. O., and Buchdahl, R. (1977). *Macromolecules* **10**, 384.

Scott, J. F. (1974). *Rev. Modern. Phys.* **46**, 83.

Slichter, C. P. (1963). "Principles of Magnetic Resonance." Harper, New York.

Slichter, C. P., and Ailion, D. (1964). *Phys. Rev. A* **135**, 1099.

Slichter, C. P., and Ailion, D. (1965). *Phys. Rev. A* **137**, 235.

Slichter, C. P., and Holton, W. C. (1961). *Phys. Rev.* **122**, 1701.

Smith, J. A. S. (ed.) (1974). "Advances in Nuclear Quadruple Resonances," vol. 1. Heyden, London.

Solomon, I. (1959). *C. R. Acad. Sci.* **248**, 92.

Solomon, I., and Ezratly, J. (1962). *Phys. Rev.* **127**, 78.

Stanley, H. E. (1971). "Introduction to Phase Transitions and Critical Phenomena." Oxford Univ. Press, London and New York.

Starunov, V. S., and Fabelinski, J. L. (1974). *Zh. Eksp. Teor. Fiz.* **66**, 1740. [*Soviet Phys. JETP* **39**, 858 (1974).]

Steele, W. A. (1969). *In* "Transport Phenomena in Fluids," (H. J. M. Hanley, ed.). Dekker, New York.

Steele, W. A. (1976). *Adv. Chem. Phys.* **34**, 1.

Steele, W. A. (1980). *In* "Vibrational Relaxation of Molecular Liquids and Solids," (S. Bratos and R. M. Pick, eds.). Plenum, New York.

Steele, W. A. (1981). *Molecular Phys.* **43**, 141.

Stegeman, G. I. A., and Stoicheff, B. P. (1968). *Phys. Rev. Lett.* **21**, 202.

St. Pierre, A. G., and Steele, W. A. (1972). **57**, 5478.

Terhune, R. W., Maker, P. D., and Savage, C. M. (1965). *Phys. Rev. Lett.* **14**, 681.

Tolles, W. M., Nibler, J. W., McDonald, J. R., and Harvey, A. B. (1977). *Appl. Spectrosc.* **31**, 253.

Uhlenbeck, G. E., and Ford, G. W. (1963). "Lectures in Statistical Mechanics." Amer. Math. Soc., Providence, Rhode Island.

Vacher, R., and Boyer, L. (1972). *Phys. Rev. B* **6**, 639.

Van Hove, L. (1954). *Phys. Rev.* **95**, 249.

Wang, S. C. (1929). *Phys. Rev.* **34**, 243.

Wang, C. H. (1973). *J. Chem. Phys.* **59**, 225.

Wang, C. H. (1977). *Molecular Phys.* **33**, 207.

Wang, C. H. (1980). *Molecular Phys.* **41**, 541.

Wang, C. H. (1982). *J. Mol. Sci.* (Wahan, China) **2**, 43.

Wang, C. H., and Fischer, E. W. (1984). *J. Chem. Phys.* (in press).

Wang, C. H., and Liu, Q.-L. (1983). *J. Chem. Phys.* **78**, 4363.

Wang, C. H., and Ramshaw, J. D. (1972). *Phys. Rev. B* **6**, 3253.

Wang, C. H., Ma, R. J., Fytas, G., and Dorfmüller, Th., (1983). *J. Chem. Phys.* 78, 5863.

Wang, C. H., Ma, R. J., and Liu, Q.-L. (1984). *J. Chem. Phys.* **80**, 617.

Wang, C. H., Fytas, G., and Fischer, E. W. (1984). *J. Chem. Phys.* (in press).

Ward, J. F. (1965). *Rev. Modern Phys.* **34**, 1.

Waugh, J. S., and Wang, C. H. (1967). *Phys. Rev.* **162**, 209.

Waugh, J. S., Huber, L. M., and Haeberlen, U. (1968a). *Phys. Rev. Lett.* **20**, 180.

Waugh, J. S., Wang, C. H., Huber, L. M., and Vold, R. L. (1968b). *J. Chem. Phys.* **48**, 662.

Wax, N. (ed.) (1954). "Selected Papers on Noise and Stochastic Processes." Dover, New York.

Weitz, E., and Flynn, G. (1979). *Annu. Rev. Phys. Chem.* **25**, 275.

Wilson, E. B., Jr., Decius, J. C., and Cross, P. C. (1955). "Molecular Vibrations: The Theory of Infrared and Raman Vibrational Spectra." McGraw-Hill, New York.

Woodbury, E. J., and Ng, W. K. (1962). *Proc. IRE* **50**, 2367.

Wright, R. B., Schwartz, M., and Wang, C. H. (1973). *J. Chem. Phys.* **58**, 5125.

Wynne, J. T. (1972). *Phys. Rev. Lett.* **29**, 650.

Yajima, T., and Takatsuji, M. (1964). *J. Phys. Soc. Japan* **19**, 2343.

Yariv, A. (1971). "Introduction to Optical Electronics." Holt, New York.

Yacoby, Y. (1976). *Phys. Rev. B13*, 4132 (1976).

Zwanzig, R. (1961). *Lect. Theor. Phys.* **3**.

Zwanzig, R. (1965). *Annu. Rev. Phys. Chem.* **16**, 67.

Index

M

N

O

P

R